LAW AND THE GOVERNANCE OF RENEWABLE RESOURCES

STUDIES FROM NORTHERN EUROPE AND AFRICA

Edited by

Erling Berge and Nils Christian Stenseth

Supported by

The Norwegian University of Science and Technology
The Norwegian Man and Biosphere Committee
The Agricultural University of Norway
The Research Council of Norway

A Publication of the
International Center for Self-Governance

ICS PRESS
Institute for Contemporary Studies
Oakland, California

© 1998 Erling Berge and Nils Christian Stenseth

All rights reserved. No part of this book may be used or reproduced in any manner without express written permission, except in the case of brief quotations in critical articles and reviews.

This book is a publication of the Institute for Contemporary Studies, a nonprofit, nonpartisan public policy research organization. The analyses, conclusions, and opinions expressed in ICS Press publications are those of the authors and not necessarily those of the Institute or of its officers, its directors, or others associated with, or funding, its work.

Inquiries, book orders, and catalog requests should be addressed to ICS Press, 1611 Telegraph Avenue, Suite 902, Oakland, CA 94612. Tel. (510) 238-5010; Fax (510) 238-8440; Internet www.icspress.com. For book orders and catalog requests, call toll-free in the United States: (800) 326-0263, outside the United States: (717) 325-5686.

Special thanks to the Norwegian Research Council for their contribution to this book.

Project manager: Mary Jennings
Project editor: Jan Ponyicsanyi
Composition: Sarah McFarland
Cover design: Denise Goldman

Book set in Palatino by Sarah McFarland and printed and bound in Canada by Hignell Book Printing, 488 Burnell Street, Winnipeg, Manitoba.

0 9 8 7 6 5 4 3 2 1

ISBN# 1-55815-504-X

Library of Congress Cataloging-in-Publication Data

Law and the governance of renewable resources : studies from Northern Europe and Africa.
   p.   cm.
   Includes bibliographical references.
   ISBN 1-55815-504-X
   1. Renewable natural resources--Law and legislation--Europe, Northern. 2. Renewable natural resources--Law and legislation--Africa.
K3478.L39   1998   98-26344
346.04'4--DC21   CIP
Printed in Canada

Contents

List of Tables v
List of Figures vii
Note from the Publisher ix
Foreword xi
Acknowledgments xiii

Culture, Property Rights Regimes, and Resource Utilization 1
Erling Berge

Part 1 Theory from Law and Social Science

1 Institutional Analysis, Design Principles, and Threats to 27
   Sustainable Community Governance and Management of
   Commons
   Elinor Ostrom

2 The Economic Rationale of Communal Resources 55
   Thráinn Eggertsson

3 Distributional and Political Issues in Modifying Traditional 75
   Common-Property Institutions
   Gary Libecap

4 Human Rights and Resource Management—An Overview 93
   Hans Christian Bugge

Part 2 Norwegian Law and Common Property

5 The Legal Language of Common Property Rights 123
   Erling Berge

6 Legal Rights Regarding Range Lands in Norway with 129
   Emphasis on Plurality User Situations
   Thor Falkanger

7 Common Property in Norway's Rural Areas 141
   Hans Sevatdal

Part 3 Saami Reindeer Herding in Northern Norway, Sweden, and Finland

8 The History of Rights to the Resources in Swedish and 171
   Finnish Lapland
   Kaisa Korpijaakko-Labba

9   Kautokeino 1960: Pastoral Praxis   181
    Robert Paine

10 The Legal Status of Rights to the Resources of Finnmark with   205
    Reference to Previous Regulations of the Use of Nonprivate
    Resources
    Torgeir Austenå and Gudmund Sandvik

11 The Legal Status of Rights to Resources in Swedish Lapland   221
    Bertil Bengtsson

12 The Legal Status of Rights to Resources in Finnish Lapland   233
    Heikki Hyvärinen

13 The Proposal of the Norwegian Government Commission   245
    on the Rights of the Saami to Land and Water in Finnmark
    Torgeir Austenå

Part 4 The Fisheries of the Barents Sea

14 The Legal Status of Rights to Resources in the Barents Sea   257
    Geir Ulfstein

15 Managing the Barents Sea Fisheries: Impacts at National and   265
    International Levels
    Alf Håkon Hoel

16 Management under Scarcity: The Case of the Norwegian   283
    Cod Fisheries
    Bjørn Sagdahl

Part 5 Comparisons and Conclusions

17 The Namibian Fisheries Resource and the Role of Statutory   313
    Law, Regulations, and Enforcement of Law in Its Utilization
    Carl-Hermann Schlettwein and Pierre Roux

18 State Law Versus Village Law: Law as Exclusion Principle   339
    under Customary Tenure Regimes among the Fulani of Mali
    Trond Vedeld

19 The Analytical Importance of Property Rights to Northern   371
    Resources
    Audun Sandberg

Contributors   401

List of Tables

.1	Types of Property Rights Regimes	19
5.1	Types of Profits	125
9.1	The Annual Cycle of Herd Knowledge	200
16.1	Quotas and Catches of Norwegian-Arctic Cod 1977-1991	292
18.1	Range Land and Crop Tenure in the Niger Delta	346
18.2	Bundles of Rights to Common Pool Range Lands and Flooded Crop Land Associated with Status-Positions in Fulani Villages	347
18.3	Mismatch Between Constitutional Rules and Rules-in-Use as Perceived by Group of Appropriators	349

List of Figures

.1a	Resource Distribution in the Norwegian and Barents Sea: Distribution of Cod during Winter	4
.1b	Resource Distribution in the Norwegian and Barents Sea: Migrations of Norwegian Spring Spawning Herring in 1995	5
.1c	Administrative Boundaries in the Barents Sea	6
.2a	Borders of Reindeer Herding Areas of Norway	8
.2b	Reindeer Herding Areas in Finnmark	9
.2c	Dialect Families of Traditional Saami Society	10
.3	Cultural Filters in the Oakerson (1992) Framework for Analyzing the Commons	14
1.1	A Framework for Institutional Analysis	28
1.2	Linking Levels of Analysis	30
9.1	The Scale of Reindeer Pastoralism in Finnmark 1961-1962	182
9.2	Kautokeino Reindeer Ranges; West, Middle, and East Ranges	183
9.3	The Sii'da as Reindeer Management Unit	184
9.4	The Sii'da and Herd Management of Commensurate Proportions	185
9.5	Summer Herds: Njar'ga, Suolo, and Nanne	187
9.6	Njar'ga Pastures of Gow'dojottelit	188
9.7	Winter Herds	193
9.8	Spring Herds (1961)	195
9.9	Annual Cycles of Herd Knowledge	201
19.1	Bundles of Rights Associated with Ownership Positions	390

Note from the Publisher

Law and the Governance of Renewable Resources is ICS Press's eighth book on the issues of the environment and common-pool resources. It is our intention to continue publishing in this important area of research and policy development.

Common-pool resources are interesting in their own right. Whether it be fisheries, water basins, or common range lands, effective governance and management of these resources are the keys to long-term use.

The importance of this book is that it forces us to focus on the critical role played by law in the process of governance, not only of renewable resources but also of other spheres of our lives. Law is a vital enabling condition for governance, and is particularly important in "modern" societies, where we think of the governing process as giving us law. Yet in the older traditions of self-governance, enabling acts were the constitutional frameworks within which communities developed their governing institutions and capacities. In fact, these very development processes many times created customary and formal laws.

One of ICS Press's main objectives is to advance the knowledge and understanding of self-governing processes. In the case of renewable resources, we find fascinating examples of how critical are issues of governance to the livelihood of the people who depend on these resources. This raises an interesting question—is necessity one of the key elements in building self-governing solutions to the significant problems facing any society?—and leads us to ask whether societies lose the sense of necessity to govern their own lives when they become wealthy through modern industrial practices.

This book enables us to understand the central role that law plays in creating self-governing solutions to issues of common resources. It will deepen and extend the debate about how we understand these issues and how policy makers begin to fashion solutions to such ongoing policy issues. If we want to apply law and self-governing approaches to the use of renewable resources, we will also have to rethink the role that public administrators play in the development of the capacities of communities—no matter how diverse—to craft self-governing solutions. The issue will be, how do we craft laws and institutions that create powerful incentives for bureaucrats to become civil servants?

We are pleased to again be publishing an important book on common-pool resources—a book that we hope will further develop a firm intellectual foundation for environmental policy and resource use.

Robert B. Hawkins, Jr.
President and CEO
Institute for Contemporary Studies

Foreword

Government and cooperation are in all things the laws of life; anarchy and competition the laws of death.

—John Ruskin

Law is the foundation of resource use. Reflecting the political, cultural, and economic values of society, law creates behavioral incentives that determine whether resources are sustained over time or exhausted before their time.

The law creates the institutional environment of resource governance—the structure within which we make decisions about use, coordinate the behavior of users, and monitor resource outcomes. The institutional environment also contains the rules and procedures by which these actions are taken. And so, through structure and rules, institutions shape resource governance. Whether that governance is effective or ineffective depends on the degree to which it attends to ecological integrity, economic productivity, and social equity.

In the past, ecological, economic, and social factors were critically important to governance because people were directly dependent on the ecological systems within which they lived. Survival depended not only on the ability to extract the productive riches of resources but also on the ability to restrain from over-extraction so that resources could renew themselves. As a result of this dependence, resource governance took on a fundamental importance. Especially in areas where resources were not abundant, balance, continuity, and fairness were underlying governance themes.

Over time, the global processes of industrialization and market expansion have meant, for many, a lessening of direct dependence on local resources. But those who work in natural resource systems are still economically and socially dependent on resource productivity, and even those who are distant from direct use are indirectly affected by resource health. All citizens maintain a stake in renewable resources, but the nature of the stake is more diverse than in the past.

Today the old themes of balance, continuance, and fairness continue to thread through resource governance. Balance must be struck in allocations between different user groups and between present and future. Continuance, the expression of duty to the future, is articulated as sustainability. Fairness plays out in the distributional justice of allocations, and in the legitimacy of rules and procedures for decision-making.

It is also the case that today the health of many renewable resources is fragile. Use has evolved from subsistence to intense exploitation, enhanced by technological innovations. Population pressures have intensified. Competition is fierce, and values about appropriate uses have changed. Compliance with rules is often low. In the face of contemporary pressures, how will renewable resources be sustained? How will ecological integrity, economic productivity, and social equity be integrated?

Many of the answers to these questions lie in the past. Working from the context of the present resource problems, it is to long-standing governance systems that we increasingly look to find the elements of effective governance. And it is often in the regions of climatic extremes—where there is little room for error—that the keys to good governance are found.

This volume contains many of the keys. The contributions of an impressive list of resource scholars place the practice of resource governance against a theoretical background of institutional design, economic rationale, distributional justice, and human rights. The focus is on northern regions, where renewable resources like fish, reindeer, and range lands are important to both the existence and the quality of life. Cases from southern regions are also included for comparison.

The unique collection provides rare and rich detail on the law and practice of resource use that is extremely useful to the understanding of effective governance. Many of the cases are new to the international literature of resource management, and it is gratifying to know that a wider readership can now be exposed to the complexities of these governance systems and the lessons they contain.

The message of this book is that sustaining resources is an old problem, one that has always been embedded in time and context, in governance and in laws. The times and contexts of renewable resources are changing rapidly. And so too is the dynamic interaction between the law, governance, economics, culture, and politics. All are affected by the history of renewable resources, and all contribute to their future.

Susan S. Hanna
Professor of Marine Economics
Oregon State University

Acknowledgments

This book grew out of the conference "Common Property Regimes: Law and the Management of Nonprivate Resources" (February 16-21, 1993). The conference was initiated by the Norwegian MAB-Committee and organized by the Department of Land Use Planning and Center for Sustainable Development at the Agricultural University of Norway. We are grateful for the efforts of Dr. Ragnar Øygard, director of the Center for Sustainable Development, and his executive officers, Colin Murphy and Anne Utvær. Their efforts brought the conference funding and made it real.

During the critical period after the conference, Derek Ott, then a PhD student at Rutgers University, assisted the editors in collecting the various papers, corresponding with authors, getting permission to publish their papers, and even typing papers not available on diskette. He also did valuable work in reading preliminary versions of papers and suggesting improvements. We appreciate his efforts. The proceedings of the conference were made available in two volumes (Berge 1993; Berge and Ott 1993). Based on the papers presented at the conference and on commissioned papers two preliminary versions of this edited volume appeared in 1994 (Berge, Ott, and Stenseth, eds. 1994) and 1995 (Berge and Stenseth, eds. 1995).

During 1996 and 1997 the encouragement of Elisabeth Case of Washington University, St. Louis; the interest of the people at the International Center for Self-Governance, Institute for Contemporary Studies, San Francisco; and a publishing grant from the Research Council of Norway (grant no. 117369/521) led to further additions and revisions.

The editors appreciate the permission to use maps made available for the book by the Boazodoallohálddahus (the reindeer herding administration), Alta; and the Institute of Marine Research, Bergen. The prolonged efforts of Kjetil Fridèn of NTNU, Trondheim, led to the compilation of the maps and other figures for the book. We are grateful for his efforts.

The book was made possible by the effort of many people and the financial support of the Agricultural University of Norway; the Norwegian University of Science and Technology, Trondheim; the

Research Council of Norway; the Regional Council for Northern Norway and Namdalen; the Royal Norwegian Ministry of Foreign Affairs; the Royal Norwegian Ministry of Agriculture; and the Royal Norwegian Ministry of Environment. The editors are grateful to all who assisted in making this publication possible.

Erling Berge, Norwegian University of Science and Technology
Nils Christian Stenseth, University of Oslo

February 1998

REFERENCES

Berge, Erling, ed. 1993. "Proceedings of a Conference on Common Property Regimes: Law and the Management of Nonprivate Resources." Lofoten, 16-21, February 1993. Volume 1. Ås: Department of Land Use Planning, The Agricultural University of Norway.

Berge, Erling and Derek Ott, eds. 1993."Proceedings of a Conference on Common Property Regimes: Law and the Management of Nonprivate Resources." Lofoten, 16-21, February 1993. Volume 2. Ås: Department of Land Use Planning, The Agricultural University of Norway.

Berge, Erling, Derek Ott, and Nils Chr. Stenseth, eds. 1994. "Law and the Management of Divisible and Non-excludable Renewable Resources." Ås: Department of Land Use and Landscape Planning, Agricultural University of Norway.

Berge, Erling and Nils Chr. Stenseth, eds. 1995. "Law and the Management of Renewable Resources," ISS report no. 46. Department of Sociology and Political Science, University of Trondheim.

Erling Berge   Introduction

Culture, Property Rights Regimes, and Resource Utilization

The fish population in the Barents Sea is a valuable resource in jeopardy. The range lands of Finnmark are also a valuable resource in jeopardy. Around the world one finds a surplus of resources in jeopardy. Property rights to these resources vary—some are privately owned, some are owned by the state, some are owned in common by a group of appropriators, and some are for all practical purposes no one's property. In a world with a rapidly growing population and with technological development rapidly increasing the efficiency of resource appropriation, rights to resources, and sensible resource management have become issues of prime importance. The problems involved in governing resources, such as the fish in the Barents Sea and the range lands in Finnmark, reach the agendas of political decison makers more frequently than ever before.

Somehow public interest in the mismanagement of private resources is not quite as much in fashion as the mismanagement of the nonprivate: state-owned resources, those owned jointly, and in particular those without recognized owners. The public view here seems to be that the reason for mismanagement is that the resources are not privately owned. Only with the care and forethought of an interested individual will it be possible to manage resources balancing the goals of maximum return in the present against the goal of as good or better returns for future generations. The culturally shared model of resource management seems to be one of a farmer taking care of his farm in order to leave it to his children in as good a condition as he himself got it from his father. This cultural idea got its scientific expression in the model of the "tragedy of the commons" (Hardin

1968). Much of the debate around resource managment has revolved around the problems raised by Hardin, and most of it is rooted in a failure to understand the complexity of property rights with a concomitant conflation of common property resources and open access resources. Indeed, "it would be difficult to find an idea as misunderstood as 'commons' or 'common property'" (Bromley 1992a, 3).

The relevant part of the culture of a population of resource appropriators are the opinions, attitudes, beliefs, norms, and values concerning who can legitimately appropriate which benefits from the resource. If somebody believes he has a right to utilize a resource in a particular way, and everybody who knows about it concurs, it does not matter what the legal code says or what the state wants. For so long as all people act in good faith, their definition of the situation will be the reality of the situation. The problems arise at the point where someone contests the right to any particular resource utilization (for example, because it infringes on what this person believes to be his right). If the conflict is solved locally without recourse to the formal legal system, we are within the bounds of a traditional management regime. But if the problem escalates, the local resource users have to face the possibility that the state and its representatives may bring a new definition of the situation into the negotiations. And where will the civil servants find their definition of the situation when they need to decide the outcomes of such disputes? From where will they gain insight into what can be considered proper resource utilization? And how will they go about curtailing usage incompatible with the principles of justice their definition of the situation implies? (An example is their choice of classification of a resource as common property for the inhabitants of a community, private property for a citizen, or state property.)

Property rights can be described as an asymmetrical relation between an owner (or owners) and the non-owners. The relation is specified by listing the claim rights, privileges, legal powers, and immunities of the owner and the concomitant duties, lack of rights, liabilities, and lack of legal powers of the non-owners (Hohfeld 1913, 1917). These rights and duties are enforced either by a state or by the members of a culture. A consistent system of property rights, including the specification of who can be owner, is called a property rights regime. Usually regimes are classified as private property, state property, common property, and non-property.[1] In particular the study of the so-called "common property" has given important contributions to the knowledge about how the distribution of rights and duties regarding the usage of natural resources affects the economic surplus from the resource and the distribution of welfare among people depending on it.[2]

This introduction to the chapters that follow will begin to explore the interrelations of law, culture, and property rights regimes. But before that analysis begins, something more must to be said about the specific problems at the heart of the discussion.

Problems of Managing Fishing

In the northeast Atlantic and Barents Sea there is a vast resource of fish, an ecosystem where big fish eat small ones and the small ones eat the even smaller sea life, where seals and whales compete with birds and men to harvest from the abundance, and where seasonal and long-term changes interact with stochastic factors of ocean currents and climate to affect the volume and distribution of biomass across species as well as geography. In the competition for the harvest, man has increased his power tremendously and rapidly during the last few decades. The possibility of depleting the ecosystem far beyond the point of profitable harvesting and possibly into no recovery has become real. How can we avoid it?

The problem has several dimensions—international as well as national. The interests of Russia, Norway, Iceland, Greenland, and the Faroes are more or less directly involved. Norway and Russia, and Norway and Denmark (Greenland) are involved in disputes about the border between their extended economic zones (EEZ) (see figure 1c). In the middle of the Norwegian Sea and between Svalbard and Novaja Semlja there are regions not now within any nation's administration. And throughout these various jurisdictions the fish migrate back and forth (see figures 1a and 1b). Within each jurisdiction there are problems of legitimacy and justice in the consequences of regulations as well as repercussions throughout the ecosystem of the regulatory policies being chosen. The problem is to improve our understanding of how the regulations simultaneously affect both the viability of the ecosystem and the quality of the social system organizing the appropriators.

Problems of Range Land Management

Finnmark and Finnmarksvidda are the main habitats of the reindeer herders of the Saami people and their herds (but they are found as far south as Røros and as far east as the Kola peninsula). Throughout known history there have been long-term swings in the availability of some of the critical resources ensuring the survival of the reindeer during critical times. When the critical resource did not suffice, some of the animals starved, the herds were depleted, and conditions improved.

But both the Saami society and the Norwegian society have changed. Modern society has encroached on the habitat along the margins, modern technology has made it possible to follow the herds more closely (but this also necessitated larger herds to pay for the

FIGURE 1a
Resource Distribution in the Norwegian and Barents Sea: Distribution of Cod during Winter

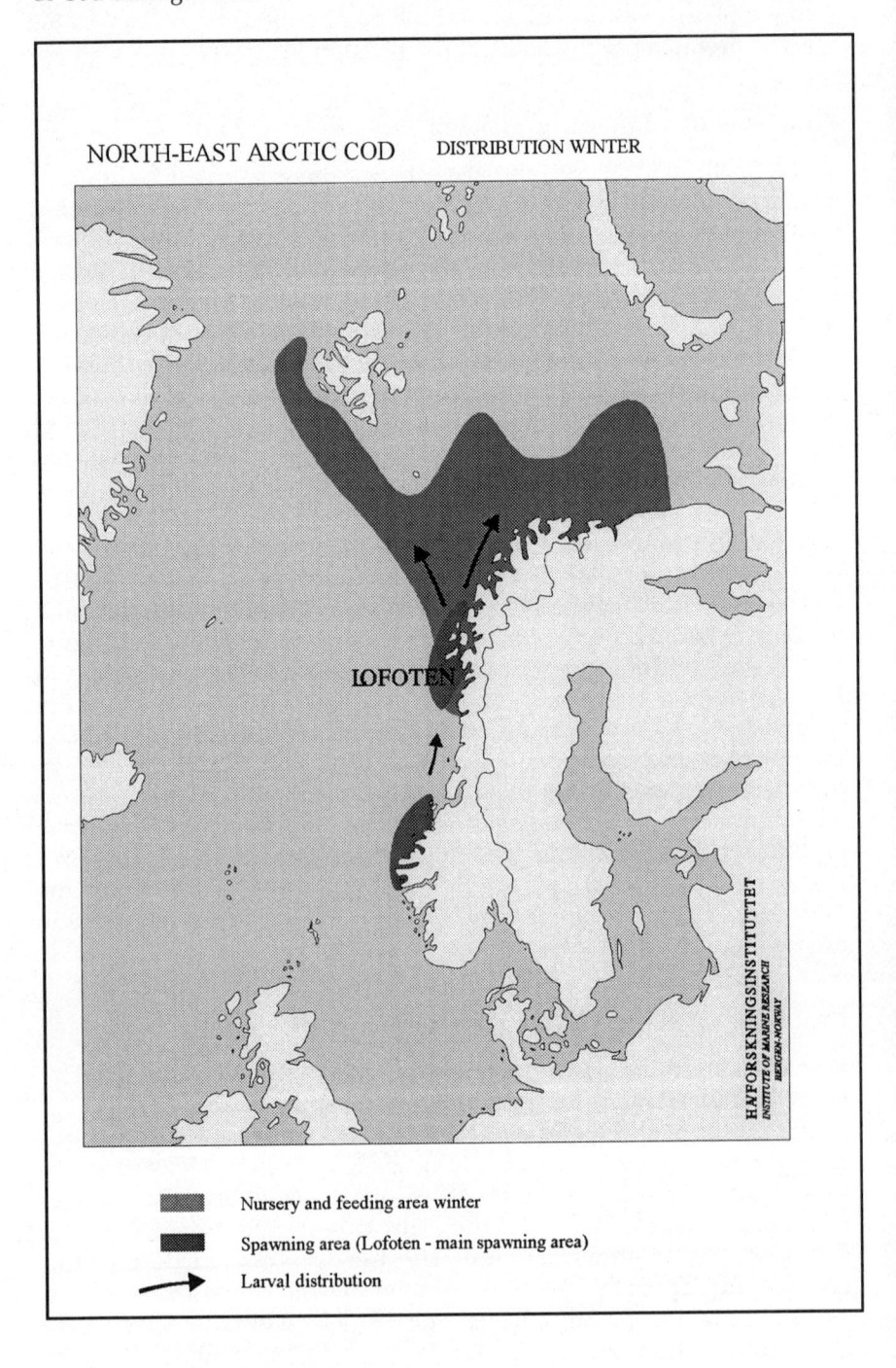

FIGURE 1b
Resource Distribution in the Norwegian and Barents Sea: Migrations of Norwegian Spring Spawning Herring in 1995

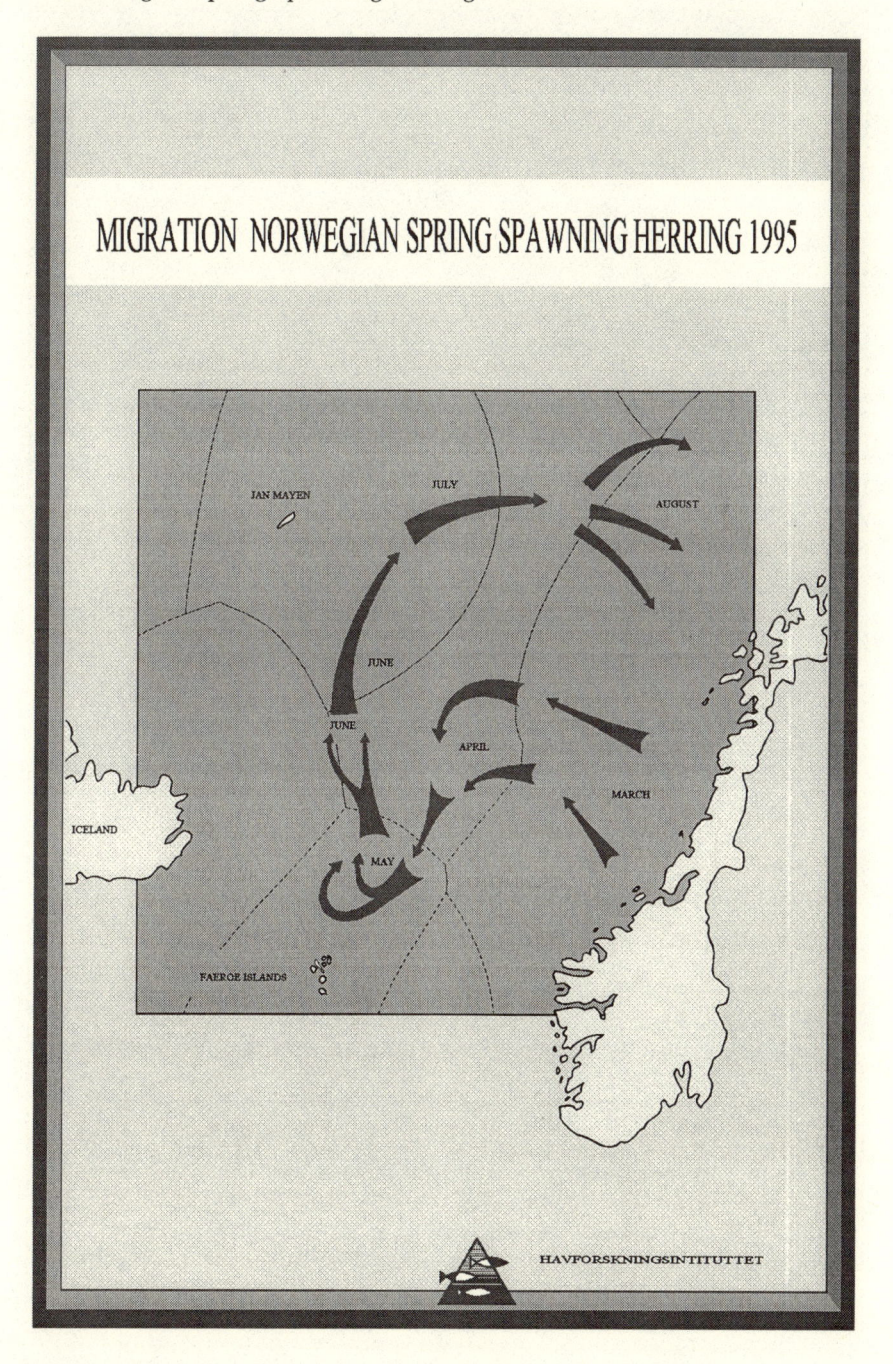

FIGURE 1c
Administrative Boundaries in the Barents Sea

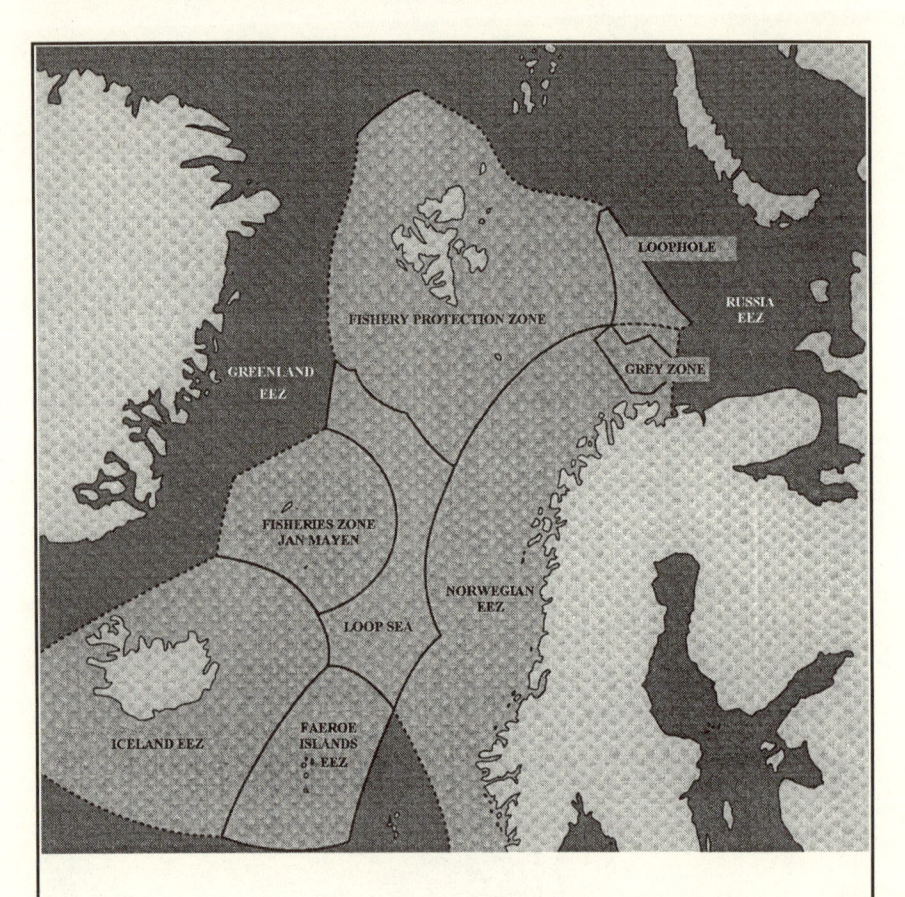

(EEZ=Extended Economic Zone according to the Law of the Sea)
- Loophole and Loop Sea areas are currently outside the jurisdiction of any state.
- Gray Zone—an area in dispute in the negotiations for a boundary between Norway and Russia.
- Fishery Protection Zone—the area around Svalbard claimed by Norway to come under Norwegian administration according to the Svalbard Treaty.
- Fisheries Zone Jan Mayen—the EEZ around Jan Mayan, a Norwegian island.

technology), and new households have added more herds. The overcrowding is visible, at least in the small regions of limited resources which most of the herds depend on during critical periods in the spring. And its consequences are showing up in the conflicts among herd owners and their anxiety about the future. Is there anything the reindeer herders can do to regain control of their future?

The problem has several dimensions. The Saami population are a separate people within the Norwegian state, as well as within Sweden, Finland, and Russia (see figure 2c). They enjoy the rights of citizenship like every other citizen. But their status as an aboriginal people also gives them special protection according to the UN Covenant on Civil and Political Rights (article 27). And the ILO Convention of 1989 concerning indiginous and tribal peoples in independent countries indicates that rights of ownership and possession of the land these peoples traditionally have occupied ought to be recognized.

The precise content of these rights is so far unresolved, and they vary among the countries. Many of the unresolved problems are tied to the problems of governing the range lands. Both Saami and Norwegians acknowledge that Saami culture and national identity to some degree is tied to reindeer herding as an industry.[3] But only a minority of the Saami people are actually reindeer herders. The implication of all this for the management of the range land as well as the Saami culture is unclear.

Among the Saami there are internal problems tied to the management of the range land. Since access to the range land now, at least in principle, is closed, which relations will the reindeer herders be able to maintain to the Saami excluded from the reindeer herding industry, and how will the closure affect Saami culture and identity? The list of questions could be extended, but our first goal is to understand what is happening.

What Are the Dynamics of These Problems?

We know a fair amount of what happens to the resources and how it happens. People make it happen. People do make their own history. But here as elsewhere, they have not chosen the conditions under which they make their history. And if it is the conditions which dictate what kind of history people make, we need to ask if it is possible to give some scope of choice of conditions to the fishermen of the Barents Sea and the Saami people. Is it possible for them to affect the conditions shaping their choices? In other words, is it possible to shape the institutions governing the resource utilization on the range lands in Finnmark and in the Barents Sea according to goals expressing the desired path of development for a social system? The assumption—not to say presumption—of modernity is that it is possible and that science can give the answer of how to do it.

FIGURE 2a

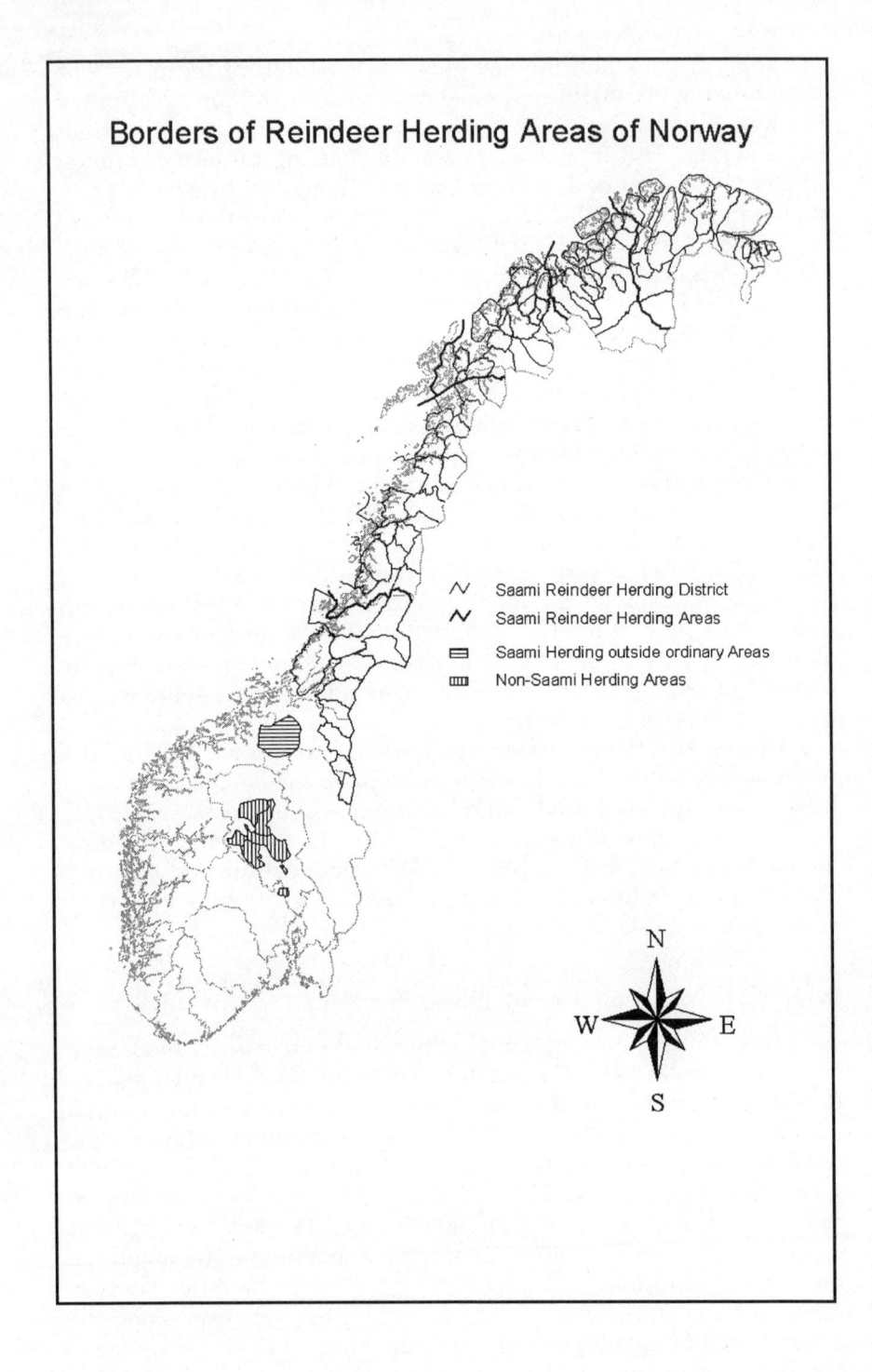

Borders of Reindeer Herding Areas of Norway

/\/ Saami Reindeer Herding District
/\/ Saami Reindeer Herding Areas
▤ Saami Herding outside ordinary Areas
▥ Non-Saami Herding Areas

FIGURE 2b

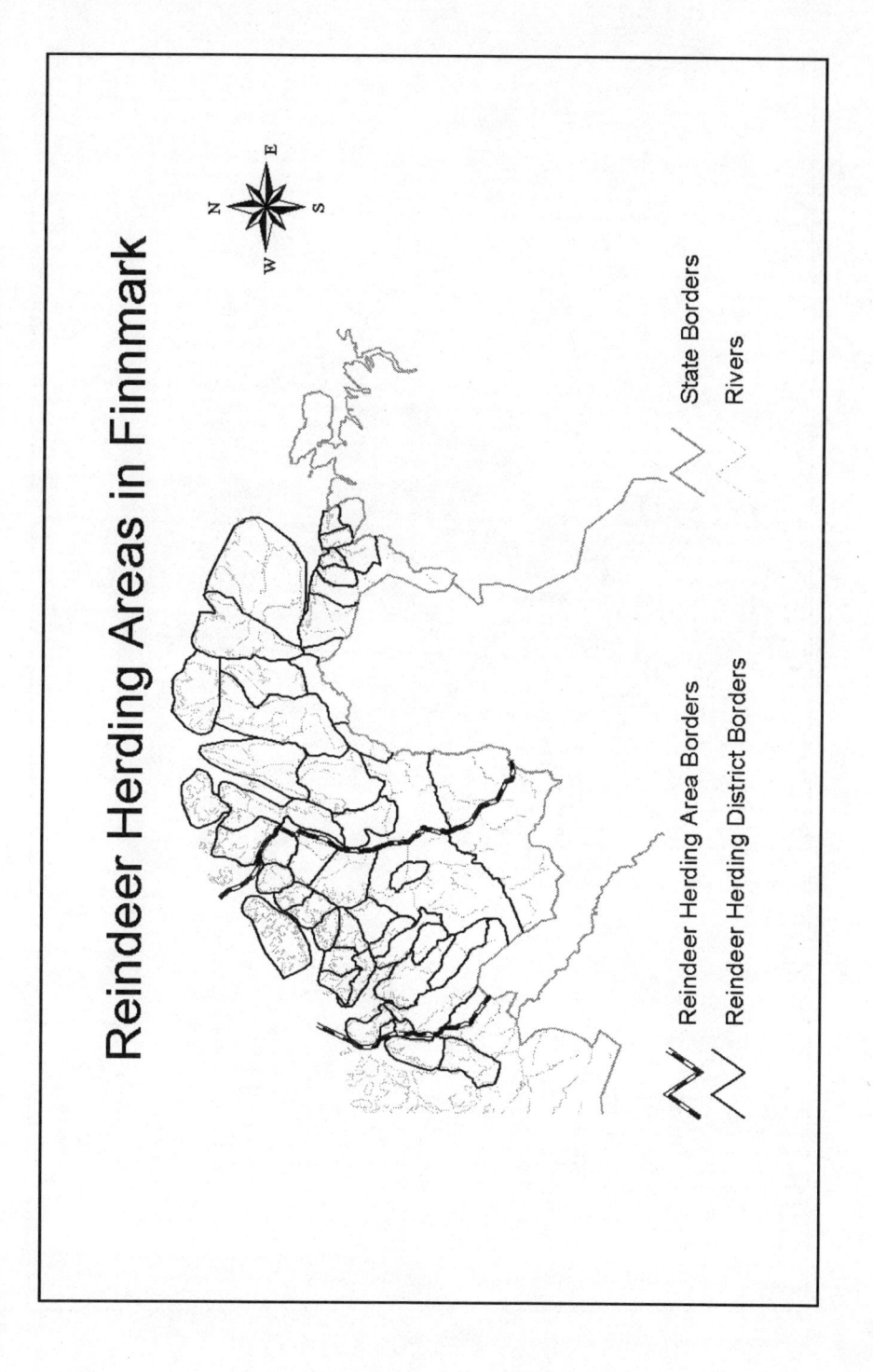

FIGURE 2c

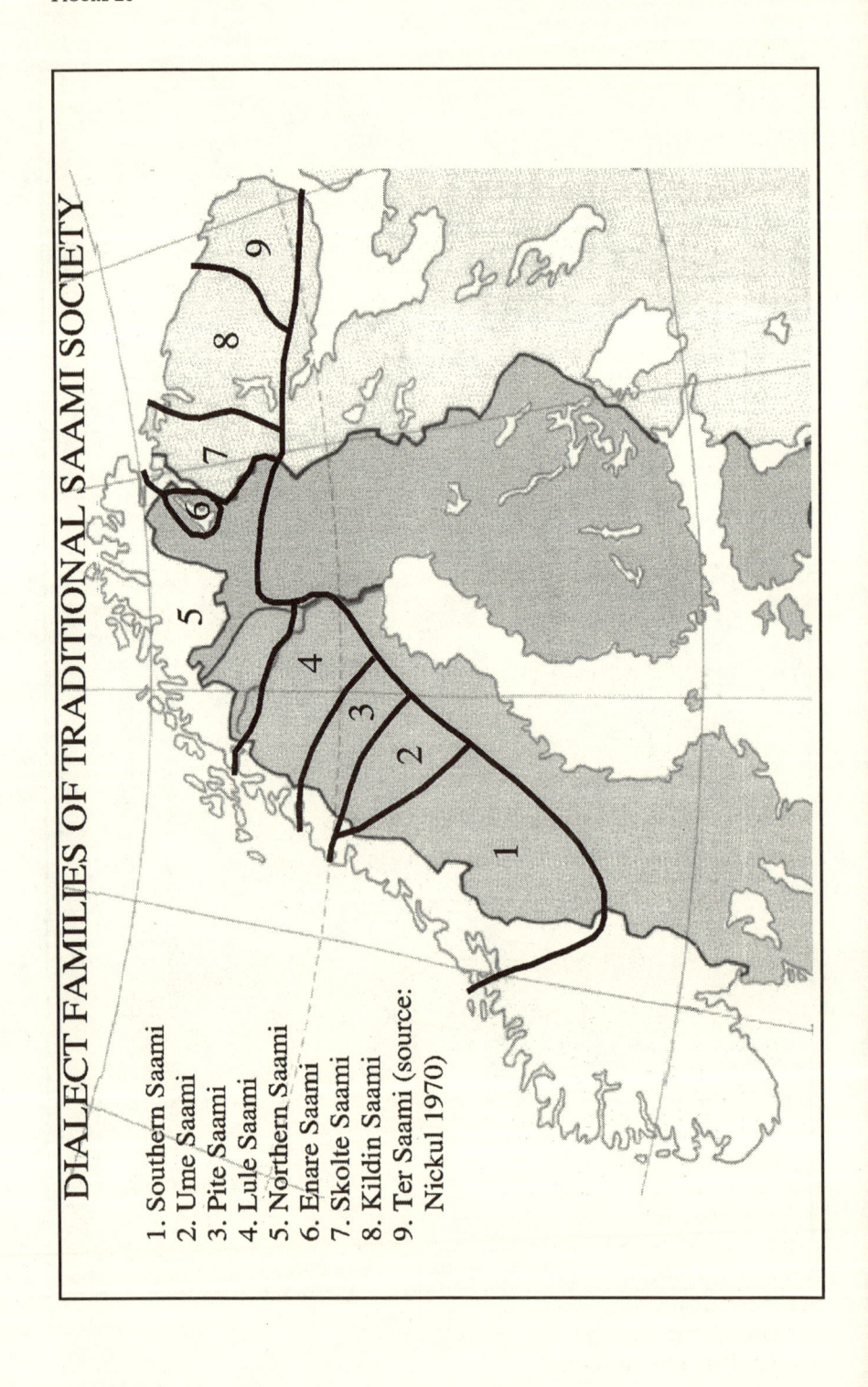

DIALECT FAMILIES OF TRADITIONAL SAAMI SOCIETY

1. Southern Saami
2. Ume Saami
3. Pite Saami
4. Lule Saami
5. Northern Saami
6. Enare Saami
7. Skolte Saami
8. Kildin Saami
9. Ter Saami (source: Nickul 1970)

The intent here is to explore the conditions which will make it possible for reindeer herders jointly to regulate the allocation of the critical resources of the range lands, and the conditions which will make it possible for all owners of fishing vessels to catch their fair share of the fish harvest with a minimum of effort and without endangering the survival of the ecosystem.

A Comparative Approach

While we know a fair amount about what happens to the ecosystems and why, we know considerably less about which conditions make people behave in a way where resources are used sustainably and even less about how to come from the present conditions to another set of conditions. To disentangle the various factors affecting resource utilization, a comparative approach has been chosen.

While the resource systems of Finnmark and the Barents Sea are similar in important ways, the social systems involved in the management of the resources are very different. Of particular importance for the comparison of the situations is that in Finnmark the resource users are an ethnic group of aboriginal status, and in the Barents sea there are international considerations both in relation to the law of the sea, the status of the Svalbard territory and the signatories of the Svalbard treaty, and in relation to the geo-political and industrial interests of Russia.

To increase the scope of comparisons one should look to other countries for contrasting cases on range land management where ethnicity is not a salient issue and for cases of management of fishing rights in a less complex international setting. But productive comparisons require a standardized theoretical language to describe the various cases. We are beginning to find this theoretical language in the rapidly developing field of theories of property rights regimes.

Theories of Property Rights Regimes

So far there is no one theory of property rights regimes. Relying on Eggertsson's (1990) study, *Economic Behavior and Institutions*, we can speak of a "naive" theory of property rights which assumes that property rights will be defined and enforced in a way that will maximize the aggregate wealth of a society. The naive theory may better be thought of as a prescription for how property rights ought to be defined and enforced by the omnipotent and totally good state, rather than as a description of how reality looks. It is not hard to find evidence disproving it as description.

One strain of theory trying to improve on the naive theory has been called the "interest group theory of property rights." (It could also have been called "political economy.") According to this theory, political clout of occupational organizations or classes will determine changes in the legal system at the margin and thus cumulatively strengthen particular groups at the expense of others and without regard to the overall efficiency of the economy.

Another development of the naive theory emphasizes the nature of transaction costs and how these shape the activity of the state in relation to property rights. The bottom line of all the theories, though, is that a property rights regime determines who can *legitimately* claim the benefits from some suitably defined resource. A property rights regime is a real world system of action which affects the distribution of the various goods defined by the society as worthy of attention.

The most significant word for a property rights regime is "legitimacy." The degree and source of legitimacy determine the kind of protection given by state and society to any particular holder of a right. The next most important phrase here is "goods worthy of attention." Property rights are not defined for everything. The more valuable something is considered to be within a culture, the more precise the property rights will be and the more elaborate the protection of them will be. A property rights regime defines and distributes a system of cultural values.

Property Rights Regimes and Culture

One might imagine that a property rights regime was determined by the characteristics of the resource it is supposed to govern. But so far, evidence indicates that cultural factors take precedence over natural factors (Godelier 1984). The role of the actual characteristics of a resource is more in the way of limiting the variation of regimes. Given such and such characteristics there are some constellations of rights and duties that will not work or will work only very poorly. At the very least the resource managers need to perceive some characteristic of a resouce before it can affect their management. This points to the primacy of language and culture in shaping the available information for managers of resources. For the members of a local community, the culture will be like a filter, selecting some aspects of resources as significant and disregarding others.

The indeterminacy of this cultural filter is also the prerequisite for the development of a unique culture. By choosing one constellation of rights and duties, one set of values will be promoted instead of another. This means, for example, that if a dominant culture—in the name of the "supreme value" of economic efficiency—dictates the choice of rights and duties, this will promote commercialization to the detriment of others' values. The ways and means of choosing can be debated. The element of power involved in multicultural settings is more easily seen. Thus the norms and rules determining the distribution of benefits can either be upheld by consensus, or by power, or by some combination.

The values of a culture as expressed in a property rights regime can be protected in a variety of ways. If something is considerd to be of great value or to be important for the daily effort to secure a decent living standard, it may be protected either through the norms and regulations promulgated directly by people in their everyday encounters, or it may be protected through acts and regulations enacted by a state on behalf of the society and promulgated by a police and court system. But just like people in their everyday encounters have to recall and interpret the rules they want to apply, statutory law has to be considered relevant and interpreted by the police and the court system. Among lawyers and law enforcement officers there will develop systems of perception of wrong doing and wrong doers as well as norms about appropriate interpretations and suitable reactions. The legal subculture is an imporant part of the forces shaping an actual property rights regime.

A property rights regime can thus be said to be affected by at least three types of cultural factors:

1. The characteristics of a resource perceived as important for the question of who can legitimately appropriate which of the various benefits yielded by the resource
2. The beliefs and norms among the actual resource appropriators about who can legitimately appropriate which of the various benefits yielded by the resource
3. Norms about justice and equity among legal authorities as expressed in acts and the interpretation of law concerning who can legitimately appropriate which of the various benefits yielded by the resource

The framework for analyzing common property institutions proposed by Oakerson (1992) is focused on interaction and decision making. But before interaction processes can start, the actors need information and they must agree on preferences. One can interpret the impact of culture as providing them with filters through which relevant data and preferences are provided.

Characteristics of Resources

The variety of characteristics of a resource may confuse us. Not many of them are of interest to the management regime. The significant aspects of a resource are its perceived qualities in relation to the goal it is assumed to contribute to fulfill. For analytical puposes, concepts such as divisibility, appropriation, sensitivity, and resilience have proven helpful.

FIGURE 3
Cultural Filters in the Oakerson (1992) Framework for Analyzing the Commons

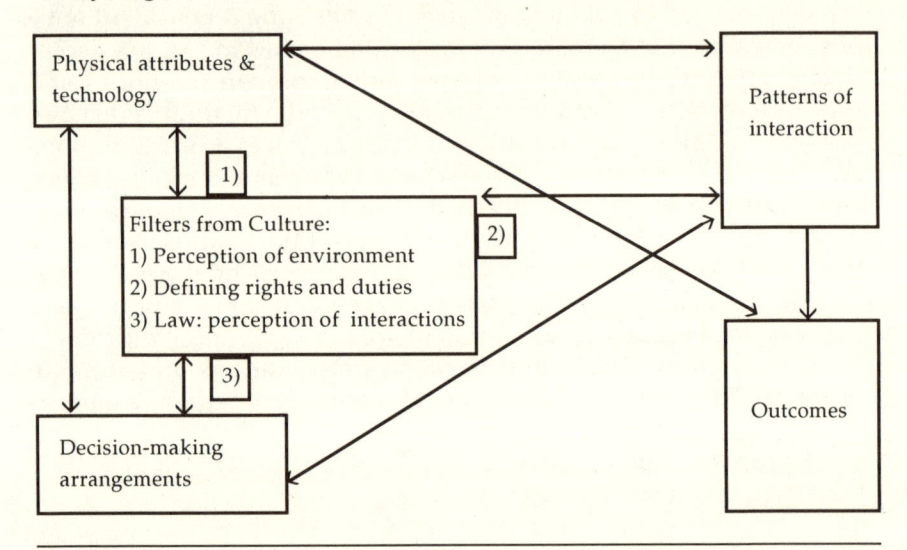

```
┌─────────────────────┐ ←─────────────────────────→ ┌─────────────┐
│ Physical attributes │                             │             │
│ & technology        │                             │ Patterns of │
│                     │                             │ interaction │
└─────────────────────┘                             │             │
   ↑   ↑   ↕ 1)                                     │             │
       ┌────────────────────────────┐  ←──────────→ │             │
       │ Filters from Culture:      │ 2)            └─────────────┘
       │ 1) Perception of environment│                     ↕
       │ 2) Defining rights and duties│
       │ 3) Law: perception of interactions│          ┌─────────────┐
       └────────────────────────────┘                 │             │
   ↕        ↕ 3)                                      │  Outcomes   │
┌─────────────────────┐                               │             │
│ Decision-making     │ ←─────────────────────────    │             │
│ arrangements        │                               └─────────────┘
└─────────────────────┘
```

Divisibility and Production

One important aspect for our concerns here is the degree of divisibility in time and/or space. Some talk about this as subtractability. The distributional problems of a society are very different for divisible goods compared to goods with important indivisibilities. If there are noticable indivisibilities in the utilization of a resource, one has to look for other ways of managing the resource than granting individual physical shares to each user if the distributional problem is to be solved.

Divisibility in consumption is, however, not the same as divisibility in production. Also the divisibility of the productive ecology needs to be considered. The range lands of many pastoral societies will typically be indivisible in important ways. Optimal use of the range land will usually imply access to seasonal pastures as well as transport corridors between the various pastures. And if there is a stochastic component (for example, as in where the rain falls), the pastures must be large enough to exploit this component. The possibility for dividing the pasture equitably may not exist. For the fish resource similar considerations exist. In a multispecies ecosystem such as the Barents Sea and northeast Atlantic, single species or parts of the ocean cannot be managed separately from the rest. Ecologically determined indivisibilities must be reflected in the property rights system to ensure sustainable resource management.

While the existence of indivisibilities is important, one should also note the many ways it is possible to divide resources like land. In most traditional[4] property rights systems there are different rules for regulating access to different types of resources (for example, arable land, trees, water, or pastoral land). There also are legal systems with different rules for fee simple rights, usufruct rights, management rights, and the rights of *cestui que trust*. In some sense it seems appropriate to talk of resource-specific property rights regimes.

Appropriation and Consumption

Other important characteristics of a resource are those who are perceived by the members of a society to affect the procedures of appropriation (for example, maximum sustainable yield or externalities in consumption or appropriation). The process of appropriation and/or consumption of a resource may create externalities (various types of crowding or queuing phenomena), the nature of which needs to be taken into consideration in the property rights regime in order to maintain the stream of benefits. For renewable resources there exist upper bounds on the volume of extraction from the resource which must be observed if the resource shall maintain its ability to provide benefits in the future. These bounds are determined by two qualities—sensitivity and resilience.

Sensitivity and Resilience

Sensitivity is the ability of an ecosystem to resist natural forces of degradation following some human interference. The resilience of an ecosystem is its ability to restore its productive capability after human interference. The volume and method of extraction from a resource must be tailored to the system's sensitivity and resilience.

Social Change and Problems of Property Rights

Both population growth and technological change will affect resource mangement (Jodha 1989).[5] From the point of view of property rights regimes, changes are likely to arise in relation to exclusion from and inclusion in the group of appropriators, inheritance of rights, long-term interests in use, transfer of rights, decisions on the joint use of resources, and new usage of a resource.

The Problem of Exclusion and Inclusion

The question of inclusion or exclusion regarding the group of people allowed access to a resource is fundamental. How is membership in the group acquired and how is it maintained? A particular instance of this problem is inheritance.

Inheritance of Rights. How can a resource user ensure that his or her heirs will be able to enjoy the same quantity and quality of a resource? This problem gets more complex in situations with common property compared to individual property. How should the interests of co-owners and descendants be accommodated? The problem has led to the distinction between ownership in common (descendants inherit) and joint ownership (co-owners inherit). The role of inheritance can be tied to the problem of how to secure long-term interests in the resource utilization.

Long-term interests in the utilization. A property rights regime powerfully shapes the time horizon of the actors utilizing the resource. The security of tenure (of any kind of rights) and how it is protected forms the possibility for long-term investment in a resource.

Transfer of rights. The problems of membership in a group of appropriators are closely related to problems of transferring rights, privileges, powers, and immunities (partly or totally) among group members for periods of time (or forever). If transfer is possible, the question is what kinds of restrictions are put on the transaction.

Decision rules for resource utilization. Decisions on joint use of a resource require meta-rules about how to decide on joint use. The existence or not of procedures for establishing or changing the meta-rules is an important aspect of a property rights regime.

New usage of a resource and the role of the remainder. An interesting aspect of all property rights regimes may be described as the problem of the remainder. If different actors control different resources within an ecosystem and their positively described rights are recognized, who controls the remainder (that which is left when everything positively described is accounted for)? The owner of the remainder will be the one to profit from new opportunities as they arise in relation to the resource.

Problems like these, as shaped and interpreted within a culture in development, have been a driving force in the development of the law of property in the Western world (Berman 1983).

Law and Legal Culture

Within the legal infrastructure and machinery of enforcement of the state, one finds norms about justice and equity expressed in law and interpretations of law concerning who can legitimately appropriate which benefit from the resource. The legal regulation is expressed in two ways. It is expressed in the form of acts, and it is expressed in the judgements by the courts of law where the interpretation of acts and traditions establish a legal canon. The legal subculture has, of course, links to the common culture of the people, but these may at times appear weak (Watson 1991, 221-244).

Enforcement of Rights

The role of public opinion and the use of cultural means of enforcing rights (various ways of applying informal sanctions) are important aspects of a property rights regime. But such means are never sufficient in complex societies (and one may doubt if they ever were, even in tribal societies). One problem for holders of claim rights, privileges, powers, and immunities is to defend their rights. Property rights are legitimate if public opinion says so and if some social power—the state or some other central or local institution—recognizes the rights holder and is prepared to enforce his rights. An important part of a property rights regime is the remedies granted rights holders who feel they have been wronged.

One important distinction in the legal culture is the division of public and private interests. Does the violation of a regulation affect only the private interests of a citizen, or does it also affect the public interest? In those cases where it is a violation of public interests, the legal culture of property rights will encompass the system of enforcement since the resources and traditions of this system determine which violations will be investigated and brought to court.

Describing Property Rights Regimes

As a baseline for comparing property rights regimes, each regime must be given a precise and standard description.

A Precise Description of the Property Relation

For the group of actors (persons or groups of persons) allowed access to a resource, the following points ought to be considered (see Hohfeld 1913, 1917):

1. What specific claim rights does membership in such a group entail, and how are they exercised and defended against nonmembers?
2. Which privileges does a claim rights holder enjoy regarding the resource? Under which specific conditions can they be enjoyed, and what happens to anyone trying to interfere with the enjoyment?

3. Which powers (to create new types of property relations regarding the resource) does a claim rights holder have? What are the liabilities of the nonmembers?
4. Which immunities will a claim rights holder have (legitimate, customary, and/or legal protection) regarding someone trying to usurp his powers, and how are they protected?

A Precise Description of Decision Rules

For the rights defining a property relation, one needs to know if a rule can be described as a convention among the local population or if its origin is some legitimate decision of a recognized system-responsible actor. For any system-responsible actor one needs to know the rules governing the decision on property rights rules.

Types of Property Rights Regimes

A property right is a *legitimate* rule of appropriation for a well-defined stream of benefits from some recognized resource.[6] This suggests that it may be interesting to distinguish between different streams of benefits from the same physical resource. Property rights regimes are usually divided into state, common, and private property rights—regimes, sometimes with the absence of property rights—the open-access regime—added on as a fourth type. I think one ought to be more specific than this and talk about the property rights regime for a specific stream of benefits from a resource. The resource-specific property rights regime consists of all the rules and procedures which determine who can legitimately appropriate any particular stream of benefits from a resource.

The major types of regimes seem to be determined according to the number of appropriators, as well as those who may legitimately claim an interest in the distribution of a particular stream of benefits from the resource. The relevant distinction according to the number of appropriators seems to be size—whether the actor is a single individual, a recognized group, or all members of a society. The term "group resource" signifies any resource where more than one independent decision maker, but not all members of a society, can claim legitimate rights to appropriate (part of) the particular stream of benefits from the resource.

One may also distinguish between private and nonprivate resources according to who may legitimately claim an interest in the distribution of any benefit from the resource. A nonprivate resource is characterized by a stream of benefits where both those who have rights of appropriation and other members of the society have legitimate interests in decisions on the distribution of benefits. No

TABLE 1
Types of Property Rights Regimes

	Private Regimes	Nonprivate Regimes
•Legitimate unit for appropriation is	•Legitimate interests in decisions on use are mainly with *appropriators* because of few externalities.	•Legitimate interests in decisions on use are mainly with *society* because of many externalities.
Individual (the legal person)	Ordinary	State
A group of individuals (contractually defined)	Common	Joint
All members of society (symbolically represented by a monarch or government)	Sovereign	Public

one except the appropriators can claim legitimate interest in the distribution of the stream of benefits for a private resource.

It seems that the degree of public interest in the distribution of benefits varies with the nature and extent of externalities created either by the process of appropriation or by the process of consumption. If such externalities are perceived to be few or of little importance, the legitimate interests in the utilization are mainly private.

Concluding Remarks

Both Finnmarksvidda and the Barents Sea belong to a class of resources which in most countries today is considered as some kind of common resource. Exactly what this entails varies enormously from situations where the resource for all practical purposes is no one's property, to situations where the resource is managed by a corporation as if it is ordinary private property.

Patterns of resource use tested by history and guarded by tradition will usually be sustainable. Today it is recognized that circumstances (population, technology, organizations, legal codes, cultural procedures, etc.) may be changing so rapidly that the sustainability of the prevalent pattern of utilization—whether traditionally enforced or enforced by a state—is an open question. Pressures to change unsustainble resource utilization will be mounting as soon as enough people perceive the problem. But there is much to be done between perceiving the problem and gaining knowledge about how to solve it, and the path to action often takes problematic twists. Still, the problem of sustainable resource use is one of the big questions for world society. To divide it into problems of early detection, knowledge about consequences of policy interventions, and strategies for implemetation of changes may make it slightly less formidable. In the present context we shall be concerned with causes and consequences of various patterns of resource usage.

According to the concepts used here, one way to change a pattern of utilization is to change the structure of property rights to the resources. But before one can start the task of designing modifications to a property rights regime, the existing system of rights—both those recognized in a legal code and enforced by the state and those recognized in a culture and enforced by traditional means—must be known in detail. The problem we want to confront is thus to understand how the various parts of a system act in concert to produce the observed sustainable or unsustainable pattern of utilization to see if the regimes of utilization can be changed in a direction approaching a more sustainable pattern of utilization.

NOTES

1. For the absence of any property rights regime, see Bromley (1989).
2. See, for example, Baland and Platteau (1996); Bromley, ed. (1992); Eggertsson (1990); Libecap (1989); McCay and Acheson, eds. (1987); Ostrom (1990); Ostrom et al. (1994).
3. The right to hold reindeer within the publicly defined reindeer herding areas of Norway is restricted to Norwegian citizens of the Saami people, and since July 1, 1979, it also depends on either being active as a reindeer herder on that date or having proof that at least the father or mother or one grandparent of the person was an active reindeer herder. The right is independent of the ownership of ground within the ten reindeer herding districts defined by an act from 1894 (see figure 2a).
4. "Traditional" used here to characterize cultures, property rights systems, or other institutions, will mainly refer to the degree of presence of a state or a civil service concerned with the public interest. The absence of a civil service means that ordered society rests on conventions, norms, and

sanctions which members of the society are able to agree upon and enforce as part of their everyday activities. It usually implies small-scale societies.

5. The interrelations of population, resources, technology, and social change is a huge topic; see for example Stinchcombe (1983); Hawley (1986); Teitelbaum and Winter, eds. (1988); McNicoll and Cain, eds. (1989); Boserup (1990); and Jansson, Hammer, Folke, and Costanza, eds. (1994).

6. By a natural or environmental resource we shall mean any physically bounded and identifiable entity recognized as a resource by some legitimate social actor.

REFERENCES

Baland, J. M. and J. P. Platteau. 1996. *Halting Degradation of Natural Resources: Is There a Role for Rural Communities?* Oxford: FAO/Clarendon Press.
Berman, Harold J. 1983. *Law and Revolution.* Cambridge: Harvard University Press.
Boserup, Esther. 1990. *Economic and Demographic Relationships in Development.* Baltimore: The Johns Hopkins Press.
Bromley, Daniel W. 1989. *Economic Interests and Institutions: The Conceptual Foundations of Public Policy.* Oxford: Basil Blackwell.
Bromley, Daniel W. 1992a. "The Commons, Property, and Common-Property Regimes," 3-16, in *Making the Commons Work,* edited by D. Bromley.
Bromley, Daniel W. ed. 1992b. *Making the Commons Work.* San Francisco: ICS Press.
Eggertsson, Thráinn. 1990. *Economic Behaviour and Institutions.* Cambridge: Cambridge University Press.
Godelier, Maurice. 1984. Territory and Property in Some Pre-Capitalist Societies, 27-70, in *The Mental and the Material,* edited by M. Godelier. 1986. London: Verso.
Hardin, Garret. 1968. "The Tragedy of the Commons." *Science,* 162. 1243-1248.
Hawley, Amos. 1986. *Human Ecology.* Chicago: The University of Chicago Press.
Hohfeld, W. N. 1913. "Some Fundamental Legal Conceptions as Applied in Judicial Reasoning." *Yale Law Journal* 23. 16-59.
Hohfeld, W. N. 1917. "Fundamental Legal Conceptions as Applied in Judicial Reasoning." *Yale Law Journal* 26. 710-770.
Jansson, A., M. Hammer, C. Folke, and R. Costanza, eds. 1994. *Investing in Natural Capital.* Washington D.C.: Island Press.
Jodha, N. S. 1989. "Depletion of Common Property Resources in India: Micro-Level Evidence," 261-283, in "Rural Development and Population: Institutions and Policy," edited by G. McNicoll and M. Cain. *Population and Development Review* 15. Supplement.

Libecap, Gary D. 1989. *Contracting for Property Rights*. Cambridge: Cambridge University Press.
McCay, Bonnie J. and James M. Acheson, eds. 1987. *The Question of the Commons: The Culture and Ecology of Communal Resources*. Tucson: University of Arizona Press.
McNicoll, Geoffrey and Mead Cain, eds. 1989. "Rural Development and Population: Institutions and Policy," *Population and Development Review* 15. Supplement.
Nickul, Karl. 1977. *The Lappish Nation: Citizen of Four Countries*. Bloomington: Indiana University.
Ostrom, Elinor. 1990. *Governing the Commons: The Evolution of Institutions for Collective Action*. Cambridge: Cambridge University Press.
Ostrom, Elinor, Roy Gardner, and James Walker. 1994. *Rules, Games, and Common-Pool Resources*. Ann Arbor: The University of Michigan Press.
Oakerson, Ronald J. 1992. "Analyzing the Commons: A Framework," 41-59, in *Making the Commons Work*, edited by D. Bromley.
Stinchcombe, Arthur L. 1983. *Economic Sociology*. Orlando: Academic Press.
Teitelbaum, Michael S. and Jay M. Winter, eds. 1988. "Population and Resources in Western Intellectual Traditions." *Population and Development Review* 14. Supplement.
Watson, Alan. 1991. *Roman Law and Comparative Law*. Athens: The University of Georgia Press.

PART ONE

Theory from Law and Social Science

Introduction

A theory about the governance of renewable resources must necessarily cross several disciplines. The four chapters in this section are written by a political scientist, two economists, and a lawyer. The papers span neo-classical economic theory by way of empirically grounded conclusions from comparative political theory to human rights.

The first chapter by Elinor Ostrom underlines the intimate connection between law and social science. The analysis summarizes the state of the art on what characterizes long enduring self-governed common property institutions, emphasizing the importance of rule making and the different implications of rule making at various levels. At the end of the paper she discusses the possible processes which may lead to the breakdown and failure of well-designed institutions after centuries of successful operation.

Thráinn Eggertsson, in chapter 2, presents a short introduction to economic theory relevant for the choice of management regime to a resource "when several independent producers jointly draw inputs from a natural resource which they share and to which they hold exclusive rights." If wealth maximization is the goal, it is argued that under certain specified circumstances, common or joint property rights systems may be the form of property rights which provides the best advantage for minimizing the aggregate cost of production, governance, and exclusion. And it is suggested that the efficiency of common property rights depends not only on economic factors, but also on the nature of political and social institutions.

Gary Libecap discusses the incentives for changing existing property rights regimes and the problems encountered in the modifications of them in chapter 3. In particular, considerations of distributional justice among the resource appropriators and the incentives among politicians, bureaucrats, and new resource appropriators are highlighted. It is concluded that as traditional appropriators turn to outside politicians and administrative agencies to address resource use problems, new objectives and interests are added. The broader array of competing political objectives for resource assignment and use may not lead to policies that advance the interests of traditional users or significantly protect the resource. Accordingly, caution is necessary in calling for public policy intervention, and once a path of regulatory change is taken, the distributional concerns of the various parties involved must be considered if collective action is to be successful in safeguarding the resource and the traditional societies that depend upon it.

Finally, in chapter 4 by Hans Chr. Bugge we turn to a little addressed topic: the interrelations of human rights and access to resources. The study gives an overview of the various aspects of human rights in relation to resource management in international law. The conclusion is that so far human rights have had a rather limited role in law on resource management. But this seems to be changing as environmental degradation and destruction of natural resources threaten the lives and well-being of millions of people, making the fulfillment of many basic human rights more and more difficult. The problems may become more severe for the next generation. Scarce resources may become one of the most important causes for social unrest, conflicts and wars, social injustice, and the suppression of human rights. Therefore, governance of natural resources and human rights issues should be added to the international discussion in this field to a much greater extent.

ELINOR OSTROM   CHAPTER 1

Institutional Analysis, Design Principles, and Threats to Sustainable Community Governance and Management of Commons

The Institutional Analysis and Development Framework

The Institutional Analysis and Development (IAD) framework is an evolving method for identifying and analyzing how attributes of a physical world interact with those of the general cultural setting and the specific rules-in-use to affect the incentives facing individuals in particular situations and the likely outcomes to result (Kiser and Ostrom 1982; Oakerson 1992; E. Ostrom 1986; V. Ostrom 1991; V. Ostrom, Feeny, and Picht 1993; Thomson 1992).[1] The IAD framework links the characteristics of a physical world (such as forests) with those of the general cultural setting (the villages and harvesters that use forests), the specific rules that affect the incentives individuals face in particular situations (how forest products can be harvested, utilized, and maintained), the outcomes of these interactions (regeneration or deforestation), and the evaluative criteria applied to these patterns and outcomes (efficiency, equity, sustainability). Common pool resources (CPRs) share two characteristics of a physical world: (1) it is costly to develop institutions to exclude potential beneficiaries from them and (2) the resource units harvested by one individual are not available to others. Recent research projects have applied this framework to develop a database on common pool resources (particularly irrigation systems and inshore fisheries) located in different regions of the world (Tang 1991, 1992; Schlager 1990; Schlager and Ostrom 1992, 1993; E. Ostrom, Benjamin, and Shivakoti 1992). After somewhat more than a year's developmental work, we have now designed a new database to record information about forest resources and institutions in many different countries (E. Ostrom et al. 1993).

Analysis of human actions and consequences frequently starts with a focal arena as shown in Figure 1. Examples include action situations where individuals decide when and how much to harvest of forest products from different locations, whether to establish a forest users association, or whether to fence off a particular part of a forest to prevent animals from foraging within. What arena is analyzed depends on the questions of interest to the analyst. The analyst wanting to examine recurrent structures of situations must, however, find ways of separating one situation from another for the purpose of analysis. Further, individuals who participate in many situations must also know the difference among them. The actions that can be taken

FIGURE 1
A Framework for Institutional Analysis

Focal Area

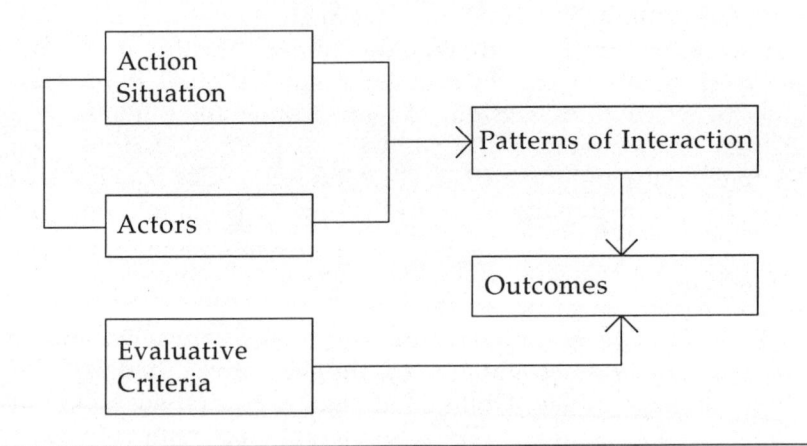

```
    ┌──────────────┐
 ┌──│   Action     │──┐
 │  │  Situation   │  │
 │  └──────────────┘  │
 │                    └──▶ Patterns of Interaction
 │                                    │
 │  ┌──────────────┐                  │
 └──│   Actors     │──┐               ▼
    └──────────────┘  │          ┌──────────┐
                      │          │ Outcomes │
                      │          └──────────┘
    ┌──────────────┐  │                ▲
    │  Evaluative  │──┘                │
    │   Criteria   │───────────────────┘
    └──────────────┘
```

in harvesting timber are different from those that can be taken in harvesting thatch or those that are involved in selling either timber or thatch. An individual who is repeatedly mixed up about what situation he or she is in, is not normally considered as competent to take independent actions.

What is distinctive about the IAD framework, when contrasted to frameworks that are closely tied to a single scientific discipline, is that all action situations are viewed as being composed of the same set of elements. Thus, while harvesting or marketing timber or thatch differ in many important ways, these diverse situations can all be described by identifying and analyzing how particular elements constituting the situations under analysis lead to the patterns observed. These elements include identifying:

- Who are the participants?
- What are the positions they hold?
- What actions can they take?
- What information do they possess?
- What outcomes can occur?
- How are actions and outcomes linked?
- What benefits and costs are assigned to actions and outcomes?

These elements are themselves relatively complex. Many different action situations can be constructed from them. At the same time that the IAD framework stresses a universality of working parts, it enables analysts to examine unique combinations of these parts. The array of potential outcomes that can be analyzed and evaluative criteria, such as equity, efficiency, sustainability, and adaptability, is also very large. Further, these elements are themselves constituted by a deeper layer of attributes about a physical and material setting, the community within which a situation occurs, and the specific rules-in-use that affect the structure of the situation.

Action situations are perceived to be nested within at least three relevant tiers of action (Kiser and Ostrom 1982) (see Figure 2). *Operational rules* directly affect day-to-day decisions made by the participants in any setting. *Operational-level actions* occur whenever individuals directly affect variables in the world by doing such things as harvesting products, worshipping at a forest shrine, planting seeds, building fences, patrolling the borders of a forest, or feeding leaves to their animals. *Collective-choice rules* affect operational activities and results through their effects in determining who is eligible and the specific rules to be used in changing operational rules. *Collective-choice actions* occur whenever individuals decide about operational activities.

Thus, the actions taken at an annual meeting of a forest users association to keep a forest closed for the harvesting of a particular product except for a specified time is a collective-choice action. *Constitutional-choice rules* affect operational activities and their effects in determining who is eligible and the rules to be used in crafting the set of collective-choice rules that in turn affect the set of operational rules. *Constitutional-choice actions* occur whenever individuals decide about how collective-choice actions will be made. Consequently, the decision of a forest users association to create an executive committee that will meet once a month to make decisions about joint activities to be undertaken is a constitutional-choice action. Constitutional choices are frequently made without recognition that they are indeed creating a future structure to make future rules about an operational level.

At each level of analysis there may be one or more arenas in which the types of decisions made at that level will occur. The elements of an action situation and of an actor are used to construct these arenas at all three levels. The concept of an "arena" does not imply a formal setting, but can include such formal settings as legislatures, governmental bureaucracies, and courts. Policy making regarding the

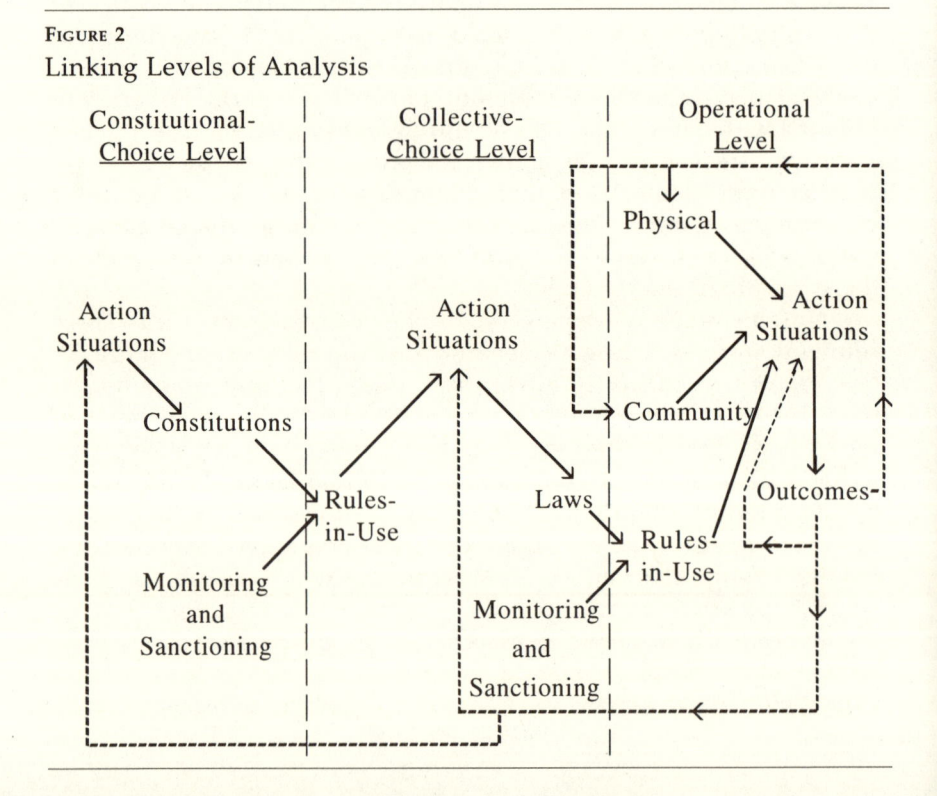

FIGURE 2
Linking Levels of Analysis

rules that will be used to regulate operational-level action situations is usually carried out in one or more collective-choice arenas as well as being enforced at an operational level. Dilemmas are not limited to an operational level of analysis. They frequently occur at the collective-choice and constitutional levels of analysis.

Design Principles of Sustainable Community-Governed Commons

The IAD framework has been an underlying foundation for all of our empirical studies of common-pool resources and common-property regimes. One line of inquiry that we have pursued over time is the study of long-lasting resource systems that are self-governed by the users. Many of these systems have been studied in depth by perceptive scholars such as Robert Netting, Thráinn Eggertsson, Gary Libecap, Daniel Bromley, Margaret McKean, Fikret Berkes, and David Feeny. The resources involved vary from irrigation systems to mountain grazing lands and both inshore and ocean fisheries. The most notable similarity among these systems is the sheer perseverance of these resource systems and institutions. The institutions can be considered robust in that the rules have been devised and modified over time according to a set of collective-choice and constitutional-choice rules (Shepsle 1989). In other words, these systems have been sustainable over very long periods of time. Most of the environments studied are complex, uncertain, and interdependent environments where individuals continuously face substantial incentives to behave opportunistically. The puzzle that was addressed in *Governing the Commons* (Ostrom 1990) is how did the individuals using these systems sustain them over such long periods of time.

The specific rules-in-use differ markedly from one case to the next.[2] Given the great variation in specific rules-in-use, the sustainability of these resources and their institutions cannot be explained by the presence or absence of particular rules. Part of the explanation that can be offered for the sustainability of these systems is based on the fact that the particular rules do differ. By differing, the particular rules take into account specific attributes of the related physical systems, cultural views of the world, and the economic and political relationships that exist in the setting. Without different rules, appropriators could not take advantage of the positive features of a local CPR or avoid potential pitfalls that could occur in one setting but not others.

A set of seven design principles appears to characterize most of the robust CPR institutions. An eighth principle characterizes the larger, more complex cases. A "design principle" is defined as a conception used either consciously or unconsciously by those constituting and reconstituting a continuing association of individuals about a general organizing principle. Let us discuss each of these design principles.[3]

Clearly Defined Boundaries

Individuals or households with rights to withdraw resource units from the CPR and the boundaries of the CPR itself are clearly defined.

Defining the boundaries of the CPR and of those authorized to use it can be thought of as a "first step" in organizing for collective action. So long as the boundaries of the resource and/or the individuals who can use the resource remain uncertain, no one knows what they are managing or for whom. Without defining the boundaries of the CPR and closing it to "outsiders," local appropriators face the risk that any benefits they produce by their efforts will be reaped by others who do not contribute to these efforts. At the least, those who invest in the CPR may not receive as high a return as they expected. At the worst, the actions of others could destroy the resource itself. Thus, for any appropriators to have a minimal interest in coordinating patterns of appropriation and provision, some set of appropriators have to be able to exclude others from access and appropriation rights. If there are substantial numbers of potential appropriators and the demand for the resource units are high, the destructive potential of all users freely withdrawing from a CPR could push the discount rate used by appropriators toward 100 percent. The higher the discount rate, the closer the situation is to that of a one-shot dilemma where the dominant strategy of all participants is to overuse the CPR.

Congruence between Appropriation and Provision Rules and Local Conditions

Appropriation rules restricting time, place, technology, and/or quantity of resource units are related to local conditions and to provision rules requiring labor, materials, and/or money.

Unless the number of individuals authorized to use a CPR is so small that their harvesting patterns do not adversely affect one another, at least some rules related to how much, when, and how different products can be harvested are usually designed by those using the resource. Well-tailored appropriation and provision rules help to account for the perseverance of the CPRs themselves. Uniform rules established for an entire nation or large region of a nation rarely can take into account the specific attributes of a resource that are used in designing rules-in-use in a particular location. In long-surviving irrigation systems, for example, subtly different rules are used in each system for assessing water fees used to pay for water guards and for maintenance activities, but in all instances those who receive the highest proportion of the water also pay approximately the highest proportion of the fees. No single set of rules defined for all irrigation systems in a region would satisfy the particular problems in managing each of these broadly similar, but distinctly different, systems.

Collective-Choice Arrangements

Most individuals affected by operational rules can participate in modifying operational rules.

CPR institutions that use this principle are able to tailor better rules to local circumstances, since the individuals who directly interact with one another and with the physical world can modify the rules over time so as to better fit them to the specific characteristic of their setting. Appropriators who designed CPR institutions that are characterized by the first three principles—clearly defined boundaries, good-fitting rules, and appropriator participation in collective choice—should be able to devise a good set of rules if they keep the costs of changing rules relatively low.

The presence of good rules, however, does not account for appropriators following them. Nor is the fact that the appropriators themselves designed and initially agreed to the operational rules an adequate explanation for centuries of compliance by individuals who were not originally involved in the initial agreement. It is not even an adequate explanation for the continued commitment of those who were part of the initial agreement. Agreeing to follow rules *ex ante* is an easy "commitment" to make. Actually following rules *ex post*, when strong temptations are present, is the significant accomplishment.

The problem of gaining compliance to rules—no matter what their origin—is frequently assumed away by analysts positing all-knowing and all-powerful *external* authorities that enforce agreements. In many long-enduring CPRs, no external authority has sufficient presence to play any role in the day-to-day enforcement of the rules-in-use. Thus, external enforcement cannot be used to explain high levels of compliance. In all of the long-enduring cases, active investments in monitoring and sanctioning activities are very apparent. These lead us to consider the fourth and fifth design principles.

Monitoring

Monitors who actively audit CPR conditions and appropriator behavior are accountable to the appropriators and/or are the appropriators themselves.

Graduated Sanctions

Appropriators who violate operational rules are likely to receive graduated sanctions (depending on the seriousness and context of the offense) from other appropriators, from officials accountable to these appropriators, or from both.

In long-enduring institutions, monitoring and sanctioning are undertaken primarily by the participants themselves. The initial sanctions used in these systems are also surprisingly low. Even though it is frequently presumed that participants will not spend the time and effort to monitor and sanction each other's performance, substantial evidence has been presented that they do both in these settings.

To explain the investment in monitoring and sanctioning activities that occurs in these robust, self-governing, CPR institutions, the term "quasi-voluntary compliance" used by Margaret Levi (1988, chapter 3) is very useful. She uses the term "quasi-voluntary compliance" to describe taxpayer behavior in regimes where most everyone pays taxes. Paying taxes is *voluntary* in the sense that individuals *choose* to comply in many situations where they are not being directly coerced. On the other hand, it is "*quasi*-voluntary because the noncompliant are subject to coercion—if they are caught" (Levi 1988, 52). Levi stresses the *contingent* nature of a commitment to comply with rules that is possible in a repeated setting. Strategic actors are willing to comply with a set of rules, Levi argues, when:

1. They perceive that the collective objective is achieved; and
2. They perceive that others also comply.

In Levi's theory, enforcement is normally provided by an external ruler even though her theory does not preclude other enforcers. To explain commitment in many of the cases of sustainable community-governed CPRs, external enforcement is largely irrelevant. External enforcers may not travel to a remote village other than in extremely unusual circumstances. CPR appropriators create their own internal enforcement to (1) deter those who are tempted to break rules, and thereby (2) assure quasi-voluntary compliers that others also comply. The Chisasibi Cree, for example, have devised a complex set of entry and authority rules related to the coastal and islaurine fish stocks of James Bay as well as the beaver stock located in their defined hunting territory. Fikret Berkes (1987) describes why these resource systems and the rules used to regulate them have survived and prospered for so long: "Effective social mechanisms ensure adherence to rules which exist by virtue of mutual consent within the community. People who violate these rules suffer not only a loss of favour from the animals (important in the Cree ideology of hunting) but also social disgrace" (Berkes 1987, 87).

The costs of monitoring are kept relatively low in many long-enduring CPRs as a result of the rules-in-use. Rotation rules used in irrigation systems and in some inshore fisheries place the two actors most concerned with cheating in direct contact with one another. The irrigator who nears the end of a rotation turn would like to extend the time of his turn (and thus, the amount of water obtained). The next irrigator in the rotation system waits nearby for him to finish, and would even like to start early. The presence of the first irrigator deters the second from an early start, and the presence of the second irrigator deters the first from a late ending. Monitoring is a by-product of their own strong motivations to use their water rotation turn to the fullest extent. The fishing site rotation system used in Alanya (Berkes 1992) has the same characteristic that cheaters are observed at low cost by those who most want to deter another cheater at that particular time and location. Many of the ways that work-teams are organized in the Swiss and Japanese mountain commons also have the result that monitoring is a natural by-product of using the commons.

The costs and benefits of monitoring a set of rules are not independent of the particular set of rules adopted. Nor are they uniform in all CPR settings. When appropriators design at least some of their own rules, they can learn from experience to craft enforceable rather than unenforceable rules. This means paying attention to the costs of monitoring and enforcing as well as the benefits that those who monitor and enforce the rules obtain. A frequently unrecognized "private" benefit of monitoring in settings where information is costly is obtaining the information necessary to adopt a contingent strategy. If an appropriator who monitors finds someone who has violated a rule, the benefits of this discovery are shared by all using the CPR, as well as providing the discoverer a signal about compliance rates. If the monitor does *not* find a violator, it has previously been presumed that private costs are involved without any benefit to the individual or the group. If information is not freely available about compliance rates, then an individual who monitors obtains valuable information from monitoring.

By monitoring the behavior of others, the appropriator-monitor learns about the level of quasi-voluntary compliance in the CPR. If no one is discovered breaking rules, the appropriator-monitor learns that others comply and no one is being taken for a sucker. It is then safe for the appropriator-monitor to continue to follow a strategy of quasi-voluntary compliance. If the appropriator-monitor discovers rule infractions, it is possible to learn about the particular circumstances surrounding the infraction, to participate in deciding the appropriate level of sanctioning, and then to decide about continued compliance or not. If an appropriator-monitor finds an offender, who normally follows rules but happens to face a severe problem, the experience confirms what everyone already knows. There will always be times and places where those who are basically committed to following a set of rules succumb to strong temptations to break them.

A real threat to the continuance of quasi-voluntary compliance can occur, however, if an appropriator-monitor discovers individuals who break the rules repeatedly. If this occurs, one would expect the appropriator-monitor to escalate the sanctions imposed in an effort to halt future rule breaking by such offenders and any others who might start to follow suit. In any case, the appropriator-monitor has up-to-date information about compliance and sanctioning behavior on which to make future decisions about personal compliance.

Let us also look at the situation through the eyes of someone who breaks the rules and is discovered by a local guard (who will eventually tell everyone) or another appropriator (who also is likely to tell everyone). Being apprehended by a local monitor when the temptation to break the rules becomes too great has three results: (1) it stops the infraction from continuing and may return contraband harvest to others; (2) it conveys information to the offender that someone else in a similar situation is likely to be caught, thus increasing confidence in the level of quasi-voluntary compliance; and (3) a punishment in the form of a fine plus loss of reputation for reliability is imposed.

The fourth and fifth design principles—monitoring and graduated sanctions—thus take their place as part of the configuration of principles that work together to enable appropriators to constitute and reconstitute robust CPR institutions. Let me summarize my argument to this point. When CPR appropriators design their own operational rules (Design Principle 3) to be enforced by individuals who are local appropriators or accountable to them (Design Principle 4) using graduated sanctions (Design Principle 5) that define who has rights to withdraw from the CPR (Design Principle 1) and that effectively restrict appropriation activities given local conditions (Design Principle 2), the commitment and monitoring problem are solved in an interrelated manner. Individuals who think a set of rules will be effective in producing higher joint benefits and that monitoring (including their own) will protect them against being a sucker, are willing to make a contingent self-commitment of the following type: I commit myself to follow the set of rules we have devised in all instances except dire emergencies if the rest of those affected make a similar commitment and act accordingly. Once appropriators have made contingent self-commitments, they are then motivated to monitor other people's behavior, at least from time to time, in order to assure themselves that others are following the rules most of the time. Contingent self-commitments and mutual monitoring reinforce one another, especially in CPRs where rules tend to reduce monitoring costs.

Conflict-Resolution Mechanisms

Appropriators and their officials have rapid access to low-cost, local arenas to resolve conflict among appropriators or between appropriators and officials.

In field settings, applying rules always involves discretion and can frequently lead to conflict. Even such a simple rule as "Each irrigator must send one individual for one day to help clean the irrigation canals before the rainy season begins" can be interpreted quite differently by different individuals. Who is or is not an "individual" according to this rule? Does sending a child below ten or an adult above seventy to do heavy physical work meet this rule? Is working for four hours or six hours a "day" of work? Does cleaning the canal immediately next to one's own farm qualify for this community obligation? For individuals who are seeking ways to slide past or subvert rules, there are always ways that they can "interpret" the rule so that they can argue they meet it while subverting the intent. Even individuals who intend to follow the spirit of a rule can make errors. What happens if someone forgets about labor day and does not show? Or, what happens if the only able-bodied worker is sick, or unavoidably in another location?

If individuals are going to follow rules over a long period of time, some mechanism for discussing and resolving what is or is not a rule infraction is quite necessary to the continuance of rule conformance itself. If some individuals are allowed to free ride by sending less-valuable workers to a required labor day, others will consider themselves to be suckers if they send their strongest workers who could be used to produce private goods rather than communal benefits. Over time, only children and old people will be sent to do work that requires strong adults and the system breaks down. If individuals who make an honest mistake or face personal problems that prevent them from following a rule cannot find mechanisms to make up their lack of performance in an acceptable way, rules can be viewed as unfair and conformance rates decline.

While the presence of conflict-resolution mechanisms does not guarantee that appropriators are able to maintain enduring institutions, it is difficult to imagine how any complex system of rules could be maintained over time without such mechanisms. In the cases described above, these mechanisms are sometimes quite informal and those who are selected as leaders are also the basic resolvers of conflict.

Minimal Recognition of Rights to Organize

The rights of appropriators to devise their own institutions are not challenged by external governmental authorities.

Appropriators frequently devise their own rules without having created formal, governmental jurisdictions for this purpose. In many inshore fisheries, for example, local fishers devise extensive rules defining who can use a fishing ground and what kind of equipment can be used. So long as external governmental officials give at least minimal recognition to the legitimacy of such rules, the fishers themselves may be able to enforce the rules themselves. But if external governmental officials presume that only they can make authority rules, then it is difficult for local appropriators to sustain a rule-governed CPR over the long run. At any point when someone wishes to break the rules created by the fishers, they can go to the external government and get local rules overturned.

Audun Sandberg (1997) provides an insightful analysis of what happens when the individuals using common-pool resources for many centuries do *not* have recognized authority to create their own rules. The formal rules for the northern Norwegian commons were first written as law in the eleventh century and have remained unchanged until 1993 and thus represented "more than 1000 years of unbroken traditions of oral and codified Common Law" (Sandberg 1993, 14). The rules, however, specified very generalized rights only and did not recognize any local governance responsibilities. Since most commons, and especially the northern commons came to be conceptualized as the king's commons, it was easy to conceptualize that the king was the only law giver with authority to change laws over time. Through a long process that started with the Protestant Reformation and accelerated around 1750, this has eventually led to a conception in government that all forests and mountains in northern Norway that are not private property and which would in other countries be considered a commons, are considered state property (Sandberg, 1993, 19). The further effort of the state to then ration access to forests, grazing areas, fisheries, and other common-pool resources to those engaged in full-time specialized employment has had an unintended effect of being disruptive to the mixed economic way of life of many Northerners who were part-time farmers, part-time fishers, part-time foresters, and part-time herders (see chapter 19). Converting this sustainable way of life into a modern system including heavy reliance on transfer payments to specialized farming, fishing, and reindeer ranching was probably not fully expected by anyone. Now, however, the economic and social base has been weakened substantially enough that simply assigning local authority to make rules related to the use of common-pool resources would probably not be a sufficient way out of a major dilemma.

Nested Enterprises

Appropriation, provision, monitoring, enforcement, conflict resolution, and governance activities are organized in multiple layers of nested enterprises.

In larger systems, it is quite difficult to devise rules that are well-matched to all aspects of the provision and appropriation of that system at one level of organization. The rules appropriate for allocating water among three major branches of an irrigation system, for example, may not be appropriate for allocating water among farmers along a single distributary channel. Consequently, among long-enduring self-governed CPRs, smaller-scale organizations tend to be nested in ever-larger organizations. It is not at all unusual to find a larger, farmer-governed irrigation system, for example, with five layers of organization each with its own distinct set of rules.

Threats to Sustainable Community-Governed Commons

The study of community-governed and managed commons provides evidence of immense diversity of physical settings and institutional rules relatively well-matched to the local setting. It is important to recognize, however, that not all community-governed CPRs cope effectively with the array of problems they face over time. Some efforts at self-governance fail before resource users even get organized. Others fail within a few years. Others survive for long periods of time but are destroyed as a result of a variety of conditions. One source of failure is institutions that are not characterized by many of the design principles. Earlier studies have shown that small-scale CPRs that are characterized by only a small number of these design principles are more likely to fail than those characterized by a larger number of them.

However, even institutions that are characterized by the design principles fail. Thus, we need to speculate about other threats to community governance that arise from observations in the field, theoretical conjectures, and empirical findings of scholars studying small-scale CPRs or related situations. The reader is cautioned that the next two sections are far more speculative in nature than the first two sections.[4] It is important, however, to share speculations so that further research and analysis can be directed toward improving the knowledge claims of some speculations and reducing our confidence in others. Here is a list of eight threats to sustainable community governance of small-scale CPRs that I have come across in different contexts:

1. Blueprint thinking
2. Overreliance on simple voting rules as the primary decision mechanism for making all collective choices
3. Rapid changes in technology, human, animal or plant populations; in factor availability; in substitution of relative importance of monetary transactions; and in heterogeneity of participants

4. Transmission failures from one generation to the next of the operational principles on which community governance is based
5. Turning to external sources of help too frequently
6. International aid that does not take account of indigenous knowledge and institutions
7. Corruption and other forms of opportunistic behavior
8. Lack of large-scale institutional arrangements related to reliable information collection, aggregation, and dissemination; fair and low-cost conflict resolution mechanisms; educational and extension facilities; and facilities for helping when natural disasters or other major problems occur at a local level

Let us briefly discuss each of these.

Blueprint Thinking

Blueprint thinking occurs whenever policy makers, donors, citizens, or scholars propose uniform solutions to a wide variety of problems that are clustered under a single name based on one or more successful exemplars. David Korten called this the "blueprint approach" and made a devastating critique of its prevalence in development work at the end of the 1970s. As Korten describes it, "Researchers are supposed to provide data from pilot projects and other studies which will allow the planners to choose the most effective project design for achieving a given development outcome and to reduce it to a blueprint for implementation. Administrators of the implementing organization are supposed to execute the project plan faithfully, much as a contractor would follow construction blueprints, specifications, and schedules. An evaluation researcher is supposed to measure actual changes in the target population and report actual versus planned changes to the planner at the end of the project cycle so that the blueprints can be revised" (Korten 1980, 496). Korten's critique is just as relevant in the 1990s as it was more than a decade ago.

Even advocates of community governance fall into the trap of blueprint thinking. Whenever a policy is adopted that calls for the creation of large numbers of farmer organizations in a short period of time, there is a potential threat of blueprint thinking. Nirmal Sengupta, for example, describes the efforts of the Sone Command Area Development Agency in India to defend itself against questions raised in 1978 by policy makers as to why one part of its objectives was not being met—"that pertaining to the formation of irrigation associations" (Sengupta 1991, 242–43). The Agency then turned to "the Cooperative Department to frame model bylaws for the irrigation-specific cooperatives called Chak Societies" (Sengupta 1991, 243). The model bylaws contained forty-two major clauses and several minor clauses, but failed to address how irrigation cooperatives might be similar to or different from cooperatives established for other purposes. In the next year, twenty-two Chak Societies were initiated in the Sone Command Area. But, few of them performed in the way that policy makers thought they should, and the whole idea of registering irrigation associations using the model bylaws was dropped. The only way to get a large number of organizations set up in a hurry is to have an organizational charter and constitution written for all units. Then, one can simply call meetings and have people sign up. Such efforts result in large numbers of paper organizations and little else.

Overreliance on Simple Voting Rules

Closely related to blueprint thinking is the presumption that certain voting rules—either simply majority or unanimity—are the only rules that should be used in making collective decisions. The problem that users face is gaining general understanding of and agreement to a set of rules—not simply having a short discussion and a *pro forma* vote. The extensive theoretical and empirical studies growing out of social choice theory have demonstrated repeatedly that if the members of a community are strongly divided on an issue, it is extremely unusual to find any rule that enables them to achieve a final decision that is stable and can be considered to reflect the preferences of those affected. Substituting a simple majority vote for a series of long discussions and extensive efforts to come close to a consensus before making decisions that commit a self-governing community may lead to those in leadership positions simply arranging agendas so that they win in the short run. But as soon as rules are seen as being imposed by a majority vote rather than being generally agreed upon, the costs of monitoring and enforcement are much higher. The group has lost quasi-voluntary compliance and must invest more heavily in enforcement to gain compliance.

Similarly, reliance on unanimity prior to major changes may also challenge the long-term viability of a self-governing society. Once formal unanimity is adopted, only one person needs to hold out to delay decisions or impose high costs on most everyone else. The adaptability of a self-governed system may be too rapid if only simple majority votes are relied upon and too slow if only unanimity is used.

Rapid Exogenous Changes

All rapid changes in technology; in human, animal, or plant populations; in factor availability; in substitution of relative importance of monetary transactions; or in the heterogeneity of participants are a threat to the continuance of any self-organized system, whether it is a firm in a competitive market or a community-governed CPR. Individuals who have adapted an effective way of coping with a particular technological, economic, or social environment may be able to adjust to slow changes in one or several variables if substantial feedback is provided about the consequences of these changes for the long-term sustainability of the resource and/ or the set of institutions used for governing that resource. They may even be able to adjust to changes in these variables that occur at a moderate rate. The faster that key variables change and the more variables that change at the same time, the more demanding is the problem of adaptation to new circumstances. These kinds of threats are difficult for all organizations. Those that rely to a greater extent on quasi-voluntary compliance are, however, more threatened than those who are able to coerce contributions (Bromley and Chapagain 1984; Goodland, Ledec, and Webb 1989).

Ottar Brox (1990) provides a vivid illustration of what happened in the northern regions of Norway when technology, population density, and other factors changed rapidly. As he points out, traditional northern Norwegian fisheries were *seasonal fisheries*. "Large oceanic fish populations migrate during phases in their life or yearly cycles, and occur within reach of coastal fishermen only during short seasons" (Brox 1990, 231). Using traditional harvesting techniques, "coastal fishermen did not have the boats, gear and preservational techniques necessary to follow the fish populations continually. . . " (Brox 1990, 231). This had the consequence that it was almost impossible to destroy the fishery. Nor were the part-time farmers and part-time fishers able to reap most of the resource rent from fishing until the Norwegian Raw Fish Act of 1938, which empowered fishermen with the right to negotiate legally enforceable landing prices. Fishers who for many centuries could not themselves reap the rents from a migratory fishery now could do so, and could do so in an era of fast-changing technology making it possible to capture and store ever-greater quantities of fish. Further, other fishermen from other countries after the Second World War had the technology and capital to substantially increase effort dramatically above that which could be devoted prior to this era. A fishery that had survived, and even flourished, during many centuries of part-time fishing, rapidly became a threatened resource without adequate institutional means to respond to the changed incentives facing all of the participants.

Transmission Failures

Rapid change of population or culture may lead to a circumstance in which the general principles involved in the design of effective community-governed institutions are not transmitted from one generation to another. When individuals substitute rote reliance on formal rules for an understanding of why particular formal rules are used, they can make arguments for how to interpret the formal rules that undercut the viability of community organization. Relating this back to voting rules, for example, the charter or constitution of a community organization may specify that simple majority rule will be used in making decisions about future projects and how the costs and benefits of these projects will be divided. If the founders of such an organization recognize the importance of gaining general agreement, they will rarely push forward on a large project that is supported by only a minimal winning coalition. In such an instance, there are almost as many community members in opposition as those who support the project. But, if over time, the principle of gaining general agreement to future projects prior to implementation is not conveyed and accepted by those who later take on leadership responsibilities, then decisions receiving only minimal support may be pushed forward. Leaders of communities who rely on minimal winning coalitions for too many decisions may find themselves having to rely on patronage, coercion, or corruption to keep themselves in power rather than on a foundation of general agreement.

Similarly, if those who are required to devote particular resources or refrain from particular actions see these "rules" as obstacles to be overcome, rather than as the written representation of general underlying principles of organization, they may push for interpretations of rules that lead to their general weakening. If each household tries to find every legal way to minimize the amount of labor contributed to the maintenance of a farmer-governed irrigation system, for example, eventually the cumulative effect is an insufficient maintenance effort and the unraveling of the contingent contributions of all. If one family tries to make a favorable interpretation of how much labor they should contribute, given the land they own, others come to know that this family is interpreting rules in a manner that is highly favorable to them. Others who would be favored by such an interpretation begin to use it as well. The total quantity of labor contributed declines. Unless there is a community discussion about the underlying principles that can be used in interpreting rules, practices may evolve that cannot be sustained over time. Then, the danger exists that the unraveling continues unabated until the community organization falls apart.

Turning to External Sources of Help Too Frequently

A threat to long-term sustainability can be the availability of funds from external authorities or donors that appear to be "easy money." These funds can undercut the capabilities of a local institution to sustain itself over time.[5] This is particularly salient in regard to farmer-governed irrigation systems.[6] Monetary resources for constructing, operating, and maintaining irrigation systems are frequently contributed by the taxpayers of the nation in which the irrigation system is located or the taxpayers of those nations providing economic assistance funds. When these funds are used, the financial connection between supply and use is nonexistent. Whether the resources so mobilized are directly invested in the construction and operation of irrigation systems or are diverted for individual use by politicians or contractors depends on the professionalism of those involved and on active efforts to monitor and sanction diversions of resources. When the farmers themselves are involved in the construction and operation of irrigation systems, they provide low-cost monitoring of how resources for these activities are used. This is lost when the users are not involved in construction or operation. Expensive auditing systems are then needed, but are rarely supplied. Consequently, a considerable portion of the mobilized resources is diverted to purposes other than those for which it was intended.

Further, the design of projects is oriented more toward capturing the approval of those who fund new construction than toward providing systems that solve the problems facing present and future users. To convince politicians that large chunks of a national budget should be devoted to the construction of irrigation projects, planners attempt to design projects that are "politically attractive." This means that politicians who support such expenditures can claim that the voters' funds are being used to invest in projects that will greatly expand the amount of food available and lower the cost of living.

International Aid that Ignores Indigenous Knowledge and Institutions

To convince external funding agencies that major irrigation projects should be funded through loans or grants, the evaluative criteria used by these agencies in selecting projects has to play a prominent role in the design of projects. Projects designed by engineers who lack experience as farmers or training as institutional analysts are frequently oriented toward winning political support or international funding. This orientation does not lead to the construction of projects that serve most users (as in small-scale farmers) effectively or encourage the investment of users in their long-term sustenance. Inefficiencies occur at almost every stage. At the same time, this inefficient process leads to the construction of projects that generate substantial profits for large landholders and strong political support for a government.

Processes that encourage looking to external sources of funding make it difficult to build upon indigenous knowledge and institutions.

A central part of the message asking for external funds is that what has been accomplished locally has failed and massive external technical knowledge and funds are needed to achieve "development." In some cases, no recognition is made at all of prior institutional arrangements. This has three adverse consequences: (1) property rights that resource users had slowly achieved under earlier regimes are swept away and the poor lose substantial assets, (2) those who have lost prior investments are less willing to venture and further investments, and (3) a general downgrading of the status of indigenous knowledge and institutions.

Corruption and Other Forms of Opportunistic Behavior

All types of opportunistic behavior are encouraged, rather than discouraged, by (1) the availability of massive funds to subsidize the construction and operation of large-scale irrigation projects and (2) the willingness (or even eagerness) of national leaders to subsidize water as a major input into agricultural production. Corrupt exchanges between officials and private contractors are a notorious and widespread form of opportunism; corrupt payments by farmers to irrigation officials are less well-known, but probably no less widespread. Free-riding on the part of those receiving benefits and the lack of trust between farmers and officials, as well as among farmers, are also endemic. Further, the potential rents that can be derived from free irrigation water by large-scale landowners stimulate efforts to influence public decision making as to where projects should be located and how they should be financed. Politicians, for their part, win political support by strategic decisions concerning who will receive or continue to receive artificially created economic rents.

Robert Bates explains many of the characteristics of African agricultural policies by arguing that major "inefficiencies persist *because* they are politically useful; economic inefficiencies afford governments means of retaining political power" (Bates 1987, 128). Part of Bates's argument relates to the artificial control exercised over the prices paid for agricultural products, a topic that is not addressed in this study. The other part of Bates's argument relates to the artificial lowering of input prices. When they lower the price of inputs, private sources furnish lesser quantities, users demand greater quantities, and the result is excess demand. One consequence is that the inputs acquire new value; the administratively created shortage creates an economic premium for those who acquire them. Another is that, at the mandated price, the market cannot allocate the inputs; they are in short supply. Rather than being allocated through a pricing system, they must be rationed. Those in charge of the regulated market thereby acquire the capacity to exercise discretion and to confer the resources upon those whose favor they desire. Public programs which distribute farm credit, tractor-hire services, seeds, and fertilizers, and which bestow access to government managed irrigation schemes and public land, thus become instruments of political organization in the countryside of Africa (Bates 1987, 130).

Thus, there is an added dimension to rent seeking in many developing countries. The losses that the general consumer and taxpayer accrue from rent-seeking activities are one dimension. The second aspect of rent seeking in highly centralized economies is the acquisition of resources needed to accumulate and retain political power. All forms of opportunistic behavior, therefore, are exacerbated in an environment in which an abundance of funds is available for the construction of new and frequently large-scale irrigation projects that provide subsidized water. This is exactly the political and financial milieu that irrigation suppliers have faced during the past forty years in most developing countries. Developed countries have made vast amounts of money available to developing countries through bilateral and multilateral loans and aid agreements.

Lack of Large-Scale Supportive Institutions

While smaller-scale, community-governed resource institutions may be far more effective in achieving many aspects of sustainable development than centralized government, the absence of supportive, large-scale institutional arrangements may be just as much a threat to long-term sustenance as the presence of preemptive large-scale governmental agencies. Obtaining reliable information about the effects of different uses of resource systems and resource conditions is an activity that is essential to long-term sustainability. If all local communities were to have to develop all of their own scientific information about the physical settings in which they were located, few would have the resources to accomplish this.

Let me use the example of the important role that the U.S. Geological Survey has played in the development of more effective, local groundwater institutions in some parts of the United States. What is important to stress is that the Geological Survey does not construct engineering works or do anything other than obtain and disseminate accurate information about hydrologic and geologic structures within the United States. When a local set of water users wants to obtain better information about a local groundwater basin, they can contract with the Geological Survey to conduct an intensive study in their area. Water producers would pay a portion of the cost of such a survey. The Geological Survey would pay the other portion. The information contained in such a survey is then public information available to all interested parties. The Geological Survey employs a highly professional staff who rely on the most recent scientific techniques for determining the structure and condition of groundwater basins. Local water producers obtain the very best available information from an agency that is not trying to push any particular future project that the agency is interested in conducting. Many countries, such as India, that do have large and sometimes dominating state agencies, do *not* have agencies that provide public access to high-quality information about resource conditions and consequences. Recent efforts to open up groundwater exploration in India may lead to the massive destruction of groundwater basins rather than a firm basis for long-term growth.

Similarly, the lack of a low-cost, fair method for resolving those conflicts that spill out beyond the bounds of a local community is also a threat to long-run sustainability. All groups face internal conflicts or intergroup conflicts that can destroy the fundamental trust and reciprocity on which so much effective governance is based. If the only kind of conflict-resolution mechanisms available are either so costly or so biased that most self-governed CPRs cannot make use of them, these conflicts can themselves destroy even very robust institutional arrangements.

Coping Methods for Dealing with Threats to Sustainability

There are no surefire mechanisms for addressing all of the above threats. There are three methods that I would like to discuss here because they are not frequently mentioned as being important ways of increasing the effectiveness of self-governed institutions. They are: (1) the creation of associations of community-governed entities, (2) comparative institutional research that provides a more effective knowledge base about design and operating principles, and (3) developing more effective high school and college courses on local governance. There are, of course, many other coping mechanisms, including those adopted by local institutions, that have survived for long periods of time and are the subject of the first section of this chapter. Thus, I focus here only on three mechanisms that are not frequently thought of in relationship to the problem of sustaining self-governing institutions related to common-pool resources.

Creating Associations of Community-Governance Entities

Those who think local participation is important in the process of developing sustainable resources and more effective governance of resources are frequently committed to doing a good deal of "community organization." All too frequently, this type of organization is conceptualized as fostering a large number of community groups at the same level. If community organization is fostered by nongovernmental organizations (NGOs) who then provide staff assistance and some external resources, the organizations may flourish as long as the NGOs remain interested, but wither on the vine when the NGOs turn to other types of projects. A technique that draws on our knowledge of how self-governed institutions operate is helping to create associations of community organizations. As discussed above, most large-scale user-governed resource institutions are composed of several layers of nested organizations.

When community organizations are brought together in federations, they can provide one another some of the back-up that NGOs provide to single-layer community organizations. While no single community-governed organization may be able to fund information collection that is unbiased and of real value to the organization, a federation of such organizations may be able to amass the funds to do so. Simply having a newsletter that shares information about what has worked and why it has worked in some settings helps others learn from each other's trial and error methods. Having an annual meeting that brings people together to discuss their common problems and ways of tackling them greatly expands that repertoire of techniques for coping with threats that any one group can muster on its own. Such organizations can also encourage farmer-to-farmer training efforts that have proved to be highly successful in enhancing farmer-governed irrigation systems in Nepal.[7]

Rigorous Comparative Institutional Research

In addition to the type of exchange of information that those involved in self-governing entities can undertake on their own, it is important to find ways of undertaking rigorous, over-time comparative research that controls for the many confounding variables that simultaneously affect performance. In the field of medicine, folk medicine has frequently been based on unknown foundations that turned out to be relatively sound. But some folk medicine continued for centuries, doing more harm to patients than good. The commons that are governed by users and the institutions they use are complex and sometimes difficult to understand. It is important to blend knowledge and information obtained in many different ways as we try to build a more effective knowledge base about what works and why.

Developing Better Curricula on Local Governance

Western textbooks on governance used to focus as much on local as national governance arrangements. During the past half-century, introductory textbooks on American government have moved from a 50–50 split between national and local government to a 95 to 5 split.

The textbooks used in the West have strongly influenced the textbooks used in developing countries. Consequently, many public officials learn nothing in high school and college about how local communities can govern themselves effectively or about the threats to local self-governance. Instead, a presumption is made that governance is what is done in national capitals and what goes on in villages is outmoded if not completely useless. Thus, the last recommendation that I will make at this juncture is to bring more materials on self-governing communities into the curriculum that is offered in high schools, in professional schools, and in colleges.

NOTES

1. An earlier version of this paper was presented at the Workshop on "Community Management and Common Property of Coastal Fisheries and Upland Resources in Asia and the Pacific: Concepts, Methods, and Experiences," sponsored by the International Development Research Centre and International Center for Living Aquatic Resources Management, held at Silang, Philippines, June 20–23, 1993. Writing of this paper is supported by the Decentralization: Finance and Management Project, funding by the U.S. Agency for International Development to Associates in Rural Development, Burlington, Vermont; the Metropolitan Studies Program of the Maxwell School of Citizenship and Public Affairs at Syracuse University; and the Workshop in Political Theory and Policy Analysis at Indiana University.

2. The wide variation of specific rules tailored to local circumstances in Norway and other northern regions is well-documented by the papers included in this volume. See the chapters by Eggertsson, Falkanger, Sevatdal, Sagdahl, Korpijaakko-Labba, and Austenå and Sandvik.

3. The next section draws in part on *Governing the Commons* (E. Ostrom 1990, chapter 3).

4. The threats pointed out here, however, are closely related to the work in sociology on "unintended consequences" (Merton 1936; Boudon 1982). Sieber (1981) provides an excellent overview of a number of diverse efforts to remedy social ills that have made them even worse. Sorensen and Auster (1989) and Baert (1991) both point out how prevalent are interventions that generate substantial, if not overwhelming, reverse effects.

5. The problem of local units becoming dependent on external funding is not limited to the funding provided by international aid agencies. Sieber (1981) reviews some of the reverse effects created by domestic U.S. policy. The supposed aim of Nixon's "New Federalism" reform was to increase the autonomy of local units and strengthen the overall federal system. A study by Hudson (1980) reveals that the policy has an opposite effect in some cities

such as El Paso. "El Paso is now more dependent, politically and economically, on federal grants than it was prior to the New Federalism and local autonomy is significantly reduced" (Hudson 1980, 900, quoted in Sieber 1981, 186).

6. This and the next two sections draw on *Crafting Institutions for Self-Governing Irrigation Systems* (E. Ostrom 1992).

7. See Yoder (1991), Pradhan and Yoder (1989), and Water and Energy Commission Secretariat (1990) for descriptions of a highly innovative and successful program of assisting farmers to design their *own* institutional rules rather than imposing a set of model bylaws on them.

REFERENCES

Baert, Patrick. 1991. "Unintended Consequences: A Typology and Examples." *International Sociology* 6(2):201–10.
Bates, Robert H. 1987. *Essays on the Political Economy of Rural Africa.* Berkeley: University of California Press.
Berkes, Fikret. 1987. "Common Property Resource Management and Cree Indian Fisheries in Subarctic Canada." In *The Question of the Commons*, edited by Bonnie McCay and James Acheson, 66–91. Tucson: University of Arizona Press.
Berkes, Fikret. 1992. "Success and Failure in Marine Coastal Fisheries of Turkey." In *Making the Commons Work: Theory, Practice, and Policy*, edited by Daniel W. Bromley et al., 161–82. San Francisco: Institute for Contemporary Studies Press.
Boudon, Raymond. 1982. *The Unintended Consequences of Social Action.* London: Macmillan.
Bromley, Daniel W., et al., eds. 1992. *Making the Commons Work: Theory, Practice, and Policy.* San Francisco: Institute for Contemporary Studies Press.
Bromley, Daniel W. and D. P. Chapagain. 1984. "The Village Against the Center: Resource Depletion in South Asia." *American Journal of Agricultural Economics* 66:868–73.
Brox, Ottar. 1990. "The Common Property Theory: Epistemological Status and Analytical Utility." *Human Organization* 49(3):227–35.
Brox, Ottar. 1996. "Recent Attempts at Regulating the Harvesting of Norwegian Arctic Cod." *GeoJournal* 39(2):203–210.
Chambers, Robert. 1988. *Managing Canal Irrigation: Practical Analysis from South Asia.* Cambridge: Cambridge University Press.
Feeny, David H. 1988a. "Agricultural Expansion and Forest Depletion in Thailand, 1900–1975." In *World Deforestation in the Twentieth Century*, edited by J. F. Richards and R. P. Tucker, 112–43. Durham, N.C.: Duke University Press.
Feeny, David H. 1988b. "The Demand for and Supply of Institutional Arrangements." In *Rethinking Institutional Analysis and Development:*

Issues, Alternatives, and Choices, edited by Vincent Ostrom, David Feeny, and Hartmut Picht, 159–209. San Francisco: Institute for Contemporary Studies Press.

Feeny, David H. 1992. "Where Do We Go from Here? Implications for the Research Agenda." In *Making the Commons Work: Theory, Practice, and Policy,* edited by Daniel W. Bromley et al., 267–92. San Francisco: Institute for Contemporary Studies Press.

Gardner, Roy and Elinor Ostrom. 1991. "Rules and Games." *Public Choice* 70(2), May: 121–49.

Gardner, Roy, Elinor Ostrom, and James Walker. 1990. "The Nature of Common-Pool Resource Problems." *Rationality and Society* 2(3), July: 335–58.

Goodland, Robert, George Ledec, and Maryla Webb. 1989. "Meeting Environmental Concerns Caused by Common-Property Mis-management in Economic Development Projects." In *Common Property Resources: Ecology & Community-Based Sustainable Development,* edited by Fikret Berkes, 148–63. London: Belhaven Press.

Hudson, W. E. 1980. "The New Federalism Paradox." *Policy Studies Journal* 8:900–06.

Kiser, Larry L. and Elinor Ostrom. 1982. "The Three Worlds of Action: A Metatheoretical Synthesis of Institutional Approaches." In *Strategies of Political Inquiry,* edited by Elinor Ostrom, 179–222. Beverly Hills, Calif.: Sage.

Korten, David C. 1980. "Community Organization and Rural Development: A Learning Process Approach." *Public Administration Review* 40(5), Sept./Oct.: 480–511.

Levi, Margaret. 1988. *Of Rule and Revenue.* Berkeley: University of California Press.

Merton, Robert K. 1936. "The Unanticipated Consequences of Purposive Social Action." *American Sociological Review* 1:894–904.

Netting, Robert McC. 1981. *Balancing on an Alp.* New York: Cambridge University Press.

Oakerson, Ronald J. 1992. "Analyzing the Commons: A Framework." In *Making the Commons Work: Theory, Practice, and Policy,* edited by Daniel W. Bromley et al., 41–59. San Francisco, Calif.: Institute for Contemporary Studies Press.

Ostrom, Elinor. 1986. "An Agenda for the Study of Institutions." *Public Choice* 48:3–25.

Ostrom, Elinor. 1990. *Governing the Commons: The Evolution of Institutions for Collective Action.* New York: Cambridge University Press.

Ostrom, Elinor. 1992. *Crafting Institutions for Self-Governing Irrigation Systems.* San Francisco: Institute for Contemporary Studies Press.

Ostrom, Elinor, Paul Benjamin, and Ganesh Shivakoti. 1992. *Institutions, Incentives, and Irrigation in Nepal.* Volume 1. Bloomington: Indiana University, Workshop in Political Theory and Policy Analysis.

Ostrom, Elinor, Sharon Huckfeldt, Charles Schweik, and Mary Beth Wertime. 1993. "A Relational Archive for Natural Resources Governance and Management." Working paper. Bloomington: Indiana University, Workshop in Political Theory and Policy Analysis.

Ostrom, Vincent. 1991. *The Meaning of American Federalism: Constituting a Self-Governing Society*. San Francisco: Institute for Contemporary Studies Press.

Ostrom, Vincent, David Feeny, and Hartmut Picht, eds. 1993. *Rethinking Institutional Analysis and Development: Issues, Alternatives, and Choices*. Second edition. San Francisco: Institute for Contemporary Studies Press.

Pradhan, Naresh C. and Robert Yoder. 1989. *Improving Irrigation System Management through Farmer-to-Farmer Training: Examples from Nepal*. IIMI Working Paper no. 12. Colombo, Sri Lanka: International Irrigation Management Institute.

Sandberg, Audun. 1993. "Entrenchment of State Property Rights to Northern Forests, Berries and Pastures." Paper presented at the Mini-Conference on Institutional Analysis and Development, held at the Workshop in Political Theory and Policy Analysis, Indiana University, Bloomington, Indiana, December 11–13, 1993.

Schlager, Edella. 1990. "Model Specification and Policy Analysis: The Governance of Coastal Fisheries." Ph.D. dissertation. Department of Political Science, Indiana University.

Schlager, Edella and Elinor Ostrom. 1992. "Property-Rights Regimes and Natural Resources: A Conceptual Analysis." *Land Economics* 68(3), Aug.: 249–62.

Schlager, Edella and Elinor Ostrom. 1993. "Property-Rights Regimes and Coastal Fisheries: An Empirical Analysis." In *The Political Economy of Customs and Culture: Informal Solutions to the Commons Problem*, edited by Terry L. Anderson and Randy T. Simmons, 13–41. Lanham, Md.: Rowman & Littlefield.

Sengupta, Nirmal. 1991. *Managing Common Property: Irrigation in India and the Philippines*. New Delhi: Sage.

Shepsle, Kenneth A. 1989. "Studying Institutions: Some Lessons from the Rational Choice Approach." *Journal of Theoretical Politics* 1:131–49.

Sieber, Sam D. 1981. *Fatal Remedies: The Ironies of Social Intervention*. New York: Plenum Press.

Sorensen, Anthony D. and Martin L. Auster. 1989. "Fatal Remedies: The Sources of Ineffectiveness in Planning." *Town Planning Review* 60(1):29–44.

Tang, Shui Yan. 1991. "Institutional Arrangements and the Management of Common-Pool Resources." *Public Administration Review* 51(1), Jan./Feb.: 42–51.

Tang, Shui Yan. 1992. *Institutions and Collective Action: Self-Governance in Irrigation*. San Francisco: Institute for Contemporary Studies Press.

Thomson, James T. 1992. *A Framework for Analyzing Institutional Incentives in Community Forestry*. Rome, Italy: Food and Agriculture Organization of the United Nations, Forestry Department, Via delle Terme di Caracalla.

Thomson, James T., David Feeny, and Ronald J. Oakerson. 1992. "Institutional Dynamics: The Evolution and Dissolution of Common-Property Resource Management." In *Making the Commons Work: Theory, Practice, and Policy*, edited by Daniel W. Bromley et al., 129–60. San Francisco, Calif.: Institute for Contemporary Studies Press.

Water and Energy Commission Secretariat, Nepal and International Irrigation Management Institute (WECS/IIMI). 1990. *Assistance to Farmer-Managed Irrigation Systems: Results, Lessons, and Recommendations from an Action-Research Project*. Colombo, Sri Lanka: International Irrigation Management Institute.

Yoder, Robert. 1991. "Peer Training as a Way to Motivate Institutional Change in Farmer-Managed Irrigation Systems." In *Workshop on Democracy and Governance Proceedings*. Burlington, Vt.: Associates in Rural Development.

Thráinn Eggertsson   Chapter 2

The Economic Rationale of Communal Resources

Introduction

We are concerned in this study with the logic of economic organization when several independent producers jointly draw inputs from a natural resource which they share and to which they hold exclusive rights. We use the term communal property to refer to this arrangement and distinguish it from situations where exclusive rights to a resource do not exist and access to the asset is open.[1]

The structure of organization is a chief concern of the economics of institutions which attribute forms of organization not only to economies of scale, but to problems of information and costly enforcement of contracts. In explaining the emergence of property rights and alternative forms of organization, most economists have followed an approach that has been called the "naive model."[2] The naive model explains the structure of institutions and organizations in terms of the demand for these arrangements by rational individuals, who are constrained by information and other transaction costs and seek arrangements that maximize the joint value of their assets. The approach is naive because it does not seek to explain the supply of property rights, which is the domain of social and political organizations.[3] Our initial discussion is based on versions of the naive model, but we go on to consider the role of social and political organizations in shaping property rights.

It is sometimes argued that communal property regimes must deal with categorically different problems of organization than other regimes of exclusive rights. We maintain that all forms of exclusive property rights involve essentially the same measurement and policing problems, and that the appropriate structure of rights depends on technology, physical characteristics of the resources, relative prices, and social and political institutions. In terms of the criterion of wealth maximization, communal property is the optimal arrangement in some situations, but unsuitable under other circumstances. Further, the condition of open access is associated with all forms of exclusive rights, including individual property, and arises because the marginal benefit of enforcing full control over all attributes of a valuable asset tends to fall short of the marginal cost (Barzel 1989).

We begin by discussing why one expects to find communal property regimes in some and not in other situations and proceed to look at the variables that push resource regimes in the direction of communal ownership. These issues are examined in terms of the naive model, as social and political institutions are assumed to be exogenous.[4] We then introduce the wealth effect and examine how the struggle over distribution can affect the structure of communal property regimes. The next step is to consider the supply of exclusive property rights. We continue by examining some of the factors that may undermine communal property and finally conclude with a few thoughts about property rights in the Saami range lands in Finnmark, Norway.

Economic Factors and the Choice of an Exclusive Resource Regime

Imagine a group of individuals (households or firms) that contemplates the utilization of a contiguous natural resource such as range land, a forest, or a fishery. The individuals are capable of collective action (but collective action requires the use of scarce resources); the objective of the group is to maximize the joint value of their resources (the choice is not constrained by individual wealth effects); and the enforcement of property rights is entirely with the group and its members (although exogenous social norms, customs, and conventions affect the cost of enforcement). The users can choose from a large menu of regimes, each characterized by several dimensions.

The number of independent users that share the entire resource or portions of it is a key dimension of a resource regime. At one extreme we have open access, when the community decides not to incur the cost of excluding outsiders and regulates the behavior of users. Then there are various regimes of exclusive rights ranging from the sharing of the entire resource by the group (communal property) to individual holdings (individual or private property). In between communal and individual property are intermediate communes which are subgroups of two or more individuals who share property.

Another dimension of regimes concerns the contractual relations among people who share a resource. For instance, it is conceivable that the individual producers could minimize costs by merging into one firm that would become the sole user of the resource. This dimension was explored by Coase (1937) in his study of the nature of the firm. Yet another margin concerns the degree of precision and detail that the community decides to give the rules for operating the resource. The costs of explicit rules are balanced against the benefits of limiting potential disputes over uncertain rights. Libecap (1978) uses nineteenth century data from Nevada's richest mining area to test and find support for the thesis that property rights will be made more precise as resources become more valuable. Finally, we note that the community must determine the extent of the rights to use, earn income from, and transfer or dispose of the resource.

Let us consider more closely the factors that are supposed to push a resource regime in the direction of exclusive rights, which is an issue that has been given considerable attention in the literature. In a pioneering article, Demsetz (1967) explains the introduction of exclusive rights in land among Indian hunters in the eastern part of Canada in terms of the cost and benefits of internalizing externalities from non-exclusive use of the resource. In this case the driving force behind the formation of exclusive rights was a sharp increase in output demand which induced the Indians to divide open hunting regions into smaller exclusive hunting territories.

Demsetz's approach is employed by Anderson and Hill (1975), who explicitly include the cost of exclusion to explain the evolution of exclusive rights to the utilization of land, water, and cattle on the Great Plains of the American Midwest during the second half of the nineteenth century. Field (1986, 1989) has refined the Demsetz approach on two margins: first, by explicitly considering both the cost of excluding outsiders, exclusion cost, and the cost of controlling the propensity to excessive use when a resource is shared, governance cost; and second, by considering a continuum of communality, ranging from individual ownership through a series of intermediate communes of increasing size to a commune of the whole.

A brief description of the Field model may help us to highlight critical variables that affect the relative efficiency of communal property. In the model, it is assumed that the community will select the arrangement that maximizes the aggregate net returns from a natural resource, such as range land. There are two corner solutions, individual property and communal property, and the internal solutions involve sets of intermediate communes of different sizes. The resource consists of units of homogeneous quality; the individual producers are also homogeneous; and their production functions are identical. All inputs other than the natural resource are privately owned. The creation of value is based on three activities:

1. The transformation of inputs into outputs that is described by a transformation function corresponding to the conventional production function.
2. The exclusion of intruders by monitoring, fencing, and other means. Successful exclusion is rewarded by greater output at each level of input use, which implies that the transformation function shifts up.[5] The exclusion function can be seen as a production function which depends both on the technology of exclusion and social institutions and organizations.
3. The policing of insiders to limit excessive utilization when two or more individuals share (a portion of) the resource. We refer to this activity as "governance." The group decides on the level of utilization that maximizes the value of the resource and assigns user rights to each individual. The relationship between inputs and level of control achieved is described by the governance function, and the cost of governance is balanced against the resulting increase in net income.

Various assumptions can be made about the nature of the exclusion and governance functions. In the Field model exclusion costs depend directly on the length of the borders, which are at a maximum when the resource is divided into individual properties and at a minimum when the resource is one property shared by the whole group. Internal governance problems arise when two or more individuals share a property, and governance costs rise directly with the number of joint users on each plot and peak when the resource is one property. If there were no governance costs and exclusion costs, the division of the resource would be determined by the economies of scale in the transformation function.[6] Below we assume that there are constant returns to scale and focus attention on the role of governance and exclusion in determining the degree of exclusivity.

Consider again the complex optimization problem confronting the community of users. Net income depends not only on the allocation of inputs in conventional production (transformation), but also on the use of inputs in exclusion and governance; furthermore, both governance cost and exclusion cost are influenced by the division of the resource into properties.[7] Many small intermediate communes imply relatively low governance costs but high exclusion costs, and few large intermediate communes have relatively large governance costs and low exclusion costs. In sum, the degree of exclusivity depends on a trade-off between governance and exclusion costs, other things equal.

Economic Forces Supporting Communal Property Regimes

The higher the exclusion costs relative to governance costs, the more likely will a community that strives to maximize wealth select large communal arrangements. Therefore, in order to understand the economic logic of communal property, we must examine the factors influencing the levels of the cost functions for governance and exclusion.

Exclusion depends on technology, the physical characteristics of the resource, relative prices (including the prices of inputs in the exclusion function and the output price), and on the social institutions that constrain the players. In extreme cases, and given the state of technology, the physical characteristics of a resource can make it prohibitively costly to divide it into exclusive sub-units, which leads to the corner solution of a single communal property. Exclusion costs are also influenced by the size of the area required for individual operations. For instance, in arid or infertile regions the typical individual may demand a large geographic area for grazing her flocks or need to vary the pastures with the seasons or climatic changes. When the cost of monitoring or fencing individual properties is high, communal regimes become an attractive alternative, as does the reliance on natural boundaries, when possible.[8]

The relative prices of inputs in the exclusion function are an important factor influencing the choice of communal property regimes, for instance in communities where the price of timber and other material for fences is high. Also, an increase in output price creates new incentives for outsiders to intrude and makes it more costly to maintain any level of exclusion. The technology of exclusion is an important determinant of exclusion costs, and primitive exclusion technology increases the relative effectiveness of communal arrangements. When there are important economies of scale in exclusion (particularly in operating a system of individual properties), a small community of users may favor communal property for (some of) its natural resources. Although it is not self-evident, the political integration of a country may bring scale economies in exclusion and increase the attractiveness of individual property. The interaction between transformation technology and exclusion technology should also be noted. The cost of exclusion depends on what is produced and how it is produced, and the choice of output and production methods is not independent of exclusion (and governance) cost. Furthermore, a change in transformation technology (or a change in relative prices) can affect the choice of regime. For instance, a new transformation technology in agriculture can make the production of fodder on individual plots the optimal alternative and eliminate the dependence on pastures; or new fishing technology may introduce foreign and domestic vessels (and open access) in a fishery that used to be the communal property of coastal fishermen.

It is important to realize that a continued upward shift in exclusion costs, with a constant governance function, first pushes a system toward communal ownership but eventually, as the upward drift continues, places the resource in the public domain. In many instances, communal property is the only practical alternative to open access, and because of their proximity the two arrangements are often confused.

Governance costs depend on social institutions, technology, relative prices, and the physical characteristics of the resource and its environment, just as exclusion costs do. Low governance costs for large groups of users encourage communal property. It has been argued, for instance by Runge (1992) and Bromley (1992), that poverty is the cause of communal property because the arrangement is frequently found in poor communities. Runge states that "low levels of income imply that formalized private-property institutions are outside the village-level budget for resource management." And Bromley adds, "In fact, as Runge reminds us, low-valued resources are more likely to be managed under common [communal] property for the simple reason that there is insufficient economic surplus to support the more expensive private-property regime. I make the same point elsewhere."[9] We prefer different reasoning. The statement that poor communities cannot afford exclusive rights may apply to the purchase of expensive consumer goods, but not to the choice of property regimes. In fact, poor communities can afford only regimes that maximize the net output from their natural resources, the difference between gross output and costs. The observation that communal property regimes are found relatively frequently in developing countries is to be explained in terms of the available technology in transformation, exclusion and governance, relative prices, and social institutions. Many low-income communities rely on a mixture of individual and communal rights: for instance, the livestock, farmland, tools, and housing are often the property of individual economic units (households) while grazing land remains communal property.[10] That communal property regimes are found in wealthy communities as well, such as Switzerland with its celebrated Alpine pastures, also undermines the poverty argument.[11]

The Wealth Effect and Communal Property Regimes

In our discussion so far, we have ignored the individual wealth effects of introducing alternative property rights regimes. Even though it has been assumed that new property rights regimes are only chosen if they increase aggregate wealth (or minimize unavoidable losses), it must be recognized that all changes in property rights involve winners and losers. Therefore, the losers have an incentive to prevent changes that are expected to worsen their (relative) wealth position, unless they are guaranteed compensation, which is often impractical. When side-payments are impractical, the outcome depends on the power of the losers relative to the winners, which is partly determined by the community's political structure.

Consider again the previous case of a community of users choosing a resource regime. The community now confronts a new constraint: each individual has the power to veto all proposals that change the status quo, and no rational (and selfish) individual will agree to a new regime that makes him or her worse off than before. Let us assume that the current situation is one of open access with excess utilization of the resource. The group does not maximize the net economic yield from the resource, but the current yield is sustainable and the resource not in immediate danger of destruction.[12]

The group is faced with a dilemma. Their calculations show that a change from open access to communal property (rather than to individual property or intermediate communes) would increase the total wealth of the community, but some individual members could easily lose from the change in regimes. As side-payments are ruled out by high transaction costs, the introduction of exclusive rights hinges on the community's ability to constrain the communal regime in such a way as to make sure that no individual will lose from the change.

Roberts (1990), using a straightforward graphic analysis of supply and demand, has analyzed the situation above (see also Roberts 1993). First, it is easy to show that the introduction of a (Pigouvian) tax, for limiting the use of the resource to the efficient level, makes all previous users worse off, unless the revenue from the tax is returned to them (Roberts 1990, 5; Weitzman 1974). However, the tax revenue does more than cover the consumers' surplus lost by the users, when the price of entry is raised.

As the use of a tax for aligning social marginal costs and benefits is information intensive, a system of marketable coupons is more practical in a world of costly information. With marketable coupons the community would establish the efficient total level of use for the resource and apply some formula to issue coupons to previous users, giving each a share in the total. Again, if the coupons are sold to the users at market price, they are worse off than before, unless the proceeds are returned to them. However, even if the coupons are given for free, the task of assigning shares to previous users in such a way that no one is made worse off becomes a complex task. Consider two individuals with equal levels of usage in the free-entry equilibrium, but individual A has a greater price elasticity of demand for the resource than individual B. If both receive the same share of coupons when communal property rights are introduced, Roberts (1990) shows that B, because of his low elasticity of demand, is made relatively better off than A. Equal treatment of the two requires that A receive a larger share of the coupons than B. When the price elasticity of demand is similar for all individuals in the group, the allocation of coupons relative to the level of prior usage or relative to some proxy for demand, such as land ownership in the case of private farmers using communal pastures, is likely to guarantee that no one is made worse off and that the relative wealth position of the individuals does not change substantially.

Finally, Roberts shows that unrestricted resale of coupons can make some individuals worse off than they were in the open-access equilibrium, particularly if the coupons are sold to outsiders who drive up the price.[13] The trouble does not arise if the allocation of coupons correctly reflects the consumers' surplus lost by each individual, but when that fails some individuals will veto unrestricted resale of coupons, even though unlimited resale maximizes the total wealth of the group.

Several scholars, such as Ostrom (1990), have emphasized that agreements on efficient communal property regimes are reached more easily in a homogeneous than a heterogeneous group. Johnson and Libecap (1982) and Libecap (1989, chapter 5) discuss how heterogeneity among fishermen limits the fisheries regulations that they can agree on.

We have discussed how the wealth effect influences the choice of property rules by a small group of producers, such as the farmers in a rural village. When resource regimes are selected by an external authority, such as a national government, the interplay of inside and outside interests, and complex procedures for making decisions, can make the story much more complex.

The Supply of Exclusive Property Rights

We now leave the naive model behind and briefly consider the supply of exclusive rights. For social scientists who employ the rational choice model, the establishment and successful operation of a system of communal property rights by rational, non-altruistic individuals poses several puzzles. The first puzzle concerns the supply of a mechanism for selecting a system of communal property. The services of individuals who provide this apparatus have the characteristics of a public good and, therefore, are likely to be supplied in inadequate quantity. Second, the choice of constitutional and operational rules for managing the resource regime is likely to involve hard bargaining over the distribution of expected gains, possibly with indeterminate results. Third, individual compliance with rules that restrict use of the resource is also a public good, and free-riding may undermine the regime when monitoring is costly.[14]

Before we go further, it is important to note that these collective action problems are not limited to communal property but shared by all attempts to establish exclusive rights. In terms of the rational choice approach, the creation of any system of exclusive rights for a community always requires some curtailment of the propensity to free-ride. All changes in property rights have wealth effects which invite bargaining over distribution, and transaction costs always make exclusive rights incomplete and cause a certain amount of waste. The decision by a group to restrict access to a resource can be represented as a contract among its members, and all contracts are incomplete because of transaction costs, according to contract economics.[15] However, the nature of the open access problems varies from one contractual structure to another.

In the case of individual property, residual rights are exercised by an owner who both has residual control and receives (under ideal conditions) the net residual benefits of her actions, which encourages the owner to make efficient decisions that maximize wealth. However, when the proprietor expands her operations beyond the unitary firm and hires agents, she must deal with incomplete contracts and shirking by the agents, which lowers the joint value of the cooperating assets. In order to limit such losses, the proprietor usually attempts to realign the incentives of her agents by monitoring and with contractual arrangements which, for instance, link their pay to the fortunes of the firm. The internal problems of the firm (opportunism, shirking, free-riding) mount as the structure becomes more complex and changes from individual proprietorship to a partnership or to a public corporation. In the public corporation it is not clear whether any party—such as the stockholders, directors, managers, or the workers—both has residual control and receives the residual income (Milgrom and Roberts 1992, 314–315). However, in all these instances various arrangements have evolved for limiting the incentives problems, including competition in the marketplace.

Communal property arrangements, just as other forms of economic organization, depend on contracts that are structured to limit transaction costs. Recently, the complexities of communal property regimes have been documented and analyzed by various scholars, of whom Ostrom (1990) is a noted example (Eggertsson 1990, 249–262).

Why do rational actors supply the institutions of communal property? How do they overcome the collective action problem? In responding to such questions about the supply of property rights (which the naive model does not consider), the theoretical literature has not converged on a single answer, but several approaches to the problem can be discerned. We will briefly consider some of these.

The collective action problem is frequently analyzed in terms of game theory, particularly as a Prisoners' Dilemma where noncooperation is the dominant strategy. Incentives to cooperate are introduced by considering not a single game but repeated games or supergames. Others claim that the problem of cooperation is best modeled by games, such as the Assurance game or the game of Chicken, which are more likely than the Prisoners' Dilemma game to lead to some cooperation, if the game is played only once.[16] In the continuous case, hybrids of games have been suggested.[17] Many studies distinguish between formal rules that are provided by political organizations and informal rules, such as customs and norms, that are not purposefully created but evolve spontaneously (North 1990).

Many scholars have bypassed the fundamental question of how to reconcile rationality on the part of the individual with rationality on the part of the group, and focus on the role of coercion in overcoming the collective action problem. These social scientists "see in collective dilemmas reasons for the existence of institutions: forms of hierarchy in which sanctions are employed to make self-interested choices consistent with the social good" (Bates 1988, 387). Hechter (1990) associates the emergence of coercive organizations in traditional societies with the joint production of private goods in situations where individual behavior is easily visible. These organizations of producers are then used to control free-riding in the supply of public goods.

The scholar can also equip his players with internal norms and values that change the structure of the payoff matrix in their games and introduce cooperation as the dominant strategy. Although not formally stated in terms of game theory, pioneering work along these lines was undertaken in the first half of the century by a number of investigators, such as Evans-Pritchard, who studied traditional societies in Africa.[18] These studies report how customary law and ideology in traditional societies contribute to the maintenance of order. Vengeance groups, collective responsibility, the institution of compensation, exogamy and relations of kinship, the system of beliefs surrounding the institution of witchcraft, and a host of other arrangements have been interpreted as raising the cost of non-compliance and promoting cooperation.

In the naive model of property rights discussed in previous sections, social and political institutions do not enter directly, but affect outcomes by shifting the exclusion and governance functions. One can speculate that certain social structures may be likely to contribute to relatively low governance costs for communal property, while other social institutions may support low exclusion costs for individual property. For instance, it is sometimes argued that the thrust of norms and customary law in many traditional societies is to restrain individualism and lower governance costs, while traditional societies often lack specialized organizations for enforcing individual ownership rights, particularly when ownership rights can be traded.[19]

The Demise of Communal Property Regimes

Communal property regimes can give way to either open access or more exclusive (individual) property rights. We now consider in what direction economic growth is likely to push a system of communal property. There is little help to be found in formal economic models such as the Field model: an increase in either of the two critical variables associated with economic growth, the demand for the resource and population, has uncertain effects on the exclusivity of the resource regime. The reason for this indeterminacy is that each variable affects both the cost of exclusion and the costs of governance in many ways. For instance, an increase in output demand—that is reflected in a higher output price—shifts up the governance cost curve and creates an incentive for smaller communes or individual property. However, an increase in output price can also affect the cost of exclusion by increasing the incentive for encroachment, which means that additional resources are required to achieve the same level of exclusion as before. The cost curve for exclusion shifts up which directs the system in the opposite direction, toward communal property.

Furthermore, it is beyond the scope of formal models to consider directly the impact on exclusivity of the numerous developments that usually accompany economic growth, such as technological change in transformation, governance and exclusion, organizational innovation, changes in the location of industry, the nature of products, and new forms of political and social organization. Economic growth with increasing population and falling transportation costs may introduce open access by overwhelming the capacity of small appropriator organizations to provide exclusion.[20] Economic growth may also bring integration and restructuring of political units and a greater capacity to manage individual properties. Further, economic growth can contribute to the breakdown of social structures in traditional societies and raise the governance cost of communal property, and with weak social structures the capacity to exclude may also be diminished.[21]

There are myriad possibilities and special cases. The impact of economic growth on communal property arrangements in particular cases has been analyzed informally by several authors. For instance, Ensminger and Rutten (1990) study how economic growth has dismantled a communal system among the Orma, who were nomadic pastoralists in a district of northeastern Kenya. The study shows how economic growth has altered the geographic location of the industry and increased the diversity of interest within the community by introducing a sub-group of sedentary livestock producers who produce for commercial markets and demand different property rights than the nomads.[22] The new heterogeneity has increased the conflict over collective decisions. Also, with economic growth the role of the appropriator organizations has diminished while the role of the national government has increased, the government seeming to favor commercial producers. The decentralized enforcement of a stateless society has been replaced by third-party specialists.[23]

The Orma story is not solely one of increased demand for the output with a resulting increase in overgrazing and encroachment, but also a story of major changes in the structure of political and social institutions. With the national government now sharing exclusion costs, the local exclusion cost curve shifts down, which increases local demand for exclusive rights and promotes a move away from communal property.

Does the nationalization of rule making, governance, and exclusion contribute to a more or less efficient utilization of natural resources? There is no definite answer to this question. On the negative side, decision makers in government are often less affected personally than an appropriator organization by decisions that waste resources. They may sacrifice local interests to national or special interests, and their remoteness suggests that they may have less information for making decisions and receive weaker feedback about the consequences of their actions than appropriator organizations. Also, as national decision makers often face softer economic constraints than appropriator organizations, they are more likely to indulge in personal preferences that are out of tune with economic reality; for instance, they may have ideological preferences for individual property or communal property. On the other hand, local users may not be able to resolve satisfactorily their bargaining over the increase in wealth that is expected to flow from changes in property rights, and a powerful outsider could possibly break the deadlock and introduce a new structure that sharply increases the value of the resource.[24]

Conclusions

We have used the criterion of wealth maximization to study the choice of regimes of exclusive rights. On the wealth criterion, optimization requires that costs be minimized. It was argued that communal property is a form of exclusive rights that, in specific circumstances, has absolute advantage in minimizing the aggregate costs of production, governance, and exclusion. We attempted to show how the relative efficiency of communal property depends not only on economic factors but on the nature of social and political institutions.

The choice of regimes of property rights is complicated by the so-called wealth effect and by the problem of collective action. We used the example of a transition from open access to communal property to illustrate why rational agents might place inefficient constraints upon communal property, such as restrictions on the resale of user rights.

We were mostly concerned with the choice of resource regimes by small appropriator organizations, but recognized that national and local governments often have a large role in specifying and enforcing resource regimes. It was also recognized that economic growth is associated with various changes in social institutions and technology, in addition to increases in demand and in population, which makes it impossible to generalize about the impact of economic growth on the viability of communal property. Finally, it must be recognized that the objective function of those who choose the structure of resource regimes may contain other elements than wealth, narrowly defined.

The case of communal grazing pastures of the nomadic Saami reindeer herders of Finnmark in northern Norway is a clear illustration of the difficulties of designing a positive theory of communal property. Prior to the large-scale involvement by the Norwegian state, a simple economic model incorporating transaction costs might have gone far to explain the structure of property rights in the reindeer industry.[25] The Saami took their herds through a sophisticated annual cycle of spring, summer, fall, and winter pastures with the sizes of communes, herds, and appropriator organizations—the Siida organizations—varying systematically over the cycle, much in the spirit of the Field model (Sara and Kristiansen 1991). Also, Saami society instituted procedures for resolving disputes on the basis of customary law, although the details of the system are apparently not known today (Sara and Kristiansen 1991, 168). The property regime appears to have been reasonably efficient. Not a single historical example of overgrazing in the Saami reindeer regions is known, although the Saami have been nomadic herders of domestic reindeer in Finnmark at least since the 1600s (Bjørklund 1991, 183).

In the modern system, the Siida organizations are no longer autonomous. Their former authority has been transferred to the national government and its agencies which regulate the industry in detail, determining, for instance, grazing districts, grazing periods, and the maximum number of reindeer that can graze in a district. The authorities can even determine the size of individual flocks (Bjørklund 1991, 185). The administrative structure of the industry is rather complex with three levels (industry, district, and subdistrict levels) not counting the Ministry of Agriculture which tops the pyramid (Kristiansen 1991, 184). To the extent that the objectives of the top decision makers can be deducted from formal declarations, they are complex and even contradictory. The agreement of 1976 between the Ministry of Agriculture and the National Association of Saami Reindeer Herders lists the following objectives:

1. To maximize the production of food from the pastures, without weakening the resource base.
2. To guarantee personal incomes in the industry that are comparable with incomes in the other sectors of the economy.
3. To guarantee secure employment and traditional residence.
4. To guarantee that the reindeer industry develop in such a way that its central role in Saami culture is preserved (Bye 1991, 175).

Over time, the Saami have become increasingly sedentary, and motor vehicles, including snow-scooters, have lowered the cost of monitoring large herds over long distances. Also, the incentives in the reindeer industry have been affected by the instruments of government policy. These instruments include various forms of subsidies, and some scholars argue that an increase of about 100 percent in the size of the reindeer herds in the period since 1976 can be explained in large part as a response to government programs (Bjørklund 1991, 186). Crowding in the communal pastures is reflected in the falling weights of the animals and signs of overgrazing (Lenvik and Trandem 1991; Johansen et al. 1991). The evidence suggests that the national government has in part replaced the former system of communal property with open access.[26]

Why do national governments introduce open access and place resources in the public domain? We can think of three possible explanations:

1. It suits the interest of the decision makers, for some reason, which implies that they are satisfied with the outcome.
2. It is an instance of the collective action problem where decisions by rational individuals bring outcomes that no one likes.
3. The decision makers either lack data to make better decisions and/or they are using the wrong model of reality to make their decisions.

All three explanations are possible, and the answer to the puzzle is essentially an empirical question that we leave to the reader.

NOTES

1. Many scholars prefer to use the term "common property" rather than "communal property" for exclusive resources that are shared. Other scholars use the term "common property" to refer to non-exclusive assets with open access, and the "commons problem" is widely understood as implying the waste associated with open access. Much confusion has been caused by two theoretical concepts sharing the same two words, which in this instance suggests that individual rights rather than sharing may be a more productive arrangement.

2. See Eggertsson (1990, 249–262).

3. Milgrom and Roberts (1992) provide an excellent survey of the modern economics of organization. The studies they examine usually assume that the players are located in a laissez-faire environment. Also, many studies ignore the wealth effects of alternative arrangements when individuals seek

to maximize the joint value of their assets. See p. 288 in Milgrom and Roberts (1992).

4. Following North (1990) we distinguish between organizations and institutions. Organizations are groups of individuals that play together according to rules that are both internal and external. The external rules, formal and informal, and their enforcement characteristics are referred to as institutions. The definition implies that the set of institutions that a player confronts depends on his location and status in society (a dictator faces a different set of institutions than her subjects). The term "property rights" refers to the power of an agent to control valuable margins of scarce assets. Society presents individuals with various rights and duties and their enforcement, but also individuals themselves privately enforce their rights. We talk, therefore, of internal and external sources of control. Individuals incur transaction costs when they enforce control to prevent either outright theft or the appropriation of value by their trading partners. From the viewpoint of a given set of actors, transaction costs are the costs of internal control, but from the aggregate or social viewpoint there is no distinction between internal and external control, and transaction costs refer to the aggregate cost of operating a regime of property rights.

5. Inputs or outputs appropriated by intruders are given zero value in the model.

6. A formal version of the Field model is found in the 1986 working paper. Firm size is not a choice variable in the formal model, which implicitly excludes the possibility that individual producers merge into large firms. See Field (1986) and Field (1989). Merger is explicitly considered by Lueck (1994) and Caputo and Lueck (1994) in an important extension of the naive model. Lueck (1994) explores the optimal use of a fixed (natural) resource. The choice variables include group size and three contractual arrangements: 1) a fixed payment contract (a firm) where a single party owns the fixed resource and hires effort from the other individuals; 2) a communal property contract where the group members supply their own inputs and equally share the fixed output; and 3) a communal property contract where the members only share access to the fixed resource. Caputo and Lueck (1994) extend the model in Lueck (1994) in various ways and compare private ownership with sharing over three possible margins: 1) output derived from the resource; 2) access to the resource; and 3) investment in the resource. Again optimization involves choosing the group size.

7. The allocation of inputs between the three activities is not optimal unless there is equality among the marginal (net) rates of return on inputs used in transformation, exclusion, and governance.

8. For instance, in the mountain pastures of Iceland the typical farmer required a large area for his or her flock of sheep, and the relative price of fences was high. The pastures were managed as communal property. See Eggertsson (1992). Note that instead of using large communal areas to meet variable weather conditions, relatively small individual plots could be instituted along with an active trade in grazing rights between individual

owners and users. However, high transaction costs could make the introduction and operation of a market in grazing rights inefficient.

9. See Runge (1992, 33). Note also p. 5 in Bromley's introduction to Bromley (1992).

10. The poverty argument for communal rights could be rescued if the introduction of individual rights required large-scale lump investments that bear fruit only in future periods. An isolated community that cannot borrow and is too close to subsistence to save is not able to make such investments. It is an empirical question whether a financial constraint is an important explanation of communal property regimes. Note that implicit in the poverty argument is the notion that communities would do better under individual property rather than communal property, if only the financial constraint were lifted.

11. See Stevenson's (1991) extensive study of the Swiss case. In his econometric investigations, Stevenson compared communal property with individual property in Alpine grazing and found that outcomes of communal regimes were inferior to those in individual regimes. Stevenson gives several theoretical and empirical reasons why his statistical results may not be correct. However, if the results are correct, the Swiss may indeed have ideological attachment to their communal arrangements and enjoy them like consumer goods.

12. Imagine that the users are restrained by costs and thus prevented from devastating the resource. The cost constraints could be due to the inelastic supply of a cooperating input, such as water on grazing land or fishing vessels in a fishery.

13. Of course, technically these individuals could be compensated for their loss.

14. For an excellent survey of the current state of the theory of collective action, see Sandler (1992).

15. For an introduction to the theory of implicit and explicit contracts and various applications of the theory, see Werin and Wijkander (1992).

16. For an excellent discussion of these issues, see Taylor (1987).

17. The pure game-theoretic approach catches the group before it forms a community and before the individuals are constrained by social institutions such as norms, conventions, and customs, which implies that the members have not developed a common language, religion, or network of family and kinship ties. It is an amusing thought to try to visualize these isolated speechless individuals gathered to select a system of property rights and play complex games with each other. However, it must be admitted that the introduction of prior rules begs the question of the origins of cooperation.

18. Gluckman (1956) has summarized and interpreted some of their findings. Bates (1983) has retold the story in the language of game theory.

19. Here we are faced with the fundamental question of whether social and political institutions lead an independent life or merely reflect technologies and economic forces. The answer is, both.

20. The term "appropriator organization" is due to Ostrom (1992).

21. The breakdown of communal (or any) property regime need not involve the formal removal of the rules that define the regime, but a weakening of their enforcement.

22. Rainfall is localized in the region and the sedentary households "solve this problem by keeping only small milking herds in the village and hiring herders to take the majority of their stock to remote and highly mobile cattle camps" (Ensminger and Rutten 1990, 23).

23. In the case of the Orma, at one point decentralized control was successfully maintained with family ownership of wells, and the control of access to water was used to regulate access to grazing (Ensminger and Rutten 1990, 3).

24. Consider the vast dissipation of oil reserves in many parts of the American Southwest that results when several independent producers share the same underground oil reservoir. According to Libecap and Wiggins (1985), asymmetric information about the value of each lease prevents independent users from agreeing on jointly operating their reservoir. An outside government could force an agreement and set general rules that require joint operations in all cases. However, positive political theory tells us that decisions by governments are plagued by information and transaction problems, and individually rational behavior by public decision makers can bring irrational outcomes. Libecap and Wiggins (1985) report that the state governments of Texas and Oklahoma failed to design rules that encouraged unitization of oil fields, whereas in Wyoming, where oil fields were mostly on federal land, the federal government designed a structure of property rights that encouraged unitization.

25. The discussion of the Saami case is based on several of the essays contained in Stenseth, Trandem, and Kristiansen (1991).

26. Open access is both an indirect and direct result of the new law for the industry. As an example of a direct effect, the law has given free access to pastures that by tradition were exclusively owned by specific individuals or groups. There are some similarities between the Norwegian government creating open access in the pastures in Finnmark and the chronic overgrazing on the Navajo Reservation as the result of the policies of the U.S. Interior Department and the Navajo Tribal Council. The policies were intended to preserve the pastoral culture of the Navajo, but in effect they legislated a common property condition for the range and forced many Navajo to leave their traditional employment of sheep raising and accept wage work or welfare (Libecap and Johnson 1980).

REFERENCES

Anderson, Terry L. and P. J. Hill. 1975. "The Evolution of Property Rights: A Study in the American West." *Journal of Law and Economics* 18, no. 1: 163–179.

Barzel, Yoram. 1989. *Economic Analysis of Property Rights*. Cambridge: Cambridge University Press.
Bates, Robert H. 1983. "The Preservation of Order in Stateless Societies: A Reinterpretation of Evans-Pritchard's The Nuer." In *Essays on the Political Economy of Rural Africa*. Cambridge: Cambridge University Press.
Bates, Robert H. 1988. "Contra Contractarianism: Some Reflections on the New Institutionalism." *Politics and Society* 16, no. 2–3: 387–401.
Bjørklund, Ivar. 1991. "Saamisk reindrift som pastoral tilpassningsform. Noen betraktninger om økonomisk modernisering og kulturell endring på Finnmarksvidda." In *Forvaltning av våre fellesressurser. Finnmarksvidda og Barentshavet i lokalt og global perspectiv*, edited by N. Stenseth et al. Oslo: Ad Notam forlag.
Bromley, Daniel W., ed. 1992. *Making the Commons Work. Theory, Practice and Policy*. San Francisco: Institute of Contemporary Studies Press.
Bye, Karstein. 1991. "Målsettinger og virkemidler i reindriftspolitikken." In *Forvaltning av våre fellesressurser. Finnmarksvidda og Barentshavet i et lokalt og global perspectiv*, edited by N. Stenseth et al. Oslo: Ad Notam forlag.
Caputo, Michael R. and Dean Lueck. 1994. "Modeling Common Property Ownership as a Dynamic Contract." *Natural Resource Modeling* 8, no. 3: 225–245.
Coase, Ronald H. 1937. "The Nature of the Firm." *Economica* 4, November: 386–405.
Demsetz, Harold. 1967. "Toward a Theory of Property Rights." *American Economic Review* 57, May, no. 2: 347–359.
Eggertsson, Thráinn. 1990. *Economic Behavior and Institutions*. Cambridge: Cambridge University Press.
Eggertsson, Thráinn. 1992. "Analyzing Institutional Successes and Failures: A Millennium of Common Mountain Pastures in Iceland." *International Review of Law and Economics* 12: 423–437.
Ensminger, Jean and Andrew Rutten. 1990. "The Political Economy of Changing Property Rights: Dismantling a Kenyan Commons." Working paper. St. Louis: Center in Political Economy, Washington University.
Field, Barry C. 1986. "Induced Changes in Property Rights Institutions." Research paper. Amherst: University of Massachusetts, Department of Agriculture.
Field, Barry C. 1989. "The Evolution of Property Rights." *Kyklos* 42, no. 3:319–345.
Gluckman, Marx. 1956. *Custom and Conflict in Africa*. Oxford: Basil Blackwell.
Hechter, Michael. 1990. "The Emergence of Cooperative Social Institutions." In *Social Institutions: Their Emergence, Maintenance and Effect*, edited by M. Hechter, et al. Berlin: De Gruyter.
Johansen, Bernt et al. 1991. "Det biologiske ressursgrundlaget for Finnmarksreinen." In *Forvaltning av våre fellesressurser. Finnmarksvidda og Barentshavet i et lokalt og global perspectiv*, edited by N. Stenseth et al. Oslo: Ad Notam forlag.

Johnson, Ronald N., and Gary D. Libecap. 1982. "Contracting Problems and Regulations: The Case of the Fishery." *American Economic Review* 72, no. 5: 1005–1022.
Kristiansen, Gørill. 1991. "Organisasjon og forvaltning i reindriften." In *Forvaltning av våre fellesressurser. Finnmarksvidda og Barentshavet i et lokalt og global perspectiv*, edited by N. Stenseth et al. Oslo: Ad Notam forlag.
Lenvik, Dag and Nina Trandem. 1991. "Forvaltning av tamrein i Nord-Norge:status og Muligheter." In *Forvaltning av våre fellesressurser. Finnmarksvidda og Barentshavet i et lokalt og global perspectiv*, edited by N. Stenseth et al. Oslo: Ad Notam forlag.
Libecap, Gary. 1978. "Economic Variables and the Development of the Law: The Case of Western Mineral Rights." *Journal of Economic History* 38, no. 2, June: 399–458.
Libecap, Gary D. 1989. *Contracting for Property Rights*. Cambridge: Cambridge University Press.
Libecap, Gary D. and Ronald N. Johnson. 1980. "Legislating Commons: The Navajo Tribal Council and the Navajo Range." *Economic Inquiry* 18, January: 69–86.
Libecap, Gary D., and Steven N. Wiggins. 1985. "The Influence of Private Contractual Failure on Regulation: The Cost of Oil Field Unitization." *Journal of Political Economy* 93, no. 4: 690–714.
Lueck, Dean. 1994. "Common Property Is an Egalitarian Share Contract." *Journal of Economic Behavior & Organization* 25: 93–108.
Milgrom, Paul and John Roberts. 1992. *Economics, Organization and Management*. New Jersey: Prentice Hall.
North, Douglass C. 1990. *Institutions, Institutional Change, and Economic Performance*. Cambridge: Cambridge University Press.
Ostrom, Elinor. 1990. *Governing the Commons. The Evolution of Institutions for Collective Action*. Cambridge: Cambridge University Press.
Ostrom, Elinor. 1992. "The Rudiments of a Theory of the Origins, Survival, and Performance of Common-Property Institutions." In *Making the Commons Work*, edited by D. Bromley. San Francisco: Institute of Contemporary Studies Press.
Roberts, Russell D. 1990. "The Tragicomedy of the Commons: Why Communities Rationally Choose 'Inefficient' Allocations of Shared Resources." Political economy working paper. St. Louis: Center in Political Economy, Washington University.
Roberts, Russell D. 1993. "To Price or Not to Price: User Preferences in Allocating Common Property Resources," In *The Political Economy of Customs and Culture. Informal Solutions to the Commons Problem* edited by T. Anderson and R. Simmons. Lanham, Maryland: Rowman and Littlefield.
Runge, C. Ford. 1992. "Common Property and Collective Action in Economic Development." In *Making the Commons Work*, edited by D. Bromley. San Francisco: Institute for Contemporary Studies Press.

Sandler, Todd. 1992. *Collective Action. Theory and Applications.* Ann Arbor: University of Michigan Press.
Sara, Aslak Nils and Gørill Kristiansen. 1991. "Reindriften i Finnmark årssyklus, driftsstrategier og forskningsutfordringer." In *Forvaltning av våre fellesressurser. Finnmarksvidda og Barentshavet i et lokalt og global perspectiv,* edited by N. Stenseth et al. Oslo: Ad Notam forlag.
Stenseth, Nils Chr., Nina Trandem, and Gørill Kristiansen, eds. 1991. *Forvaltning av våre fellesressurser. Finnmarksvidda og Barentshavet i et lokalt og global perspektiv.* Oslo: Ad Notam forlag.
Stevenson, Glenn G. 1991. *Common Property Economics. A General Theory of Land Use Applications.* Cambridge: Cambridge University Press.
Taylor, Michael. 1987. *The Possibility of Cooperation.* Cambridge: Cambridge University Press.
Weitzman, Martin L. 1974. "Free Access vs. Private Ownership as Alternative Systems for Managing Private Property." *Journal of Economic Theory* 8, June: 225–234.
Werin, Lars and Hans Wijkander, eds. 1992. *Contract Economics.* Oxford: Basil Blackwell.

GARY LIBECAP   CHAPTER 3

Distributional and Political Issues in Modifying Traditional Common-Property Institutions

Introduction

Throughout the world, indigenous property rights systems are under pressure. Local arrangements for allocating access and use of resource stocks, including inshore fisheries, grazing lands, forests, and animal herds, have historically been a durable means of maintaining the resource and preventing the dissipation of resource rents. Often, these arrangements have been common-property institutions, whereby non–group members have been denied access, but group (community) members have been granted usufruct rights to the resource. Although problems of calculating and assigning individual allotments and insuring individual compliance with harvest rules have existed under common-property conditions, so long as the group was reasonably small, homogeneous, and had shared preferences or objectives regarding the resource, serious depletion was not an issue. Long-standing equilibrium conditions emerged.

These conditions and the associated effectiveness of local common-property institutions in preventing open-access losses (Gordon 1954), unfortunately are under stress. They are vulnerable to rising resource values, new technology, new entry, and new legal codes, drafted elsewhere in the society. Advanced harvest technology and capital equipment have dramatically increased the effects individual extraction can have on the stock. Further, rising resource prices and the depletion of stocks elsewhere have invited entry by nontraditional users, who do not adhere to local harvest rules. This entry, outmigration from traditional societies, and the introduction of other cultures have weakened social cohesion within traditional groups and the effectiveness of common-property arrangements.

Indeed, unlike more formal, impersonal private property rights, traditional institutions, which are based on local information, repeat contracts, and shared preferences, are singularly ill prepared to respond to major new entry pressures. If, instead, new entry is limited and the problem is one of a breakdown of traditional rules, then a solution is the more formal definement of individual property rights within the existing structure. The shares of all current members must be renegotiated, reduced, and made more flexible if total harvest is not to increase with entry. This negotiation, as described below, however, can involve distributional issues that tax the political framework of traditional institutions. Compliance problems also will increase if new entrants do not recognize the legitimacy of the traditional commons institution. Moreover, as existing shares are reduced to lessen pressure on the resource, the incentive to cheat increases.

Modifications of traditional institutions can be through the assignment of transferable quotas for fishing or for formal grazing permits for herding. If, however, entry is more significant, then broader public policy intervention is required to define property rights, since traditional arrangements are unlikely to provide a complete framework for addressing open-access problems.[1] New rules, however, must recognize existing practices in order to be effective, but substantial adjustments may be required.

Accordingly, in the face of the potential collapse of—or at least, of severe pressure on—many of the world's indigenous common-property arrangements, calls have been made for public policy intervention to devise ways to supplement or strengthen local common-property institutions. Adding public policy, however, brings a new set of problems. A new set of actors—politicians, agency officials, other claimants, and interest groups—are added to the original users. This creates a new bargaining setting that is more complicated with less clear results for members of traditional groups. Other interests will be weighed in the political process, so that it is no longer predictable that the arrangement will benefit or be consistent with the desires of the original indigenous group.

This chapter examines a number of issues with regard to the modification of property rights: first, the incentives to change existing property rights arrangements as new conditions emerge are summarized; second, bargaining issues that are raised due to distributional concerns are highlighted; third, the incentives of politicians, bureaucrats, and new entrants, and the implications for existing resource users are introduced; and fourth, two case studies from U.S. inshore fisheries and Indian grazing practices are presented for insights into how distributional issues have affected political bargaining over open-access problems and how those issues have affected the public policy response.

Incentives to Modify Property Institutions

Property rights are the social institutions that define or delimit the range of privileges granted to individuals to specific assets, such as parcels of land or water, fish, wildlife, and mineral deposits. Although property rights institutions vary from strictly defined private property rights to common-property arrangements for specified groups, included in all rights structures are the rights to exclude nonowners or nonmembers from access, the rights to appropriate the stream of rents from use of and investment in the resource, and rules regarding the transfer of individual property rights. As such, property rights institutions critically affect incentives for decision making regarding resource use and hence, economic behavior and performance. By allocating decision-making authority, property rights also determine who are the economic actors in a society and define the distribution of wealth.

Property rights institutions exist in order to avoid the losses of open-access conditions, where there are no restrictions on access and use. Under these circumstances, the value of the resource is dissipated through excessive and wasteful use practices. These include too rapid harvest rates because individuals do not take into account the social costs of their harvest decisions. As a result, total output by all parties using the open-access resource exceeds the social wealth-maximizing level. In addition, short time horizons dominate so that the user costs of production are ignored and long-term investment is neglected. Finally, competition for control diverts labor and capital inputs from production to predatory or defensive activities, and the associated uncertainty of control limits the emergence of markets for the exchange and allocation of the resource to higher-valued uses. In the absence of some type of market signals, the resource will not flow smoothly or routinely to new uses as economic conditions change.

Traditional societies with limited resources and production opportunities have understood well the dangers of unregulated access and use of valuable resources. Generally, the very survival of the community has depended upon successfully addressing the open-access problem. Hence, an intricate and sophisticated network of rules has been assembled over time to manage locally based natural assets, which include customs or policies regarding who can have access, harvest practices and rates, transferability, and dispute resolution.

These traditional institutions, however, are now under unprecedented pressure. The old equilibriums have been upset, and in many cases new institutional arrangements are required to address the open-access problem. New mechanisms are needed to arrive at a cooperative solution, but the parties that must cooperate are a much broader and more heterogeneous group than before and the distributional issues involved in assigning access and resource rents are much more difficult.

Distributional Issues in Contracting for Property Rights

In general, individuals within a group will negotiate among themselves to modify existing property rights institutions in order to mitigate the losses of the common pool, as soon as there are net benefits of so doing. Forces that drive these adjustments in property rights include declining harvests and income, new competition from others, and production possibilities to which the old arrangement was poorly attuned. But with new entry, no longer can the negotiations for institutional change take place solely within the traditional group. The interests of new entrants, politicians, and bureaucrats must be considered. The bargaining setting is much more complex, and dispute resolution and monitoring compliance with new property rules and harvest rates become more difficult.

These problems compound existing ones about the distribution of the gains and costs from changes in property rights arrangements. While the aggregate gains from reducing open-access problems through the redefinition of property rights are unlikely to be controversial (and empirically, this seems generally to be the case), the allocation of wealth and political power inherent in any adjusted rights structure will be a source of dispute. New property rights arrangements will not only have different production effects, but they will have different distributional implications as well. Some parties will clearly be made worse off, while others may benefit. Some parties will have their traditional harvest practices and the lifestyles associated with them limited, while other parties may be denied access altogether. These distributional effects occur even when there are significant aggregate gains. Distributional negotiations and devising a management and allocation scheme that is politically acceptable becomes the center of the problem in outlining a new property rights arrangement.

In political bargaining over institutional change, the positions taken by individual parties are determined by their expected net gains from the new arrangement with respect to status quo conditions. The benefits of the status quo are a function of current property rights, which define the individual's share of aggregate production, and the productive capacity of the resource. Those who have small shares under existing arrangements or who suffer particularly from the decline in harvests are most likely to expect some benefit from adjustments in property rights arrangements. Others who have adapted well to existing open-access conditions or who have disproportionately large shares will be reticent to make major changes, particularly if there is uncertainty regarding their future shares. Each bargaining party will attempt to mold the resulting arrangement in ways that maximize his share of the aggregate returns. This maneuvering affects the timing and nature of the property rights that are adopted and the aggregate benefits that are obtained. Accordingly, in modifying common-property practices to address open-access conditions, not only does a new management scheme have to be devised to limit harvests, but a new formula for allocating access and use must be created. This is not only a key problem within the traditional group, but it becomes the key political problem once public policy is brought in to supplement traditional arrangements.

The issue, then, becomes one of devising an allocation mechanism to assign the gains and costs from institutional change in acceptable ways, while addressing the open-access problem. Because over harvest, the depletion of the stock, and other conditions associated with open-access require restrictions on future exploitation, some parties will be adversely affected by the institutional change. They may be temporarily or permanently denied access and use or have their traditional use practices dramatically changed. By compensating influential parties that might be harmed in the proposed change, a political consensus for institutional change can emerge. Those share concessions, however, necessarily alter the nature of the property rights under consideration and the size of the aggregate gains that are possible. If influential parties cannot be sufficiently compensated through share adjustments to win their support, otherwise beneficial institutional change may not occur. Even though society is made worse off by the failure to address the new open-access problem, disputes over the distribution of access and resource rents can block a cooperative solution.

In principle, it is possible to imagine a side payment scheme that would compensate those who otherwise would oppose a socially-desirable change in property rights. But empirically, the record suggests that these side payments often will either be incomplete or not forthcoming, delaying collective action. Questions arise as to who should receive payments, who should pay, the size of the compensation, and its form. All of these issues are subject to dispute. These problems, for example, have affected the timing and assignment of individual quotas in fisheries, grazing permits, crude oil production quotas, and orange shipment quotas.[2]

Questions arise as to the basis for assigning quotas or other forms of use rights. Two possibilities are to grant them on the basis of prior possession or on the basis of previous production. Previous production as a criteria for shares, however, may involve severe information problems in documentation or verification. Prior possession does not consider new entrants, and fairness issues may arise if the distribution scheme leads to a skewed assignment of property rights.[3] A uniform allocation formula is a conventional alternative because it reduces the information problems associated with verifying past production and allows for inclusion of new entrants. It also avoids more complex and politically risky distributional arrangements and addresses fairness criteria. But uniform allocations disadvantage particularly skilled or successful parties who may have adapted well to the status quo. These individuals will have reason to oppose adjustments in property rights because they bear more of the costs and receive fewer of the benefits of the new arrangement. Conflict also will arise regarding the means for entry or exchange of property rights, since these practices often will involve outsiders. Finally, strategic bargaining by key parties to increase their share in the new arrangement can block or delay agreement, if unanimity rules are required to institute change.

All things equal, the intensity of political bargaining over distributional issues and the likelihood of successful property rights change will be influenced by (1) the size of the aggregate expected gains from institutional change, (2) the number and heterogeneity of the bargaining parties, (3) the skewness of the current and proposed share distribution, and (4) information problems. The larger the expected aggregate gains, the more likely politicians can devise shares to make influential parties better off, so that institutional change can proceed. On the other hand, the larger the number of bargaining parties, the greater the number of claims that must be addressed by politicians in assigning or modifying property rights, making institutional change more difficult. Time and precedent are critical factors in determining the number and bargaining power of claimants. Past political agreements regarding property rights define a set of actors or vested interests who can create advantages for future bargaining by molding political institutions to their benefit. Previous agreements also affect bargaining by setting precedents and expectations among competing groups regarding the expected gains from collective action to change property rights. The more heterogeneous the private bargaining parties, the more difficult the formation of coalitions and a consensus on the proposed assignment of rights. Further, a very skewed existing rights arrangement leads to pressure in political contracting for a redistribution of wealth. Indeed, those parties without current property rights are motivated to lobby for redistribution even if there are no aggregate benefits from institutional change. Finally, information problems raise contracting costs by intensifying disputes over how the proposed change will affect individual parties and what share adjustments are necessary for compensation. Failure to agree on such compensating shares may convince those who do comparatively well under the current arrangement, even open-access, that they will be made worse off by the institutional change.

Public Policy and Political Bargaining to Change Property Rights

Consideration of the details of political bargaining is necessary for predicting the ultimate impact on traditional users of public policies to respond to open-access problems. Because property rights are politically determined, and especially in appeals beyond the traditional community to public policy, the definition and enforcement of property rights will occur in the political arena. The very existence of an open-access problem indicates that the informal customs and agreements which required little or no state intervention and were sufficient in the past, are now inadequate. However, lobbying politicians and other government officials for new or increased government support for modifying and protecting traditional use practices will activate other interest groups in the political process, as well as involve the additional interests of politicians and bureaucrats. With a broader array of competing interests, greater government intervention in the definition and enforcement of property rights will make bargaining more complex and require concessions from traditional users in the form of redistribution of resource rents to other influential constituents. The bargaining parties include private claimants (traditional users, plus new entrants, environmental groups, and other constituents), politicians (incumbents and aspiring office holders), and bureaucratic officials (who will administer public policy regarding the management of the resource). All have an incentive to devise a management scheme that advances their interests, and these may not be consistent with the interests of traditional users.

Politicians will play an important role in brokering any new arrangement, but they will have a different incentive structure and face a different array of costs and benefits than do the other parties involved in bargaining to change rights arrangements. For one thing, they have short time horizons. Politicians have no particular reason to be concerned about very long-term, sustainable resource uses. The demands they face are immediate, and there are no futures markets in votes. Current practices in the United States regarding the funding of social security and a lack of sustained interest in reducing the federal deficit are examples of an inherent short-term bias in political decisions. Additionally, vote-maximizing politicians must respond to many competing interests to insure reelection or the maintenance of political power. They have incentives to maintain status quo distributions, and do so by balancing competing demands for resource access and use, so that no group will get all that it wants through public policy.

This suggests that if traditional users are not well organized and are not politically influential, then the demands of other constituents, perhaps the new entrants, will prevail. Indeed, traditional groups with histories of reaching agreements and maintaining traditional common-property institutions will likely be small, and since income from traditional harvest practices is apt to be low, such groups are generally relatively poor. Education levels and experience in using the political process may be limited. The political influence of competing groups of claimants depends upon their wealth, size, and homogeneity (Stigler 1971; Peltzman 1976; Becker 1983). This suggests that traditional users will not be particularly effective lobbyists in their own behalf in the competition for resource access and use as property rights are being adjusted. In addition, with different political jurisdictions involved, as is a standard case, there will be different and competing politicians, ranging from national politicians (and if more than one country is involved, there will be multiple national politicians with competing interests at stake) to local politicians. All have different constituencies, and will make resource use decisions with their own objectives in mind. Traditional users may have ties to one group of politicians (local), but lack critical ties to national politicians.

Necessarily, agency officials, who administer statutes in devising public policy, also have a short-run bias. They must be responsive to elected officials. Although there is latitude in the devising of administrative rules and perhaps an ability of agency officials to act on their own preferences, agencies cannot stray too far from the desires of the existing electorate (Weingast and Moran 1983). Indeed, agency decisions regarding the administration of public policy are critically affected by the need to form political alliances with influential constituents for appropriations, staffing, and maintenance of regulatory mandates. Additionally, agency officials are not residual claimants. That is, they do not bear the full costs or benefits of their administrative policies, and hence have less incentive to devise policies that maximize the rental value of the resource than do actual resource users.

For these reasons, one cannot predict that public policy outcomes will necessarily be in the long-run interest of traditional users, even if legislation or the initiating call for government intervention is made in their behalf. Traditional users become but one of many competing interests at stake. In general, the greater the magnitude of the open-access problem, the more likely there will be a response from politicians to devise a new property rights arrangement. Once problems have become very severe, interest groups are more likely to form cohesively and effectively to pressure politicians for action. This suggests, however, that a political response to an open-access problem will not occur until late, after much of the damage has been done.

Given the various competing parties and potential for conflict over the allocation of property rights and the prediction that a political response is apt to be delayed, institutional change is likely to be an incremental process with modest adjustments from status quo conditions. The role of time and precedent in influencing the number of vested interests and the expected returns from collective action suggest an historical path dependence for property rights institutions.

These arguments imply that caution is in order regarding the efficacy of public policy intervention to address open-access problems faced by traditional users. The closer that solutions rely on existing practices, the more likely they will advance the welfare of current users and at the same time, protect the resource. The arguments also suggest that within traditional groups and across other competing users, negotiations to modify existing property rights arrangements will raise distributional concerns that will affect the new institutions that are put into place and their effectiveness in mitigating rent dissipation. Some of these issues are illustrated in the following empirical examples from the United States.

U.S. Inshore Fisheries

In some cases, public policy has not been very supportive of traditional (or at least, long standing) use practices. The political influence of other, competing users has been a critical factor. For example, Higgs (1982) describes the vibrant nature of the Pacific Northwest inshore salmon fishery at the turn of the century, when salmon were abundant and could be harvested at low cost due to their anadromous nature. Because salmon returned from the ocean to the streams from which they were spawned to deposit and to fertilize eggs, they could be harvested from fixed sites along streams leading from the Pacific Ocean using fish wheels and gill nets. A system of private property rights to those sites emerged along major rivers, such as the Columbia, similar to the well-developed property systems used earlier by Indians.

As early as 1892, however, there were concerns about the entry of new fishermen and the impact on the stock of the growing rise in total gear used in the fishery. Declining productivity created intense hostilities among various groups of fishermen, who were identified by the types of equipment they used. Each group blamed over-fishing and its consequences on others and attempted to have the fishing privileges of their rivals curtailed. Public policy solutions were demanded, and state legislatures were drawn into the fray. Gill netters increasingly were able to secure legislation in Oregon and Washington that placed discriminatory restrictions and taxes on the operators of fish wheels. Ultimately, the low-cost, productive fish wheels were outlawed by the two states. However, removing one group did not solve the open-access problem. Conflicts over access and harvest continued among owners of fish traps in Puget Sound, commercial purse seiners who relied on vessels, and sports fishermen. New political coalitions of fishermen formed to lobby for restrictions on their competitors. Because of their small numbers and highly visible large catches, fishermen who used fish traps were especially vulnerable. With the growing political influence of numerous sports fishermen and those commercial fishermen who used vessels, regulations eventually were adopted to forbid fish traps. By the early part of the twentieth century, these historical fishing practices disappeared.

As fishing pressure continued, new regulations were authorized by state legislatures and molded by regulatory agencies to force the interception of salmon in the ocean at much higher costs. Capitalization and labor costs increased as the number of boats and fishermen rose. As the stock of salmon declined from more intensive harvest, a principal regulatory response was to construct costly hatcheries and to shorten the fishing season in an attempt to raise aggregate catch. The progressive shortening of seasons intensified the rush of fishermen to complete their harvest early and added pressure for larger and faster vessels. Moreover, tensions among competing fishing groups continued as each sought to obtain legislation that favored it and posed constraints on its rivals. No long-term satisfactory solution has been obtained, despite continued regulatory efforts, and the value of the salmon fishery in the two states has declined.

Similar problems in satisfactorily addressing open-access problems have been encountered elsewhere, and their persistence is not due to some technological imperative or lack of scientific analysis. In examining property rights and regulation in the Texas Gulf Coast shrimp fishery, Johnson and Libecap (1982) describe the actions of fishermen unions in devising locally-based rules for limiting access and harvest. The Gulf Coast Shrimpers' and Oystermen's Association along the Mississippi coast devised rules to restrict entry and harvest. Under union rules, fishermen were permitted to sell only at or above the association's floor price. By setting a minimum price for small, immature shrimp, which generally exceeded market prices and had to be paid by local packers, the rules reduced the quantity of small shrimp demanded by the packers. Accordingly, the higher price required for small shrimp acted to redirect harvest to later in the season and thereby increase the yield of higher-valued larger shrimp. The market price per pound for larger shrimp set by the union for payment by packers was equal to the market price. Shrimp purchased by packers at less than the mandated price would not be peeled by union peelers. The union also obtained state legislation that recognized its practices and fixed minimum sizes for harvest. The analysis of harvest price data by Johnson and Libecap indicate that the union was successful in delaying harvests in Mississippi and in raising the size of the shrimp caught and marketed there, relative to neighboring Louisiana. Even so, this effort, as well as similar efforts by fishery unions on other U.S. coasts, was struck down by the U.S. Justice Department as violations of the Sherman Antitrust Act at the behest of those fishermen who were denied access by local union rules.

In the absence of locally-based arrangements, fishermen in most U.S. inshore fisheries have relied upon public policy with at best, spotty results. Over-fishing remains a common characteristic, and catch and incomes have fallen. Neither fishermen nor regulatory agencies have been able to devise very satisfactory harvest rules. Until recently, few quota arrangements were adopted. To avoid the redistribution problems associated with quota design, fishermen could agree only on across-the-board regulations such as season closures or equipment restrictions.

Disputes have arisen over the impact of harvest restrictions and on the response of the stock to regulatory practices. Due to differences in skill among fishermen, catch and income have varied sharply. In the design of institutions to reduce open-access losses, each party has been concerned with how the new arrangement will impact its share of total catch. For better fishermen, there has been the hazard that allowable catch and income under any new institutional arrangement would be less than they received under the status quo. These redistribution concerns have existed for a long time and have limited agreement on institutional change until fisheries became severely depleted with all harvests low. At that point more of the bargaining parties have been able to see their welfare improved by controls on catch, and agreement has become more likely. Unfortunately, by that time the costs of the open-access problem will have been long standing and the stock seriously depleted.

Among the competing contracting parties have been commercial fishermen of various kinds and sports fishermen. Because of their large numbers as voters, sports fishermen have been politically influential and have succeeded in promoting regulations that have often displaced commercial fishermen and any informal property rights arrangements they may have devised.

Historically, a political consensus has emerged among commercial and sports fishermen only for regulations that tended to avoid controversial distributional issues, and instead, focused on visible yield-enhancement—hatcheries, season closures, gear restrictions, and entry controls on outsiders. Until recent depletion changed bargaining stands, limited access schemes and individual quotas were a much less popular regulatory approach.[4]

Limited access schemes usually involved issuing a restricted number of fishing licenses and allowing entry only to licensees as a means of reducing overall harvest rates and pressure on the stock. With the number of licenses kept small relative to the number of fishermen who would fish under open-access conditions, and entry restricted to license holders, rents could be increased. If the licenses were considered to be a permanent assignment of access to the fishery and were transferable, they could become a valuable property right. Because of the potential wealth assignment involved, determining who would receive the initial licenses and the procedure by which they would be granted have been important problems to be resolved. Political influence based on numbers, cohesion, and wealth have been more critical determinants of who received licenses than have been other criteria, such as the impact of various fishing groups on fishery rents or past use practices. Because total rents could be increased and redistributed through restrictive licensing, some fishermen therefore could be made better off relative to their position under the status quo. Within the group receiving licenses, however, the problems of designing and enforcing intragroup controls on fishing remained, especially in the absence of local arrangements that generally have been prohibited by law.

Recently, individual, transferable quotas have become a more common response to this problem since they restrict entry and limit individual catch. But their long-run acceptance and use still have faced the concerns of fishermen. With transferable quotas, some of the equal access questions that may be politically important have been resolved. For example, markets have developed for the transfer of quotas to allow new fishermen to enter or to allow some of those who were excluded to reenter the fishery.

There has been, however, the problem of the initial assignment of quotas. If the quotas were granted to incumbent fishermen, they would receive a wealth transfer. Politicians have considered imposing taxes on the value of the license, perhaps to compensate those who were excluded from the fishery. Similarly, if the licenses were sold by the government through price discrimination schemes, the government could extract all of the rents so that fishermen were no better off under regulation. In either case, the adoption of taxes or pricing policies in limited-access schemes could reduce the welfare gain to fishermen from the new program and sharply reduce their enthusiasm for it.

There have been other issues regarding the size of the quota and whether it would vary among fishermen and across the season. Variable quotas to reflect past harvest practices and differences in skill have been considered a means of building support among successful fishermen for regulation. These practices, however, have been found to be uncommon in a variety of empirical studies, where uniform, across-the-board quotas are predominant.[5] Uniform quotas would be responsive to equity concerns, which are common political goals.[6] They, however, disadvantage more skilled fishermen. Adjusting quotas across seasons and within seasons by regulatory agencies to respond to new estimates of the condition of the stock also could be an important feature of regulation, but it introduces uncertainty for fishermen in calculating their expected gains from the adoption of a quota system. Further, uncertainty regarding the size of annual quotas, the duration of quota policies, and the nature of other regulatory actions have added to the difficulties facing fishermen in calculating individual benefits from the new arrangement relative to the status quo. Moreover, uncertain quotas could encourage fishermen to violate their allotments, raising enforcement costs and reducing the effectiveness of the policy in enhancing the growth of the stock and aggregate fishing incomes.

Nevertheless, regulatory officials and politicians have some incentive to adopt temporary quotas. A permanent quota system could sharply reduce the administrative authority of regulators and justification for agency staffing and budgets. Further, permanent quotas limit the ability of politicians to respond to changing political demands for free access to the fishery. With transferable permanent quotas, subsequent exchanges of access rights would be through market transactions and not through political assignments. Finally, there would be political pressures opposing a permanent quota system from fishermen who have their access and harvest opportunities reduced, as well as from input suppliers ranging from fishing crews to vessel and equipment manufacturers and retailers who have a stake in a less restrictive regulatory regime.[7]

This summary indicates some of the bargaining problems encountered in devising regulatory schemes to address open-access problems in fisheries. They have not been easy issues to circumvent. Moreover, public policies and judicial responses to open-access problems often have not considered (or ruled out) locally devised arrangements. This has reduced the effectiveness of regulation in protecting the resource.

Grazing Practices and Regulation on U.S. Indian Reservations

American Indians, particularly those in the Southwest, have pastoral economies. Almost all are under stress as the number of herders has increased, and over-grazing and deteriorating range quality are common results. This problem, unfortunately, has existed since the 1930s and again, no effective, long-term solution has been devised. There are conflicting goals of maintaining traditional pastoral cultures by granting tribal members access to the land in the face of rapidly increasing populations and of safeguarding the sustainability of the range resource. There seems to be no evidence that these conflicts are being resolved in a satisfactory manner, despite the passage of sixty years.

Grazing practices, the extent of overgrazing, and the quality of range land vary across the reservations (Johnson and Libecap 1980). Regulatory practices by tribal councils and outside government agencies, such as the Bureau of Indian Affairs, have had mixed effects on the resource stock and on the welfare of tribal members. Political factors, both within the tribes and in government agencies, have played a critical role in formulating regulatory policies regarding property rights and range land use. The experience indicates that one cannot be too sanguine that either tribal governments or the federal government can provide property rights arrangements that preserve the resource and advance the well-being of tribal members.

In the United States, the federal government holds title to Indian land, and formal use rights, where they exist, are granted to individuals through the Bureau of Indian Affairs (BIA) and local tribal councils. In assigning grazing rights, the BIA has emphasized the equal distribution of tribal land. In the process, it has rejected existing claims of large herders where they have been associated with overgrazing and where their holdings have been deemed unequal. Historically, large herders have established informal control of range land on many southwestern reservations through prior appropriation and continued occupancy.[8] There are economies of scale in herding, so that large herders have higher per animal returns. Absent an ability to obtain formal property rights to their land, large herders in many cases have engaged in "limit grazing" to reduce the threat of entry by other herders on their customary lands. Under limit grazing, herders stock beyond the level that would otherwise maximize rents in order to reduce the expected gains from entry.[9] Although this practice of overgrazing is an effective means of defining and enforcing customary grazing areas, it weakens plant stands and makes the range vulnerable to erosion and the introduction of unpalatable species.

Recognizing and enforcing the land claims of large herders to allow them to discontinue overgrazing practices and to encourage them to invest in the long-term quality of the land has not been politically feasible for either tribal councils or the BIA. Large herders have been viewed as better able to bear the costs of imposed stock reductions to improve range quality. More importantly, large herders have controlled a disproportionate amount of reservation land. Recognizing their claims would deny the potential claims of other tribal members, and in any event, federal policy since 1933 has been to emphasize the communal nature of Indian lands. Finally, large land holdings prevent the granting of herding privileges to additional members, and as populations have increased, the demand on popularly-elected tribal councils for herding opportunities has correspondingly risen. Hence, recognizing the land claims of large herders has been inconsistent with other political goals.

Accordingly, in many cases either uncompensated, forced redistributions of land have occurred through BIA policies with an emphasis on an equal distribution of the land (Navajo and Zuni reservations) or the claims of large herders have been tacitly admitted, but no clarification of rights has occurred (Cochiti, Santo Domingo, San Felipe, Sandia, Santa Ana, Taos, Santa Clara, Tesuque, Name, and San Juan reservations). Naturally, uncompensated redistribution has been resisted by herders and has been politically controversial. Unfortunately, while redistribution has brought about a rise in small herds on the Navajo reservation, for example, it has not resulted in range improvement practices. Indeed, with the rise of small herds and the political pressures to facilitate new entry by additional herders, total stocking has increased and property rights have become less, not more, confused. Not surprisingly, range land conditions have deteriorated (Libecap and Johnson 1980).

Conclusion

As per capita incomes rise around the world, there is greater concern about the rational use and conservation of natural resources. The professed goal is a sustainable interplay between man and the environment. Historically, traditional common-property institutions have been quite successful in small, homogeneous communities in maintaining resource stocks and the community wealth on which they are based. Recently, with rising populations, migration, new entry, and the introduction of new technology, these traditional arrangements have been placed under stress. Policy discussions have emerged regarding institutional changes away from traditional practices to more formal rights assignments to promote more sustainable resource uses. In some cases, pressures have arisen for more formally defining the use rights granted community members, which have previously been informal and vague. Indeed, a common result of rising resource values and greater competition for resource use is a demand for an increase in the specificity of property rights.[10] Other pressures have risen to both reduce the number of individuals who can exploit the resource and limit the harvest rates of those who are allowed to continue exploitation. These raise critical distributional issues that affect political support for institutional change. Resolving distributional conflicts over the redefinition of property rights, however, fundamentally changes the nature of the institution that ultimately can result, with implications for its effectiveness for managing the resource stock. Additionally, as traditional users turn (or are turned) to outside politicians and administrative agencies to address resource use problems, new objectives and interests are added. The melding of a broader array of competing political objectives for resource assignment and use may not lead to policies that advance the interests of traditional users or that significantly protect the resource. Accordingly, caution is necessary in calling for public policy intervention, and once a path of regulatory change is taken, the distributional concerns of the various parties involved must be considered if collective action is to be successful in safeguarding the resource and the traditional societies that depend upon it.

NOTES

1. Ostrom (1990) provides case studies and analyses of how locally based institutions can address or be modified to address open-access problems. Johnson and Libecap (1980, 1982) and Libecap and Johnson (1980) indicate the problems encountered in different resource settings when external agencies disregard existing property rights and resource practices.

2. See Johnson and Libecap (1982), Johnson and Libecap (1980), Libecap and Johnson (1980), Libecap and Wiggins (1985), Libecap (1989c), and Hoffman and Libecap (1992).

3. For discussion of fairness issues, see Hoffman and Spitzer (1982, 1985); Fogel (1992).

4. Similarly, Hoffman and Libecap (1992) find that orange growers could not agree on prorationing rules in Florida under citrus marketing orders because of disagreement on the impact on particular growers and shippers. No quota design could be devised that brought agreement. Hence, unlike California, Florida growers have relied instead on across-the-board shipping holidays (season closures) and uniform grade and size restrictions to limit shipments to market.

5. For example, see the regulatory case discussed by Hoffman and Libecap (1992) regarding orange marketing orders. See also Johnson and Libecap (1982).

6. See Fogel (1992).

7. For discussion of individual quota systems and their advantages and costs, see Scott (1989); Libecap (1989b); and Neher, Arnason, and Mollett (1989).

8. The notion of occupancy and beneficial use as a means of legitimizing claims is a common practice. It was the basis for U.S. homestead allocations under federal land policy in the nineteenth century, and remains the basis for land claims by squatters and others with otherwise formal title in Brazil. Failure of title holders to occupy and "use" their lands makes them vulnerable to entry by others. See Alston, Fuller, Libecap, and Mueller (1997).

9. The limit grazing model is developed in Johnson and Libecap (1980).

10. For example, see Libecap (1978).

REFERENCES

Alston, L. J., J. R. Fuller, G. D. Libecap, and B. Mueller. 1997. "Competing Claims to Land: The Sources of Violent Conflict in the Brazilian Amazon." Working paper. University of Arizona, Tucson.

Becker, G. S. 1983. "A Theory of Competition Among Pressure Groups for Political Influence." *Quarterly Journal of Economics* 68: 371–400.

Eggertsson, T. 1990. *Economic Behavior and Institutions*. New York: Cambridge University Press.

Fogel, R. W. 1992. "Egalitarianism: The Economic Revolution of the Twentieth Century." Simon Kuznets Memorial Lecture, April. Yale University.

Gordon, H. S. 1954. "The Economic Theory of a Common Property Resource: The Fishery." *Journal of Political Economy* 62: 124–142.

Higgs, R. 1982. "Legally Induced Technical Regress in the Washington Salmon Fishery." *Research in Economic History* 7: 55–86.

Hoffman, E. and M. L. Spitzer. 1982. "The Coast Theorem: Some Experimental Tests." *Journal of Law and Economics* 25: 73–98.

Hoffman, E. and M. L. Spitzer. 1985. "Entitlement, Rights, and Fairness: An Experimental Examination of Subjects' Concepts of Distributive Justice." *Journal of Legal Studies* 14: 259–297.

Hoffman, E. and G. D. Libecap. 1992. "Political Bargaining and New Deal Agricultural Policies: Citrus Marketing Orders in the 1930s." Working paper. University of Arizona.

Hoffman, E., K. McCabe, K. Shachat, and V. Smith. 1992. "Preferences, Property Rights, and Anonymity in Bargaining Games." Working paper. University of Arizona.

Johnson, R. N. and G. D. Libecap. 1980. "Agency Costs and the Assignment of Property Rights: The Case of Southwestern Indian Reservations." *Southern Economic Journal* 47: 332–346.

Johnson, R. N. and G. D. Libecap. 1982. "Contracting Problems and Regulation: The Case of the Fishery." *American Economic Review* 72: 1005–1022.

Libecap, G. D. 1978. "Economic Variables and the Development of the Law: The Case of Mineral Rights." *Journal of Economic History* 38: 338–362.

Libecap, G. D. 1986. "Property Rights in Economic History: Implications for Research." *Explorations in Economic History* 23: 227–252.

Libecap, G. D. 1989a. "Distributional Issues in Contracting for Property Rights." *Journal of Institutional and Theoretical Economics* 145: 6–24. Reprinted in *The New Institutional Economics*, 1991, edited by E. G. Furubotn and R. Richter. College Station: Texas A&M University Press. 214–232.

Libecap, G. D. 1989b. *Contracting for Property Rights*, New York: Cambridge University Press.

Libecap, G. D. 1989c. "The Political Economy of Crude Oil Cartelization in the United States, 1933–1972." *Journal of Economic History* 49: 833–855.

Libecap, G. D. and R. N. Johnson, 1980. "Legislating Commons: The Navajo Tribal Council and the Navajo Range." *Economic Inquiry* 18: 69–86.

Libecap, G. D. and S. N. Wiggins. 1984. "Contractual Responses to the Common Pool: Prorationing of Crude Oil." *American Economic Review* 74: 87–98.

Libecap, G. D. and S. N. Wiggins. 1985. "The Influence of Private Contractual Failure on Regulation: The Case of Oil Field Unitization." *Journal of Political Economy* 93: 690–714.

Neher, P. A., R. Arnason, and N. Mollett. 1989. *Rights Based Fishing*. Dordrecht: Kluwer.

Ostrom, E. 1990. *Governing the Commons: The Evolution of Institutions for Collective Action*. New York: Cambridge University Press.

Peltzman, S. 1976. "Toward a More General Theory of Regulation." *Journal of Law and Economics* 19: 211–40.

Scott, A. 1989. "Conceptual Origins of Rights Based Fishing." In *Rights Based Fishing*, edited by P. A. Neher, R. Arnason, and N. Mollet. Dordrecht: Kluwar. 11–39.

Stigler, G. 1971. "The Theory of Economic Regulation." *Bell Journal of Economics and Management Science*. 2: 3–21.

Weingast, B. and M. Moran. 1983. "Bureaucratic Discretion or Congressional Control? Regulatory Policy Making by the Federal Trade Commission," *Journal of Political Economy* 91: 765–800.

Wiggins, S. N. and G. D. Libecap. 1985. "Oil Field Unitization: Contractual Failure in the Prescence of Imperfect Information," *American Economic Review* 75: 368–385.

Hans Christian Bugge   Chapter 4

Human Rights and Resource Management—An Overview

Introduction

The Interrelationship Between Human Rights and Resource Management

Until recently, human rights and resource management would generally be perceived as rather separate legal areas. Ideas of human rights were developed at a time when resources were seen as unlimited.[1]

Today the link is more obvious. Degradation of natural resources and the environment threatens the economic base and welfare of millions of people. We now see more clearly than before how access to natural resources, and the wise management of such resources, is a condition for the fulfillment of many human rights—in particular, social, economic, and cultural rights. The scarcity of natural resources in many parts of the world raises questions of balanced exploitation and fair distribution of these resources. The relevance of this relationship becomes even more evident when we consider the conditions of future generations.

On the other hand, the fulfillment of human rights to information and participation, and political freedom, may be important to secure a wise and balanced management of resources. The massive environmental problems in the former socialist countries serve as a reminder of this. But neither is democracy a guarantee for environmental protection and sustainable use of resources. Too often, short-term economic growth and job creation is given political priority by politicians being primarily concerned with the next general election.

This study looks into some of these interrelationships between human rights and management of natural resources from a legal point of view.

The Concept of Human Rights

It may be useful first to clarify the concept of "human rights." In this study "human rights" mean the fundamental rights—for individuals or groups—expressed in international instruments in such a way that they have become international law. A "right" for individuals or groups usually means a corresponding obligation for the state to respect or fulfill it in their national law and policy. States that do not respect or fulfill these rights break international law. Through various means, a number of organizations—intergovernmental as well as nongovernmental—strive to have these rights implemented and respected everywhere. This is a developing branch of international law, where new aspects of these rights or even "new rights" are recognized through the dynamic process of adopting new binding texts or practices.

The concept of "human rights" is based on the idea that certain human values and interests are of a universal and fundamental nature. They apply to "citizen and alien, friend and enemy." As such, they should be respected and fulfilled by all states, and they should form the foundation of national legislation, preferably as constitutional norms. Internationally, the human rights are expressed in the Universal Declaration on Human Rights adopted by the United Nations in 1948, the UN's two comprehensive Human Rights Covenants of 1966,[2] important regional instruments such as the European Convention on Human Rights of 1950 and the African Charter on Human and Peoples' Rights of 1981, and in sectorial conventions at the global[3] as well as regional levels.

Some human rights represent basic limitations on the legislator. Their purpose is to protect the individual against abuse of power by the state (or the minority against the majority). They limit the state's freedom of action. Other human rights oblige the state to take certain actions, to provide services to meet certain human needs and aspirations. This distinction between "protection" and "provision" corresponds roughly to the distinction between the civil and political rights on the one hand, and the economic, social, and cultural rights on the other. These two sets of human rights are expressed separately in the two covenants.[4] It should be mentioned that the legal nature and status of the two types of human rights differ somewhat. The civil and political rights are, generally, easier to ensure and enforce through strictly legal means than the economic, social, and cultural rights.

Some human rights deal with the participation of the citizen in society. Expressions of this are found both among the civil and political rights, and among the economic, social, and cultural rights. They represent in several respects a link between these two types of rights. It will appear from this study that rights related to information and participation are highly relevant in the area of management of natural resources.

Originally, human rights were rights of the individual. But during the latest decades, the concept of human rights has gradually widened—some would say the concept has become less clear and less operational. Several international instruments establish rights to be enjoyed by "peoples," groups, etc. The legal status of such "collective rights" or "peoples' rights" raises particular problems. Attempts have been made to describe them more generally as "third generation rights," having in common an element of solidarity. Rights to peace, security, disarmament, development, and a healthy environment are examples of what is often referred to as "third generation" human rights. In such thinking, the "first and second generation rights" are understood to represent, respectively, the elements of (individual) liberty as is characteristic of traditional civil and political rights, and (social) equality, which has been said to be the typical aim of economic, social, and cultural rights.[5]

A "Human Right to Natural Resources"?

Is there a "human right" to natural resources—a right to possess and exploit such resources? Let us first consider if there is such a right as an individual right. One point of departure here is the "right to property."

The (Individual) Right to Property

The right to property is one of the "classical" human rights—mentioned in the 1789 French Declaration on Human and Citizens' Rights (article 17)[6] and in the Fifth Amendment of the United States Constitution,[7] and it is found in many of the more modern national constitutions. In international law, the right to property is expressed in very general terms in the UN Declaration (article 17). It is placed among the civil and political rights: "(1) Everyone has the right to own property alone as well as in association with others, and (2) No one shall be arbitrarily deprived of his property."[8] The Declaration is, however, not legally binding in the strict sense.[9]

When looking at the two covenants on human rights, one searches in vain for articles stating the right to private property. One may assume that this issue was too controversial to be adopted by the UN at a time when nearly half the world's population lived in communist-ruled societies where private property to means of production had been formally abolished.

We must turn to regional conventions on human rights to find expressions of a "right to property." The European Convention on Human Rights does not itself include an article on the right to property, but such an article is found in the First Protocol to the Convention, article 1: "Every natural or legal person is entitled to the peaceful enjoyment of his possessions. No one shall be deprived of his possessions except in the public interest and subject to the conditions provided for by law and by the general principles of international law. The preceding provisions shall not, however, in any way impair the right of a state to enforce such laws as it deems necessary to control the use of property in accordance with the general interest or to secure the payment of taxes or other contributions or penalties."[10] According to this article, the "right to property" is clearly limited in its real content. The state may expropriate private property for public use. And the state may control the use of property "in accordance with the general interest." Hence, the state may regulate—and even prohibit—the exploitation of (private) natural resources to protect public interests. The article does not give the individual a "right" to use or exploit natural resources.

The article has been subject to interpretations by the European Commission and Court of Human Rights. In interpreting the second and third sentence they have expressed a general principle of "reasonable balance" between the interests of the individual and the society in its application, and have judged state actions by this yardstick.[11] One consequence of this principle is a citizen's right to compensation if his property is expropriated, at least as the main rule. But national authorities have a considerable freedom in defining the level of compensation.

In Norway, the "right to property" has found its (indirect) expression in article 105 of our written constitution (Grunnloven), which states, "If the welfare of the state requires that any person shall surrender his movable or immovable property for the public use, he shall receive full compensation from the Treasury." The use of natural resources is regulated by an extensive legislation in Norway, giving central and local government the authority to restrict the use and exploitation of these resources. Agriculture, forestry, fishery, mining, and the use of watercourses are subject to extensive legislation and government control. Land use is strictly regulated through local and national planning instruments, and special nature conservation measures may restrict the use of land even further.

The main legal issue in this connection is not whether the state may regulate and restrict the private party's use of resources, but to what extent such regulations and restrictions have to be economically compensated by the state. The main rule in Norwegian law is that regulations and restrictions do not give rights to compensation. However, regulations that virtually eliminate any economically valuable use, may in extreme cases be seen as being equal to expropriation, and accordingly be subject to compensation.[12]

So, the right to property, as an individual human right, is not a right to use or exploit the resources of the property freely and without restrictions. The state may regulate strictly the use of natural resources, regardless of whether they are privately or publicly owned. The economic content of the right to property is defined through the application of the general principle of compensation in case of expropriation or—to a limited extent, depending on national law— regulation on the use. It should be added, however, that certain sectorial human rights instruments explicitly state a right for certain groups to own property—in various forms and with different legal content. One example is the provisions on rights to land and other natural resources in the ILO Conventions on indigenous and tribal people, to which I shall revert. Another example is the Convention on the Elimination of All Forms of Discrimination Against Women, which—generally speaking—establishes the right to acquire, inherit, and dispose of property for all women.

Another possible individual right linked to the use of natural resources could be the right to work. People whose living is directly linked to the use of natural resources, such as farmers, herders, and fishermen, might argue that the right to work in their case is without meaning if they do not have a right to exploit these resources. The right to work is expressed in article 6 of the International Covenant on Economic, Social, and Cultural Rights (CESCR),[13] and in the European Social Charter.[14] The general view is, however, that these provisions are too general and conditional to establish a right for the individual with legal force.[15] The mass unemployment in the world clearly illustrates the weakness of this "right." There is no clear reason to give people who are directly dependant on the exploitation of natural resources for their work a better protection than others in this respect.

The "Peoples'" Right to Dispose of Their Natural Resources

In international law, the right to dispose of natural resources has been defined as a "peoples' right." This principle is expressed in the Covenants on Civil and Political Rights (CCPR) and the Covenant on Economic, Social, and Cultural Rights. Article 1, paragraph 2 of both instruments provides that "All peoples may, for their own ends, freely dispose of their natural wealth and resources without prejudice to any obligations arising out of international economic cooperation, based upon the principle of mutual benefit and international law. In no case may a people be deprived of their own means of subsistence." In talking about "peoples," these articles differ from the other articles of the two covenants, which mainly deal with individual rights. They are directly linked to the principle of peoples' right to self-determination, expressed in article 1, paragraph 1 of the two covenants: "All peoples have the right of self-determination. By virtue of the right they freely determine their political status and freely pursue their economic, social and cultural development."

It appears that the issues of peoples' right to self-determination and right to dispose of their natural wealth came up during the preparation of the two covenants much as a reflection of the process of decolonization. Whether these principles should be included in the covenants on human rights was highly controversial. It was finally adopted by the UN only with a small majority. As an argument to include it was claimed that the right of peoples to self-determination is an indispensable condition for the full enjoyment of the human rights treated in the covenants.[16] There is, in the two covenants, no reference to the right of states in this respect. However, states' permanent sovereignty over their natural resources is a basic principle in international law. How does the right of "peoples" to natural resources as expressed in the two covenants relate to this right of states?

In cases of clear identification between state and people, this dichotomy does not represent any problem. The principles of state sovereignty over natural resources and peoples' rights to dispose of their natural resources, become in fact (and legally) identical. Its substantive content is that other states cannot exploit the resources without the consent of the state. The peoples' rights approach becomes more complex, and less clear in law, when several "peoples"—for example, different ethnic groups and minorities—together constitute a state. Is the real content of article 1 paragraph 2 of the covenants a right for such groups to exploit the natural resources of the land to which they traditionally belong—if necessary, against the will of the state? In other words, is the state sovereignty over natural resources subject to limitations in the form of consent of the "people" directly concerned? And it is absolutely limited by the sentence, "In no case may a people be deprived of its own means of subsistence"? This question has been central, and controversial, in discussions on the legal situation of indigenous peoples and other minorities—particularly their right to land and other resources.[17] One practical aspect of the right to land and other natural resources for these groups is their right to oppose the destruction of their livelihood.

The wording of article 1 itself does not give a clear answer. However, it appears clearly from the preparatory work and later discussions in appropriate forums that an ethnic minority within a state cannot claim the right to self-determination on the basis of the first paragraph of the article. This is regardless of whether the members of the ethnic group see and define themselves as a "people." The notion of peoples' self-determination is subordinated to the conception of the unity and integrity of the state. Also, there seems to be agreement in international law that the word "people" must have the same meaning in paragraph 2 as in paragraph 1 of the article.

The conclusion is that this article in itself does not protect ethnic minorities within a state against the exploitation by the state of natural resources on which they base their living.[18] And one seeks in vain for other rules to this effect in customary international law or other general human rights conventions. The issue will be further discussed below in relation to article 27 in the UN Covenant on Civil and Political Rights and the ILO Convention on indigenous peoples, which treat—directly or indirectly—the right of indigenous peoples to land and to have their natural resources protected.

The Right to Development

Another basis for a collective right to exploit natural resources might be the "right to development." It could be argued that "development" depends on adequate access to natural resources. The UN General Assembly adopted the Declaration on the Right to Development in 1986.[19] The idea of such a right was first advanced by a Senegalese jurist in 1972, and was strongly supported by the developing countries. The Declaration is of a very general nature, and many questions remain open as to the real content of this right—as well as its legal status and implications. It is an important example of the "third generation" of human rights mentioned earlier in this chapter, and has as such been controversial both legally and politically.[20] Its main message is a call for economic growth, poverty alleviation, social justice, and participation by all.[21] There is no article in the Declaration explicitly dealing with the use or management of natural resources. But there is a general provision in its article 8 which lays down an obligation for the state to ensure access to "basic resources" for all:

1. States should undertake, at the national level, all necessary measures for the realization of the right to development and shall ensure, inter alia, equality of opportunity for all in their access to basic resources, education, health services, food, housing, employment and the fair distribution of income. Effective measures should be undertaken to ensure that women have an active role in the development process. Appropriate economic and social reforms should be carried out with a view to eradicating all social injustices.
2. States should encourage popular participation in all spheres as an important factor in development and in the full realization of all human rights.

Protection of Natural Resources and Sustainable Development

State Sovereignty and Natural Resources

As already indicated, protection and conservation of the natural resource base may in many respects be as important, from a human rights point of view, as the right to use the resouces. According to international law, states have the sovereign right to exploit their own natural resources. Does this imply a sovereign right for states to overexploit and even damage their own environment and natural resources? Or has the state a general legal obligation to protect natural resources and the environment?

In treaty law, there is at present no general rule which makes it a legally binding obligation for all states to protect their own environment or manage their resources in a sustainable manner. However, expressions of such an obligation are found in a number of treaties dealing with more limited environmental issues. For example, article 192 of the 1982 Law of the Sea Convention says that "[s]tates have the obligation to protect and preserve the marine environment." This applies to the marine environment both within and outside national jurisdiction. The 1992 Convention on Biological Diversity says in its preamble that "[s]tates are responsible for conserving their biological diversity and for using their biological resources in a sustainable manner." In the substantive articles of the convention this obligation is expressed through more indirect and "soft" rules. For example, acording to article 6, each contracting party to the convention shall, "in accordance with its particular conditions and capabilities," develop national stategies, plans, or programs for the conservation and sustainable use of biological diversity, and "as far as possible and as appropriate" integrate the conservation and sustainable use of biological diversity into relevant sectoral or cross-sectoral plans, programs, and policies. And the many treaties on more specific issues establish important obligations for states to protect particular sectors of the environment, or limit pollution from certain sources.

However, in spite of these many important rules in treaty law, it is at present difficult to derive a general, legally binding obligation in international law for states to protect and conserve their natural resources, or to manage them in a sustainable way.[22] This may change in the future. International environmental law is a dynamic part of international law. The treaty law continues to develop. There are also many expressions of "soft law" that support the idea of a general obligation for states to protect their own environment. For example, the Stockholm Declaration of 1972 and the World Charter for Nature which was adopted unanimously by the UN General Assembly in 1982, contain a number of principles which clearly indicate relevant obligations of this kind, although only a few of them are addressed directly to states. The Rio Declaration[23] may be seen as a setback in this respect. But work to develop a general covenant on environmental protection and sustainable development has been going on in various nongovernmental forums over the last years, and the situation now seems ripe for intergovernmental negotiations on the issue.[24] So, a general obligation for states to protect their environment and to manage their natural resources in a sustainable way most probably emerges as binding international law sooner or later. The exact meaning of such an obligation will be as much a matter of political discussion as of legal interpretation. It will always be necessary to strike the difficult balance between protection and short-term development needs.

The sovereign right of the state to exploit its own natural resources has, however, for some time had at least one important limitation in international law: in the case of common resouces or transfrontier problems. It follows from general principles of international customary law that states must respect the interest of other states when several states share common natural resources such as an inland waterway or living resources in the sea. According to the same principle a state can not use its resources in such a way that it damages resources of other states. This has become particularly relevant in relation to transfrontier environmental problems such as transfrontier pollution. This principle was expressed in the famous principle 21 of the Declaration of the 1972 Stockholm Conference on the Human Environment, which states the following: "States have, in accordance with the Charter of the United Nations and the principles of international law, the sovereign right to exploit their own resources pursuant to their own environmental policies, and the responsibility to ensure that activities within their jurisdiction or control do not cause damage to the environment of other states or of areas beyond the limits of national jurisdiction." This principle is also expressed in several international conventions, among them the UN Convention on the Law of the Sea. It was repeated in the Rio Declaration, and in the Framework Conventions on Climate Change (preamble) and the Convention on Biological Diversity (article 3).

One of the crucial issues in future discussions of international environmental law is the possible conflict between the right for the states to exploit their resources freely, and the need for restrictions and safeguard measures in order to protect the atmosphere, the world's biodiversity, and the global biosphere in general.

The Concept of "Sustainable Development"

The issue of protection and conservation of natural resources took on a new dimension with the introduction of the concept of "sustainable development." This is the central concept and idea in the report from the World Commission on Environment and Development—the "Brundtland Report." Since the report was presented in 1987, the idea and objective of "sustainable development" has gained widespread political support. The concept is unclear in many respects and is interpreted in very different ways. The basic idea, however, is simple. Sustainable development is defined by the Brundtland commission as "[a] development that meets the needs of the present without compromising the ability of future generations to meet their own needs." It contains both the concept of needs to be met today and in the future—in particular the basic needs of the poor part of the world—and the concept of limitations: we must manage the natural resources and the environment in such a way that sufficient amounts are left to our children and grandchildren. A "sustainable development" is the combination of economic development and environmental protection. From a human rights point of view it includes both the economic, social, and cultural rights—the "right to development" in a broad sense—and the "right" for present and future generations to protection of the natural resource base. It presents us with a double moral challenge: solidarity with the poor people of today´s world, and solidarity with the next generation. "Intergenerational equity" is one legal expression of the idea behind "sustainable development."

Through the objective of sustainable development, human rights and wise management of natural resources merge. In this, it represents an interesting development from the Human Rights Covenants of 1966. Here, to use the natural resources is seen as a people´s right in itself. In the perspective of sustainable development, the emphasis is not so much on the right to exploit resources, but rather on the obligation to exploit resources sensibly—as a means of meeting basic needs for all. And it introduces an important time perspective: proper management of our natural resources today is a condition for the fulfillment of basic human rights in the future.

The objective of sustainable development got broad political support at the Rio Conference. But the conference did not contribute to clarifying the meaning and implications of the concept. Neither did it break much new ground in national and international law concerning management of natural resources. Nevertheless, the notion of "sustainable development" is now moving into legal texts. It is, in different forms, already included in several international treaties, and in national legislation. For example, in both Norway and Denmark it has been included in the object clause of important new laws concerning natural resources and the environment. Hence, it is rapidly developing from primarily a political objective to a legal concept—a "legal standard." This may have interesting consequences also from the point of view of individuals' rights, and in particular the weight of future generations' interests in legal considerations and decisions.

"Right to Environment" as a Human Right?

The Right to a Healthy Environment as a Material Right

With growing environmental problems worldwide, it has been much discussed how the law can be developed further to contribute to a more effective protection of nature and the environment. Many lawyers have argued in favor of developing a "right to a healthy environment" as a material human right. Already in Stockholm in 1972 such a view was present. Principle 1 of the Stockholm Declaration states: "Man has the fundamental right to freedom, equality and adequate conditions of life, in an environment of a quality that permits a life of dignity and well-being, and he bears a solemn responsibility to protect and improve the environment for present and future generations." During the two decades that have passed, many legal and political forums have discussed this idea further. The fact that a good and healthy environment is a condition for the fulfillment of already recognized human rights, such as the right to life, health, welfare, and home, also contributes to the process of giving environmental conditions legal relevance.

There is, however, still no general recognition of an enforceable "right to environment"—neither as an individual nor as a collective right.[25] It would be unrealistic not to acknowledge the complexities involved in this issue. It is not simple to grant legal status to such very broad—and in many ways relative—concepts and goals as a good or "healthy" environment. The notion of human rights is not necessarily the best starting point for meeting the legal challenges in the field of resource management and environmental protection. For example, its anthropocentric character overlooks nature's intrinsic value, and it provides little guidance when the interests of the present and future generations have to be weighed against each other. It may, in some respect, be more relevant to define the human *responsibilities and obligations* towards nature, and towards our grandchildren—as was also underlined in principle 1 of the Stockholm Declaration.

The Rio Conference in June 1992 did not contribute much to clarify or strengthen the legal status of environmental protection in a human rights perspective. However, we witness important developments in this field as well. One is the fact that a rapidly growing number of states include the protection of the environment and sustainable development in their constitutions. This is done in the form of an obligation for the state to protect the environment of the country or promote sustainable development, and/or as a corresponding right for the citizens. It is also significant that the European Court of Human

Rights, in two recent cases, has accepted the right to a healthy environment as included in the human rights of home and privacy.[26]

The Right to Information and Participation

Instead of pursuing the idea of developing a "right to environment" as a material human right, it may be more fruitful to look into some of the procedural rights as a means for securing proper management of natural resources: decision-making processes, participation, and information.

The public's right to information and participation is essential. If this right is sufficiently clarified and acknowledged, it may be an important key to basic changes in decision making, and to promoting sustainable development and a good management of natural resources. It has a basis in some of the human rights which have already been accepted, and it can be made sufficiently precise and practicable to be legally enforceable in most countries. In the broadest sense, this is a question of the political system in general, democracy, and right to information. Expressions of this are found in the Universal Declaration on Human Rights, articles 19 (freedom of opinion and expression), 20 (freedom of peaceful assembly and association), and 21 (the right to take part in the government of the country), as well as the corresponding articles 19, 22, and 25 in CCPR, and article 8 in the CESCR (the right to form trade unions).

The right to information is now broadly recognized as an important issue in relation to environmental protection. While the "right to a healthy environment" in a material sense may be problematic, procedural rights in this area are more easily accepted and applied, in particular (1) the right to be informed about the environmental situation in the area, and the environmental consequences of new projects (as in environmental impact statements), (2) the right to participate in decision-making processes, and (3) the right to legal and administrative remedies against decisions in this area. On this issue, the Rio Declaration on Environment and Development took a step forward. Its principle 10 states "[e]nvironmental issues are best handled with the participation of all concerned citizens, at the relevant level. At the national level, each individual shall have appropriate access to information concerning the environment that is held by public authorities, including information on hazardous materials and activities in their communities, and the opportunity to participate in decision-making processes. States shall facilitate and encourage public awareness and participation by making information widely available. Effective access to judicial and administrative proceedings, including redress and remedy, shall be provided." More recently, the UN has stated that "[a]ccess to information and public participation in decision making is fundamental to sustainable development."[27]

Trends in Norwegian Law

On May 25, 1992, the Norwegian Parliament, the Stortinget, adopted a new section in the written constitution dealing with the protection of the environment. Article 110b states that every citizen has the right to an environment which does not endanger health, and to a nature where diversity and ecology are preserved. The management of natural resources must be farsighted and balanced in order to ensure this right for future generations. To this end, there shall be a right to information about the state of the environment. The principles expressed in article 110b are formulated in terms of "rights" for individuals, combined with a directive to state authorities to issue more detailed provisions about the implementation of these principles. The history of this constitutional amendment[28] shows that parliament neither intended to adopt a general subjective right nor a mere programmatic declaration. It can thus be argued that it will have legal implications apart from those set out in implementing legislation, that is, for the interpretation and application of other rules of law, and for the exercise of executive and administrative discretion.[29]

The Rights of Indigenous Peoples to Natural Resources

The issue of ethnic minorities and indigenous peoples' right to land and other natural resources is an important and controversial issue in international law. There is a delicate balance between sovereignty and territorial integrity of states on the one hand, and the promotion and protection of minority identity on the other—with separatism as a possible extreme result. It is significant that the UN Declaration on Human Rights does not have any provision dealing with the issue, in spite of the fact that "protection of minorities represents one of the most important predecessors to modern, international human rights protection" (Nowak 1993). The most important general provision dealing with this issue in international law is article 27 of the Covenant on Civil and Political rights, which states: "[i]n those states in which ethnic, religious or linguistic minorities exist, persons belonging to such minorities shall not be denied the right, in community with the other members of their group, to enjoy their own culture, to profess and practice their own religion, or to use their own language."

It should be noted that the right described in article 27 is an individual right, but by reference to "the other members of the group" it indirectly also gives protection to the "group." The wording is cautious and vague. It leaves many questions open to interpretation. It has even been questioned whether the article applies to indigenous people at all. Today this is fully recognized. However, the article does not deal explicitly with the issue of indigenous peoples' rights to land and other natural resources. One question is whether it should be interpreted to imply a right for minorities not only to enjoy their culture in the strict sense of the word, but also a right to the material conditions for this culture, such as the natural resources on which the culture and lifestyle are based. This has been a much discussed issue in relation to the Saami population and culture in Norway.[30]

There are different views on this issue. The core of the matter is really the meaning of the expression "enjoy their own culture." "Culture" can be given a narrow meaning, such as "the arts and other manifestations of human intellectual achievement regarded collectively,"[31] or a wider, more anthropological meaning, such as "the customs, civilization, and achievements of a particular time or people,"[32] including learned patterns of behavior and the ideas that underlie behavior. This wider interpretation of the word culture inevitably leads to interpreting article 27 as also protecting the natural resources on which the culture and lifestyle of indigenous peoples is based.

The text of the article and the *travaux préparatoires* do not exclude a wide interpretation. But neither do they provide a clear legal basis for such an interpretation. The issue has been much discussed in human rights forums. The UN Committee on Human Rights has had the issue before it on several occasions in connection with national reporting. There seems to be a tendency to include questions related to material conditions of indigenous peoples when discussing reports on article 27. In several recent cases of individual complaints related to article 27 the committee has accepted that traditional economic activity—and implicitly the protection of natural resources on which it is based—may fall under article 27 if it is an essential element in the culture of an ethnic community.[33] The arguments now seem rather strong in favor of giving article 27 importance as a source of law also for the protection of natural resources, insofar as the existence and availability of these resources are a prerequisite for the traditional way of life, and thus the cultural survival of members of indigenous groups.

It should be underlined that article 27 deals with "ethnic, religious, or linguistic minorities" in general, and not only indigenous peoples. When interpreting and implementing the article, the historically founded claims of indigenous peoples must carry more weight than any claims made by other minorities, in particular colonizers or other immigrant groups ("new minorities"). Indigenous peoples are particularly close to and dependant on the nature in which they live. Hence, it is difficult to argue for a wide interpretation of article 27 for all types of minorities, although the text itself does not indicate any differentiation in this respect. It should also be noted that article 27 cannot be interpreted as forbidding any interference in the land or natural resources. The crucial issue is whether interference is so extensive that it really threatens the culture.

The rights of indigenous peoples are treated more directly and comprehensively in the ILO Convention of 1989 Concerning Indigenous and Tribal Peoples in Independent Countries. This convention is a partial revision of an ILO Convention on the same issue from 1957.[34] These are the only global conventions dealing with issues related to indigenous people. So far less than fifteen states have ratified the 1989 convention.[35] On the issue of land rights its article 14 states:

1. The rights of ownership and possession of the peoples concerned over the lands which they traditionally occupy shall be recognized. In addition, measures shall be taken in appropriate cases to safeguard the right of the peoples concerned to use lands not exclusively occupied by them, but to which they have traditionally had access for their subsistence and traditional activities. Particular attention shall be paid to the situation of nomadic peoples and shifting cultivators in this respect.
2. Governments shall take steps as necessary to identify the lands which the peoples concerned traditionally occupy, and to guarantee effective protection of their rights of ownership and possession . . .

This article concerning indigenous peoples' material right to land is supplemented by articles concerning their right to participate in decision making concerning natural resources. In general, indigenous peoples shall be consulted in all matters concerning their situation and the implementation of the convention (articles 6 and 7). Articles 15 and 16 are particularly relevant to issues related to management of land and other natural resources. Article 15 states:

1. The rights of the peoples concerned to the natural resources pertaining to their lands shall be specially safeguarded. These rights include the right of these peoples to participate in the use, management, and conservation of these resources.
2. In cases in which the state retains the ownership of mineral or sub-surface resources or rights to other resources pertaining to lands, governments shall establish or maintain procedures through which they shall consult these peoples, with a view to ascertaining whether and to what degree their interests would be prejudiced, before undertaking or permitting any program for the exploration or exploitation of such resources pertaining to their lands. The peoples concerned shall wherever possible participate in the benefits of such activities, and shall receive fair compensation for any damage which they may sustain as a result of such activities.

If relocation of people is found necessary as an exceptional measure, it shall take place only with their free and informed consent. It may be fair to say that the right to information and participation is a "compromise" solution between the acceptance of minorities´ exclusive rights over the natural resources of their traditional territory, and the principle of state sovereignty.

Since the convention is new and few countries have ratified, the precise content and implications of these articles are uncertain at this stage. The very fact that such a convention was adopted by the General Conference of the ILO indicates a growing international understanding of the need to protect the natural resources on which indigenous people depend for their cultural survival and development. In 1994 the Sub-Commission on Prevention of Discrimination and Protection of Minorities under the UN Committee on Human Rights adopted a draft United Nations declaration on the rights of indigenous peoples. It underlies the close interrelationship between natural resources, economic activity, and culture of indigenous people. Many of the articles of the declaration aim at a further strengthening of indigenous peoples' right to natural resources on which their traditional way of life is based. For example, article 26 says that "indigenous peoples have the right to own, develop, control, and use the lands and territories, including the total environment of the lands, air waters, coastal seas, sea-ice, flora and fauna and other resources which they have traditionally owned or otherwise occupied or used. This includes the right to the full recognition of their laws, traditions and customs, land-tenure systems and institutions for the development and management of resources...." Article 28 states that "indigenous peoples have the right to the conservation, restoration, and protection of the total environment and the productive capacity of their lands, territories, and resources...."[36] However, it remains to be seen whether there will be sufficient political support for such radical provisions in the further negotiations. Other recent discussions may create a certain doubt in this respect.[37]

Should We Further Pursue the Concept of Human Rights as a Means for Sustainable Resource Management?

The dynamic and positively loaded concept of human rights has created a temptation to mobilize its persuasive force for many widely different causes. The question that increasingly needs to be posed is whether it is a good policy to pursue this concept further into new fields or newly discovered problems. As the sketchy overview of this study shows, "human rights" are indeed a very mixed group of legal norms. Some imply a quite precise right for the individual to be protected against certain clearly defined actions by the state. The respect for and fulfillment of such rights do not depend much on economic conditions or cultural traditions in the state. Other human rights are more complex, vaguely formulated, and in reality much dependent on economic development and other more fundamental conditions to be fulfilled. Their legal content and status is therefore uncertain and much of their fulfillment in reality is left to each state. The possibilities for the international community to identify violations and enforce these rights are limited.

Clearly, it can be argued that a further "dissolution" of the concept of human rights may mean even less of operational rules and more of just empty symbols. This may entail less respect for the concept itself and make effective enforcement even more illusory. As it has been said, human rights become more a question of politics than of law and ethics. Many human rights lawyers therefore see a danger in broadening the human rights field and making it less and less precise as a legal concept and instrument.

On the other hand, there has been a positive—although slow—development in the field of human rights in many parts of the world. Surely, in a number of states gross violations of even the most basic civil and political human rights are still common, and in the poorest countries of the world the fulfillment of the economic, social, and cultural rights seems further away than ever before. But there are also positive trends, partly linked to the process of democratization in many developing countries, and partly to many states' growing respect for decisions and recommendations by international human rights institutions. Norway is a case in point: the European Convention on Human Rights and other human rights instruments influence policy, court decisions, and general legal development to a larger extent today than just ten years ago.[38] The fact that the human rights concept has been broadened and has become more vague does not seem to have influenced this trend in a very negative way.[39]

Surely, to define a problem or a social objective as a question of "human rights," and give it a status as such in international instruments, does not necessarily lead to a quick solution. It is not a *sesam-sesam*. It might even be counterproductive, as a time-consuming *détour*. But most likely it will contribute to a legal process, step by step, which gradually will give the objective or the consideration more weight in both international and national decision making. The idea of a "right to environment" may be refused by international lawyers and politicians as too vague and impossible to make operational and enforceable. But at the same time, we witness that during the last twenty years some fifty states have included the objective of environmental protection in their constitution—many with words similar to article 110b in the Norwegian Constitution which says that everybody has the right to a healthy environment and to a nature which ensures reproduction and diversity.

Individual and subjective rights in the traditional sense of claims which entail corresponding obligations for the state and society may have only a limited role to play as regards resource management. But it may play a role. And resource management should be an area of increasing attention in the human rights work. Environmental degradation and the destruction of natural resources are already threatening life and well-being for millions of people, thus making the fulfillment of many human rights even more difficult and remote. And looking into the future, the problems may become even more dramatic for the next generation. Scarce resources may become one of the most important causes for social unrest, conflicts and wars, social injustice, and suppression of human rights. Therefore, management of natural resources and human rights issues should to a greater extent be combined in the international discussion and work in this field. Sustainable development should be a basic idea and common denominator.

For the protection of minorities and indigenous peoples and their natural resource basis, the development and enforcement of international instruments with a "human rights label" may be a particularly important strategy. Human rights carry important moral arguments. What else could stem a rapid exploitation of natural resources and degradation of the natural and cultural diversity in a world where the economic and cultural forces of modernization and globalization seem to become more and more centralized, growth-oriented, and powerful?

But in the area of resource management and environmental protection, the idea of rights for individuals and groups is only one side of the coin. It is as much a question of limitation of the freedom of action. What is necessary is also more responsibilities and duties—and not only for individuals, but also for "legal persons" such as companies and other collectivities, ethnic and social groups, as well as for the state. At the same time, the exploitation and management of resources pose problems which cannot be solved at the national level. Thus, there is a strong need for more international law in this field, and a more effective international system for its enforcement. Most of this international law will have to be developed in international forums which have environmental protection and sustainable development as their primary mission, and which are rather far from those dealing with human rights issues. After all, it is primarily on these arenas—and in the national parliaments and in the business community worldwide—that the future of the global environment and resource management will be decided.

NOTES

1. I thank the late Dr. Juris Torkel Opsahl, Institute of Public and International Law, for advice and assistance in preparing this paper and Lise Rakner, Christian Michelsens Institute, for useful comments.

2. International Covenant on Economic, Social, and Cultural Rights (CESCR) and International Covenant on Civil and Political Rights (CCPR).

3. Such as the International Convention on the Elimination of All Forms of Racial Discrimination of 1965, the Convention on the Elimination of All Forms of Discrimination Against Women of 1979, and the Convention on the Rights of the Child of 1989.

4. Broadly speaking, the civil and political rights are the inheritance of western democratic and liberal values, while the economic, social, and political rights are more in line with the ideology of former welfare and socialist states. The division might be seen as a reflection of the ideological gap between the East and West during the years of the cold war.

5. These terms have also been presented as a reflection of the famous human rights slogan *"liberté, égalité, fraternité."* It is perhaps trivial to observe that no revolution, and least of all the French one, has succeeded in making all three concepts operational. On the other hand, historically they need not necessarily appear one after another, as "generations"; the values and interests they represent may, if properly balanced against each other, be harmonized and protected by law simultaneously, as three distinct dimensions of a legal system rather than "generations."

6. "La proprieté étant un droit inviolable et sacré, nul ne peut en être privé, si ce n'est lorsque la nécessité publique, légalement constatée, l'exige évidemment, et sous la condition d'une juste et préalable indemnité."

7. "No person shall be . . . deprived of life, liberty or property without due process of law; nor shall private property be taken for public use without just compensation." The Ten Original Amendments to the U. S. Constitution are called the Bill of Rights. They came into force in December 1791.

8. The *travaux préparatoires* of this article show that it caused much controversy. A number of different proposals were discussed before its final wording was adopted; see the article by Gudmundur Alfredsson in A. Eide et al. (1992).

9. UN Declarations and recommendations of this type are often referred to as "soft law." They are not legally binding in the strict sense of international law, like treaties. But the commitments they lay down often gradually develop from political declarations to legal obligations, working their way into international conventions and national legislation, or becoming customary law.

10. A similar provision is found in the African Charter: "The right to property shall be guaranteed. It may only be encroached upon in the interest of public need or in the general interest of the community and in accordance with the provisions of appropriate laws."

11. In 1982, Sweden became the first country found by the European Court of Human Rights to have violated this article, in the case of *Sporrong & Lönnroth v. Sweden*. For discussion see Bogdan (1986).

12. The question of compensation for restrictions on the use of private property is a central one in Norwegian law. It has also been a controversial issue. However, several decisions by the Supreme Court during the last twenty-five years—some of them in plenary—have to a large extent clarified the matter. Through these decisions, the right to compensation for regulations has been limited rather strictly.

13. The full text of the article is the following: "1.) State Parties to the present Convention recognize the right to work, which includes the right of everyone to the opportunity to gain his living by work which he freely chooses or accepts, and will take appropriate steps to safeguard this right. 2.) The steps to be taken by a State Party to the present Covenant to achieve the full realization of this right shall include technical and vocational guidance and training programmes, policies and techniques to achieve steady economic, social and cultural development and full and productive employment under conditions safeguarding fundamental political and economic freedoms to the individual."

14. Part 1, paragraph 1: "Everyone shall have the opportunity to earn his living in an occupation freely entered upon."

15. See Opsahl (1991).

16. See Nordenfelt (1987). Nordenfeldt is generally very critical of widening the concept of human rights through "collective" or "peoples'" rights.

17. The issue is thoroughly discussed in the Norwegian Public Report (NOU 1984) regarding the legal situation of the Saami people in Norway.

18. This understanding of article 1 of the two covenants was also implicit in the Action program adopted by the international conference on the rights of indigenous peoples in Geneva in 1981. It has, however, been criticized as too "statist"; see for example the debate in Crawford (1988).

19. G.A. Res. 41/120, adopted by a vote of 146 to one (the United States) with six abstentions.

20. The United States and other western countries have been particularly critical. However, at the World Conference on Human Rights in Vienna in June 1993, the right to development was approved by consensus as a "universal and inalienable right and an integral part of fundamental human rights."

21. The first Global Consultation on the Right to Development as a Human Right was held in Geneva in January 1990; see Barsh (1991).

22. See for example Birnie and Boyle (1992, p. 122).

23. Declaration on Environment and Development adopted at the UN Conference on Environment and Development in Rio de Janeiro in June 1992.

24. The issue was discussed at the special session of the UN General Assembly for the follow-up of the Rio Conference, held in New York June 23–27, 1997. But the wording of the final document from the meeting is

cautious. It states that "it is necessary to continue the progressive development and, as and when appropriate, codification of international law related to sustainable development" (see program for further implementation of Agenda 21, para 109).

25. On this issue, see Kiss and Shelton (1991) and the discussion in Shelton (1991).

26. The European Convention for the Protection of Human Rights and Fundamental Freedoms art. 8, 1 states that [e]veryone has the right to respect for his private and family life, his home and his correspondence." The first case is *Lopéz Ostra v. Spain* (41/1993/436/515). The court found that Spain had violated art. 8, by placing a plant for destruction of chemicals close to the home of the Lopéz Ostra family, without offereing the family an alternative home or compensation, in spite of the fact that members of the family fell ill. The second case is *Guerra and Others v. Italy*. The court found that Italy violated art. 8 by refusing to disclose information about how neighbors to a chemical factory were exposed to chemical emissions.

27. Programme for the further implementation of Agenda 21, para. 108.

28. See in particular Innst. S. nr. 163 (1991–92) with references, and the debate in parliament, Stortingstidende (Parliamentary Bulletin) 1992, pp. 3735–3743, and earlier in Backer (1990).

29. See Backer (1993). It is, for instance, arguable that this new provision will weaken the claim of individuals to compensation for restrictions on the use of property when these are imposed by environmental legislation (i.e., it strengthens the "Polluter Pays Principle"). Some Supreme Court decisions had, however, already drawn similar conclusions, e.g., Rt. p. 1279 and 1987, p. 80. See also Backer (1991) and Fleischer (1996).

The Supreme Court of Norway has in two recent decisions referred to article 110b as an element of interpretation when weighing environmental considerations against other considerations in individual cases.

30. See in particular Norges offenthige utredninger (Norway's Public Reports) (1984, 18) and Odelstingsproposisjon nr. 33 (1986-87) which formed the basis for the new Norwegian legislation concerning the Saami rights. The question was briefly discussed by Norway´s Supreme Court in plenary in the "Alta case," but no conclusion was drawn. See Norsk Retstidende (Rt.) 1982, p. 241. Recently, the matter has been further discussed in relation to the Saami rights to land and natural resources in Norway (see NOU 1997, 5).

31. The first definition of the word in *The Concise Oxford Dictionary* of 1991.

32. The second definition of the word in *The Concise Oxford Dictionary* of 1991.

33. *Lovelace v. Canada* (complaint no. R 6/24, 1981), *Kitok v. Sweden* (application no. 197/1985), and *Ominayak/Lubicon Lake Band v. Canada* (application no. 167/1984). In cases of individual complaints, the committee acts under the Optional Protocol of the Covenant.

34. ILO Convention (no. 107) Concerning the Protection and Integration of Indigenous and Other Tribal and Semi-Tribal Populations in Independant

Countries. The basic idea in this convention was the assimilation of indigenous people into modern society. Only a few states have ratified the convention.

35. The convention was ratified by Norway on June 20, 1991.

36. The issue of protection of natural resources of indigenous people is also present in the relevant part of Agenda 21, adopted at the Rio Conference. According to its chapter 26, which deals with the role of indigenous people and their communities, governments should empower indigenous people and their communities through measures that include the "recognition that the lands of indigenous people and their communities should be protected from activities that are environmentally unsound or that the indigenous people concerned consider to be socially and culturally inappropriate" (para 26.3 (a) (ii)). Governments should also recognize "their values, traditional knowledge and resource management practices with a view to promoting environmentally sound and sustainable developement" (para. 26.3 (a) (iii)).

37. The latest intergovernmental discussion on the human rights of indigenous peoples and other minorities took place at the World Conference on Human Rights in Vienna in June 1993. The final document from the conference—"Vienna declaration and program of action"—slightly strengthens and widens the formulation in article 27 of CCPR by stating "[t]he persons belonging to minorities have the right to enjoy their own culture, to profess and practice their own religion and to use their own language in private and in public, freely and without interference or any form of discrimination." The World Conference also "recognizes the inherent dignity and the unique contribution of indigenous people to the development and plurality of society and strongly reaffirms the commitment of the international community to their economic, social and cultural well-being and their enjoyment of the fruits of sustainable development." The Vienna Declaration in itself does not contribute very much when it comes to indigenous peoples' right to land and other resources.

38. A recent public report recommends that large parts of the international instruments on human rights should be incorporated into Norwegian law. See NOU (1993, 18).

39. The Vienna Conference may not have been a great step forward, but at least it confirmed the universal nature of human rights, and the will of all member states to enhance international cooperation to fulfill them. It should be remembered in this connection that the UN has received many new members during the last decade, and it now numbers some 200 states with a great variety of interests and priorities.

REFERENCES

Backer, Inge Lorange. 1990. "Grunnlovfesting av miljørettslige prinsipper." *Institutt for offentlig retts skriftserie* 6.

Backer, Inge Lorange. 1991. "Grunnloven og miljøet." *Juristkontakt* 7.
Backer, Inge Lorange. 1993. *Knophs Oversikt over Norges Rett*, 10th edition. Oslo: Universitetsforlaget.
Barsh, R. L. 1991. "The Right to Development as a Human Right: Results of the Global Consultation." *Human Rights Quarterly* 13, no. 3, August.
Birnie, Patricia W. and Alan E. Boyle. 1992. *International Law and the Environment*. Oxford: Clarendon Press.
Bogdan, Michael. 1986. "Äganderätten som folkrättslig skyddad mänsklig rättighet." *Raoul Wallenberg Institute Report*, no. 2. Lund.
Crawford, James, ed. 1988. *The Rights of Peoples*. Oxford: Clarendon Press.
Eide, A., G. Alfredsson, G. Melander, L. A. Rehof, and A. Rosas, eds. 1992. *The Universal Declaration of Human Rights: A Commentary*. Oslo: Scandinavian University Press.
Fleischer, Carl August. 1996. *Miljø-og ressursforvaltning. Grunnleggende forutsetninger*. Oslo: Universitetsforlaget.
Innst. S. nr. 163 (1991-1992).
Kiss, A. and D. Shelton. 1991. *International Environmental Law*. London and New York: Graham & Trotman.
Nordenfelt, Johan. 1987. "Human Rights—What They Are and What They Are Not." *Nordic Journal of International Law* 1.
"Norsk Retstidende." 1982. *Norwegian Law Bulletin* Rt., p. 241.
NOU (Norwegian Public Report) 1984:18. "Om samenes rettsstilling" (On the legal position of the Saami), Statens Forvaltningstjeneste, Oslo.
NOU (Norwegian Public Report) 1983:18. "Lovgivning om menneskerettigheter" (Legislation on Human Rights), Statens Forvaltningstjeneste, Oslo.
NOU (Norwegian Public Report) 1997:5. "Urfolks landrettigheter etter folkerett og utenlandsk rett" (Aboriginal people's rights to land according to international law and the law of foreign countries), Statens Forvaltningstjeneste, Oslo.
Nowak, Manfred. 1993. *UN Covenant on Civil and Political Rights Commentary*. Strasbourg: N. P. Engel Verlag.
Opsahl, Torkel. 1991. *Internasjonale menneskerettigheter. En foreløpig innføring*. Oslo: Institûtt for menneskerettigheter.
Ot.prp. (Propostition to the Odelstinget) nr. 33. 1986-1987. *Om lov om Sametinget og andre samiske rettsforhold (sameloven)*.
Shelton, D. 1991. "The Right to the Environment." In *The Future of Human Rights in a Changing World: Essays in Honor of Torkel Opsahl*, edited by A. Eide and J. Helgesen. Oslo: Universitetsforlaget.
Stortingstidende. 1992. p. 3735-3743.
World Commission on Environment and Development. 1987. *Our Common Future*. Oxford/New York: Oxford University Press.

PART TWO

Norwegian Law and Common Property

Introduction

Describing the property rights situation of renewable resources within a particular society requires detailed knowledge of the society and appropriate technical language. Property rights in general are part of mainstream culture in the Western world and a language for describing them precisely is easily available. Rights of common is nowadays not part of this mainstream culture. And furthermore, in societies such as Scandinavia where commons have a long history, their particular features are such that direct translations of the native legal language often will either be very imprecise or prone to misinterpretations. To avoid any ambiguity, Erling Berge has investigated the terms used to describe rights of common in English and American jurisprudence in "The Legal Language of Common Property Rights." With the concepts presented there at least the commons of Scandinavia can be given a precise description.

Thor Falkanger presents a general overview of Norwegian law on the use of range lands (or waste lands or outfields) in chapter 6. The range lands in Norway are to a considerable degree used by more than one person. Even in the cases with one owner only, the land will in most cases be used by others as well, due to specially created rights over the property, and due to the all men's rights. The situation is most complex for state-owned land where one may find conflicts over use between the owner and three groups of users: the farmers, the public, and those to whom the state has conferred rights in its capacity as owner. The rights and duties of the users are to some extent defined by contract, but in important respects directly by written or customary law. The solution of conflicts between the many users of the range lands will depend upon the combined effects of traditional private law and modern administrative law.

In chapter 7, Hans Sevatdal goes further into the issue of the various forms of common property in Norway: the state commons, the *bygd* commons, and the "farm" commons of southern Norway and the state-owned lands in Nordland, Troms, and Finnmark.

What Sevatdal calls *bygd* commons (in Norwegian designated *bygdeallmenning*) may illustrate the point above about the difficulty of translating special terms in the description of local property rights. The original meaning of *bygd* comes close to "local community." Also, current usage (outside the context of commons) of the word *bygd* would suggest some kind of local community, independent of more formally defined units such as school districts, parishes, or municipalities. Earlier in our history *bygd* would be used for the smallest administrative unit, the local law district, and later the parish. In Sweden the word would mean the same. But in conjunction with commons translations like "community commons" or "parish commons," the term will not give the right associations.

The areas burdened with rights of common have throughout our history always been tied to users from some specific local community, the *bygd*. Thus the local community became tied to a certain area recognized as "their" commons. But during the past eight hundred years the original usage of the word *bygd* has turned around in the legal language, and today the *bygd* is defined as comprising those farm enterprises which have rights of common in the area recognized in law as a "commons" (both state and *bygd* commons).

Since the commons are of ancient origin, and such factors as topography, climate, settlement patterns, and economy in Norway vary enormously, it must be expected that the commons are equally diverse. Classification into homogeneous groups is difficult. However, three main categories of common property can be identified: (1) state common land, (2) *bygd* common land, and (3) common property owned jointly by "farms" (*sameiger mellom bruk*). In addition, the state lands in the counties of Nordland and Troms must be treated separately and are different from the state lands in Finnmark.

Sevatdal's analysis reviews these three main types of commons, the types of lands they cover, the areas, the ownership situation, access and decision making involved, the rules of alienation, and the degree of collectiveness in use. It is noted that compared to the other Nordic countries, there is a remarkable lack of traditional local institutions for management of the commons. The research base is too weak to draw firm conclusions about the reasons for this. But one reason may be the slow decline of the old Norse Thing institution (mostly devoted to resolving conflicts and sometimes to coordination of collective action). The Thing institution started its slow decline (through centralization) in the middle of the seventeenth century, but was still strong in 1837 when the commune institution (local political units) was enacted and took over some of its political dimension. Since then it declined more rapidly and was removed formally between 1915 and 1927 (Næss 1991, Tretvik 1996). Sevatdal's guess is that problems related to the "collectiveness" in all three types of commons were solved within two regimes: (1) customs and tradition (evolved during the Thing regime), including standards for decent behavior, and (2) statutory law (particularly since 1863). Today state commons and *bygd* commons are very much institutionalized both at local and national levels.

REFERENCES

Næss, Hans Eyvind, ed. 1991. "For rett og rettferdighet i 400 år. Sorenskiriverne i Norge 1591-1991." ("Champions of Justice and Equity for Four Hundred Years. The Magistracy in Norway 1591-1991"), Oslo: Ministry of Justice.
Tretvik, Aud Mikkelsen. 1996. "...med mig tiltagne 2de edsvorne Lagrettemænd..." Om bønders deltakelse i lokalforvaltningen på 1700-tallet (On farmers' participation in local governance during the 1700s). Heimen, vol. 33:189-200.

ERLING BERGE   CHAPTER 5

The Legal Language of Common Property Rights

In order to describe and compare various instances of commons we need a precise descriptive language. A first approximation to this can be found in the history of land law in England and the United States (Singer 1993; Lawson and Rudden 1982; Simpson 1986). The problems of linking people to the rights and duties of ownership, and of linking rights and duties of ownership to resources, were experienced there and found their solution in ways similar to the situation in the Nordic countries. Even though common ownership of renewable resources is insignificant today in both England and the United States, the detailed legal language remains and can be put to use in the comparative study of common property rights regimes.

Divided or Shared Property Rights?

Real property may be held by more than one person in several ways. Property rights may be *divided* among many persons. One person may own the timber, another person may own the fuel wood, and a third person the pasture. Property rights may also be *shared*. The three persons owning timber, fuel wood, and pasture may share the property rights to the ground and to hunting and fishing.

Co-ownership

According to Lawson and Rudden (1982, 82–84) English property law recognizes two types of co-ownership: joint ownership and ownership in common. There are two important differences between them. One concerns what happens to the property on the death of one co-owner. *Joint ownership* implies that one joint owner's interest accrues on his or her death to the other joint owners, while *ownership in common* implies that on the death of one co-owner his or her fractional interest passes to the successors.[1] The other important difference is that ownership in common implies a specified fraction of interest in the object. Yet each owner in common, "no matter how small his fractional interest, has the right to possess the entire parcel—unless all the co-owners agree otherwise by contract" (Singer 1993, 801). Joint owners also have the right to possess the entire parcel,[2] but they are required to have equal fractional interest in the property.[3]

"Rights of Common" and "Profits"

The distinction between ownership in common and joint ownership applies to co-ownership in general. To describe a system of property rights recognized as a commons in the Nordic countries, we also need the concept of *rights of common*. The *rights of common* is a variable bundle of rights called "profits" sharing the characteristic that they allow the holder to remove something of value from another owner's property (originally *profits-à-prendre*).[4]

Lawson and Rudden (1982, 127–35) define a servitude as a relation between two units of land—the *servient tenement*, which is burdened with a duty, and the *dominant tenement*, for the benefit of which it exists. They list three types of servitudes: easements, profits-à-prendre, and restrictive covenants.

Simpson (1986, 107–114) recognizes three varieties of profits: (1) *profits appendant*[5]—the right to the resource is inalienably attached to some holding or farm unit,[6] (2) *profits appurtenant*—the right to the resource is attached to some holding, but alienable, and (3) *profits in gross*—the right to the resource belongs to some legal person in ordinary ownership.

Both Lawson and Rudden (1982, 130) and Singer (1993, 405) distinguish between profits appurtenant and profits in gross. Singer considers profits to be a subclass of easements in gross and states that profits today are considered freely alienable. Lawson and Rudden say that only profits in gross are freely alienable. Both find that some rights can run with the land.

Simpson's three kinds of profits are defined by a combination of two different variables. The first is a distinction between a person holding a right and a farm unit holding a right. The second is between the rights being alienable or inalienable. The point of these legal technicalities is obviously to let the rights of common run with the farm as part of the total resources available. For many farms the viability would depend on these rights of common. This attachment of the rights of common to some kind of recognized farming unit is important also in another way. It allows a reasonable way of limiting the use of the resource. In Norway, for example, it is the needs of the farm, not the farmer, which define the extent of the rights of common for pasture and wood resources. Thus one can say that even if it is the farmer who exercises the rights, it is the farm which reaps the benefit. This attachment of a right to a farm[7] will be called *quasi-ownership* and the farms will be labeled *quasi-owners* to distinguish them from legal persons.[8]

The three types of profits do not contain any category where the right is inalienably attached to a person in the way of citizen rights or human rights. However, the right to kill ground game is vested

TABLE 1
Types of Profits

	Rights vest inalienable	Rights vest alienable
Rights vest in land	Appendant	Appurtenant
Rights vest in person	All men's rights	In gross

SOURCE: Author

inalienably in the occupier of the land where the game is found, and the right to kill other game is usually vested in the freeholder (Lawson and Rudden 1982, 74).

In Norway and Sweden the "all men's rights" (*allemannsretten*) to such goods in the outfields as right of way, camping, and picking of berries and mushrooms can be described as an inalienable personal profit. The all men's rights have no restrictions on who can enjoy them, but of course there are clear limits on how to enjoy them.[9] Some other rights vest inalienably in persons as long as they are citizens of Norway, or are registered as living in a certain area or are members of a certain household.

NOTES

1. Today it is concluded that the joint ownership situation is ideal for the functioning of trusts and is said to apply to the management of property,

while ownership in common applies to the beneficial enjoyment of property (Lawson and Rudden 1982).

2. The right to the entire property for owners in common is often defined by the phrase, "the co-owners hold undivided shares." It is the physical object of ownership which is undivided.

3. For historical reasons the English terms are *joint tenancy* and *tenancy in common* if the object of interest is land. Here we will use *ownership in common* and *joint ownership* also if the object of interest is land.

4. The standard treatments of the law of property (Singer 1993; Lawson and Rudden 1982) do not discuss rights of common. Profits are defined as a type of easement by the law of servitudes. In their discussion of profits, Lawson and Rudden divide them into two types. One type is seen as "survivals of old manorial customary arrangements, whereby the tenants of a manor had the right, for instance, to pasture their animals on the waste of the manor." This type of profit is linked to some tenement. The other type of profit exists "in gross," i.e., it belongs to a person. Rights of common are discussed by Simpson (1986, 107–108), but he also sees them as "essentially incidental to a system of agriculture which is no longer in use in most of the country, though in hill-farming country the right to pasture sheep on moor land commons remains essential to the type of farming practiced."

5. In England, appendant profits were exclusively rights of pasture.

6. If the holding was split up, the appendant rights would also be subdivided.

7. Or in general to a unit without standing as a legal person such as a fishing vessel or a cadastral unit.

8. See Berge and Sevatdal (1995, 266–268). One may say that the right to use a resource is quasi-owned if it is inalienably attached to legal persons in their capacities of being residents in an area or citizens of a state or to estates in their capacity as cadastral units. An estate is not a legal person, but the right to use a particular resource can be inalienably attached to an estate and the use limited by the "needs" of the estate. The ability of estates to hold resources in quasi-ownership is the basis for calling them quasi-owners. The right to resources held in quasi-ownership may be extinguished, but not conveyed independently of the estate. Selling the estate implies selling those particular rights as well. If the quasi-owner ceases to exist, the resource held in quasi-ownership will either also cease to exist, or revert to the co-owners in case of joint quasi-ownership instead of any descendants of the estate. If two farm estates with rights to hunt in the commons are joined, the new estate will not have the hunting rights of both the former farm estates, only the hunting rights of one quasi-owner.

9. The principle of all men's rights as defined in Scandinavia seems to be unknown in the United States and England, but fairly common—although with variations—elsewhere in Europe (Steinsholt 1995). The struggle to keep and extend the rights of way tied to the system of footpaths and to establish a freedom to wander in England is vividly described by Marion Shoard (1987). In the United States public rights of access vary widely from region to region.

The only places public rights are assured are on the beaches below the mean high tide mark where the public has rights of navigation, fishing, and recreational uses including bathing, swimming, and other shore activities (Singer 1993, 249–258). Fishing could here be described as an inalienable personal profit.

REFERENCES

Berge, Erling and Hans Sevatdal. 1995. "Some Notes on the Terminology of Norwegian Property Rights Law in Relation to Rights of Access to a Resource." In *Law and the Management of Renewable Resources*, edited by Erling Berge and Nils Chr. Stenseth. ISS Rapport, no. 46. Trondheim: Norwegian University of Science and Technology.
Lawson, F. H. and Bernhard Rudden. 1982. *The Law of Property*, second edition. Oxford: Clarendon Press.
Shoard, Marion. 1987. *This Land Is Our Land. The Struggle for Britain's Countryside*. London: Paladin.
Simpson, A.W.B. 1986. *A History of the Land Law*, second edition. Oxford: Clarendon Press.
Singer, Joseoh William. 1993. *Property Law: Rules, Policies, and Practices*. Boston: Little, Brown and Company.
Steinsholt, Håvard. 1995. "Allemannsrett - og galt"(All Men's Rights—and Wrongs). Landbruksøknomisk Forum, no. 4, 5–14.

THOR FALKANGER   CHAPTER 6

Legal Rights Regarding Range Lands in Norway with Emphasis on Plurality User Situations

Introduction

Before dealing with the indicated topic it is necessary to define or explain some basic concepts. The central term in the title is *range lands*, which is not found in legal dictionaries, nor in the *Oxford Concise Dictionary*. Here it will be used in a somewhat loose sense, corresponding to the Norwegian word *utmark*,[1] as in land outside towns and built-up areas not being cultivated farmland. Positively designated range lands will in the present context include forests and natural grazing land, as well as barren mountain areas and glaciers. In this sense, range land is very often utilized by a number of persons—as will be explained below. This fact is apparently the basis for the introduction of the concept *nonprivate resources*,[2] indicating that there is not one person exclusively enjoying the benefits that may be derived from the range lands. From a lawyer's point of view it is questionable whether this term is adequate, due to the basic Norwegian concepts on ownership and delimited rights. A brief explanation follows:[3] Norwegian property law is based upon the concept that land is owned by one or several persons (natural persons or bodies corporate). The contents of ownership will depend upon the circumstances in the individual case—the property may, for example, be subject to leases, mortgages, or easements, considerably restricting the owner's enjoyment of the land. Thus, the actual contents of ownership are the residual rights, and these rights may be further limited or increased in the course of time (for example, restricted when the property is mortgaged, or increased when mortgages are redeemed). Persons with a right to a property without having the status of owner are considered as holders of limited rights, which are positively delimited. For example, in principle, the contents (the effects) of a lease depend upon the contract whereby it is created. Obviously, there may be cases where it is difficult to decide which of two holders should be named as owner and which should become the holder of positively delimited rights.[4] The implication hereof is that in a plurality-user situation, one or more of the participants may be considered the owner(s) while others are considered the holders of limited rights of different types. In practical terms they may appear on the same footing, but legally there is a fundamental difference between an owner and a non-owner utilizing range lands.

Clearly, the law is not static. In order to have a complete understanding of today's legal regime it is necessary to know quite a lot of the historical development of the legal rules, and also whether there are currently winds of change (and in which directions these winds might be blowing). With the given limits, however, it is necessary to restrict the description to a general overview of the present legal situation in Norway. The presentation will be general also in the sense that the specific problems related to Finnmark will be very briefly commented on, as these are the theme of chapter 10.

The Range Lands: Ways of Exploitation and Possible User Conflicts

In order to get a better understanding of the possible conflicts that can occur when there are several users of the same area, and examine the legal machinery for avoiding or solving such conflicts, it is useful to enumerate the benefits that may be derived from range lands. A list of resources follows:

1. forests
2. grazing land for cattle and sheep
3. fish and game
4. peat (for heating purposes, etc.)
5. nuts, berries, flowers
6. minerals
7. water (both for consumption and as a source of power)
8. building ground (for an expanding town, etc., and for recreational cabins)
9. recreational uses
10. inherent value environmentally, ecologically

It is apparent that there are a number of possibilities of conflict between various groups of users regarding these resources. Using waterfalls for production of electricity may reduce or eliminate fishing, or the development of a waterfall may be considered as disastrous from an ecological point of view. The potential for conflict also exists when the use of the land is on one hand only, which the last example clearly shows. In the present context, the focus will be on situations where more than one person or group of persons rightfully are utilizing range lands for various purposes. According to the Norwegian legal tradition, it is natural to distinguish between the rules generally considered as belonging to the private law sector and the public (administrative) law whereby the interests of the society as such are the basis (such as protection of agricultural land or environment).

The Private Law Regime

The starting point is that the person considered as the owner of a land area has the exclusive rights of disposition—legally and factually. He is omnipotent unless there are specific grounds for delimiting his powers. Even when disregarding public law—which will be dealt with in the following section—there are important practical exceptions to the main rule of owner supremacy.

Joint Ownership

First of all it should be noted that the owner may have a co-owner. Joint ownership regarding range lands is a very old institution in our country, and even today considerable areas are subject hereto—often in the form that the joint property is connected with ownership to certain farms. A number of farmers in a community have jointly owned grazing land, and since the rights to the grazing land cannot be sold separately, they must follow the farm. In particular, with a great number of joint-owners there are many possibilities for conflict-of-interest. Two acts are of particular importance: the Act on Joint Ownership of 1965 and the Act on Reallocation of Land of 1979.

The Act on Joint Ownership of 1965[5] defines the rights and duties of the joint owners. These rules are of a supplementary nature, that is, they are applicable when there is no express or implied agreement between the co-owners covering the same situation as a rule of the act. Some of its stipulations merit a short mentioning. Regarding the physical use of the land, the act is conservative: each of the joint owners is entitled to use the land in the customary, traditional manner. But it is added in section 3 that the land also may be used for other purposes "which are compatible with the present time and other circumstances." Within this framework, each owner is entitled to use the land (for cattle grazing, for example), but only to an extent corresponding with his part in the joint ownership. The expenses involved in the preservation of the property must be divided between the owners, basically according to their ownership interests.[6]

It is apparent that it may be advantageous for the owners to form some kind of organization, preferably with a board or a steering committee that has the powers to make decisions on behalf of all the owners. The act is, however, somewhat restrictive in these respects. Majority decisions are subject to rather extensive protection of the minority, and there is no obligation to have formal bodies.[7] It is up to the owners—or rather the majority of the owners—to decide whether there should be a board and whether there should be formal rules on such things as owner-meetings, voting procedures, and election of officers.[8] It should be added that some acts of a more specific nature also have rules on majority decisions.[9] For example, the Farming Act of 1955 gives greater powers to the majority regarding questions of farming and forestry than the 1965 act does, but the minority is protected inasmuch as it has a right of recourse to the agricultural authorities.

If an owner does not fulfill his obligations to make contributions to the upkeep of the jointly owned property, exceeds his rights to use the property, or is guilty of other infractions, the co-owners cannot easily get rid of him. Of course, injunctions and damages may be demanded in accordance with general principles of law. In addition, the act gives the rule that any owner may at any time require that the joint ownership be dissolved,[10] primarily so that the land is physically divided between the owners in conformity with their ownership shares. If physical division is not possible (technically, legally, or commercially), the property has to be sold with a distribution of the net revenue to the former owners. However, these procedures are not applicable in the instances where there, in principle, is the mentioned connection between the farms and the jointly owned range land.[11] But in such cases another remedy is available—the right to demand reallocation of land in accordance with the Act on Reallocation of Land of 1979.[12] If such a demand is accepted, the jointly owned land will be distributed so that now each piece of land will have one owner only. This reallocation may include not only the jointly owned land, but also the farms to which the jointly owned land belongs.

Delimitation of Ownership (One or Several Owners)

The powers of an owner—regardless of whether there is one or more owner of the same land area—are delimited in various respects. A partial list follows:

1. The rights to minerals with specific gravity 5 or higher do not belong to the owner of the ground. Norwegian

law adheres to the principle that the finder has the right to exploit mineral resources. The finder may use the necessary land areas against compensation to the owner.[13]

2. Of great importance are the rights which in Norwegian are called *allemannsrettigheter*—rights belonging to literally everyone, and which directly translates as "all men's rights." These rights flow from general law, dating back to times immemorial, but are now to a great extent codified.[14] The all men's rights include a number of different types of use of property belonging to others (for example, the right to walk over and stay on foreign property for a limited period of time, to camp and bathe there, and to pick wild berries and nuts).[15] However, in most instances these rights enjoy only weak protection. The owner may decide to use the land for a purpose which is not compatible with the exercise of the all men's rights. If he were to cultivate an area or build a house thereupon, then the rights of the public cease. Only in exceptional circumstances is a person, now deprived of his former use, entitled to compensation. But up to the moment when the owner changes his use of the area, the public in general may enjoy the indicated rights, and an attempt on the part of the owner to restrict the exercise of these rights is illegal. To some extent the conflicts between the owners and the public are solved or at least diminished by decisions or statements by particular administrative bodies (the Norwegian *friluftsnemnder*, or "outdoor life-councils"). In a larger sense it should be noted that there is a tendency to extend the all men's rights—typically, it is a broadly held view that the number of public recreational areas for fishing and hunting should be increased (a result of today's emphasis on recreational activities and "back to nature" attitude). At the same time, there are indications that the protection given the all men's rights is being strengthened—in particular, when the rights are of importance for making a living.[16]

3. Finally, for the sake of completeness, it must be added that an owner is entitled to create encumbrances on his land of such a nature that they remain as encumbrances also when the property has passed to another owner. The encumbrances may vary widely: leases, mortgages, rights of pre-emption, restrictive covenants, and easements are several examples. Obviously, the relationship between the owner and the holder of a limited right in the property

may give rise to difficulties. Such difficulties primarily must be solved by construction of the promise or agreement, creating the right (the encumbrance). In addition, there are a number of acts defining the rights and obligations of the parties, for example, as between owner (mortgagor) and mortgagee.[17] In most instances these acts are of a supplementary nature; they will be relevant only when no regulation—or sufficiently clear regulation—can be obtained by construing the promise or contract whereby the encumbrance was created. But in some instances an act may be of a peremptory character, such as setting a maximum time limit for the lease of fishing rights in a river—the underlying principle being that it will be harmful if important resources are permanently taken away from the property.[18]

State-Owned Land

The state may have acquired land through numerous means—an ordinary sales contract, or through expropriation. In such cases the relationship between the state as owner, persons with specific rights, and the public with its all men's rights is in principle the same as when the owner is a private person.

But the majority of state-owned land is not acquired in this manner. Roughly one third of the total area of Norway is state owned, and the dominant part hereof has been state owned for hundreds of years. One simple way of explaining this is that when Norway was populated, the state considered itself as owner of the areas which were not intensively used by the farmers. These areas have a particular status, inasmuch as the farmers in the vicinity—and to some extent also other residents—have rights over the lands. The particular status appears from the name given to such areas: *statsalmenning*, which may —with considerable hesitation—be translated to "state-owned common." In a number of cases the ownership to a common has been transferred to the farmers in a defined district, and thus we have a second category of commons, which may be called "district" or "farmer-owned commons." For the present purpose it is sufficient to say a few words on the state-owned commons.[19] The typical legal pattern is as follows:

1. The farmers have certain rights in the state-owned commons.
2. These rights are enjoyed by the farmers in the vicinity of each state-owned common. The geographical delimitation depends upon usage and tradition.
3. In order to have rights in the common, the farm must be a certain size.
4. The rights in the common may vary from instance to

instance. The actual contents depend upon usage. One important limitation should be noted: the rights may be exercised only to meet the requirements of the individual farm (for example, a right regarding wood or timber is limited to what can reasonably be used on the farm for purposes compatible with farming, such as woodfuel or building material).
5. The typical rights in the common concern wood, grazing, fishing, and hunting.
6. When the farmers have exercised their rights, the state, as owner, may utilize the property. This is of importance in two respects: first, if for example the forest yield exceeds the quantities the farmers are entitled to, the state will profit in respect of the excess. And second, if it is possible to utilize the land in a manner not contrary to the farmers' rights (as defined through usage), the state may do so. Thus the benefits derived from leasing plots for buildings fall to the state,[20] and, to give one more example, the state may develop waterfalls for production of electricity without compensation to the farmers.[21]
7. Fishing and hunting rights have gradually, to some extent, been transferred upon the public in general. Now the general rule is that hunting and fishing in state-owned commons may be exercised by anyone currently being domiciled in Norway—against certain payments, which may be differentiated so that people living in the vicinity of the common pay less. The fees paid shall cover certain expenses, and the excess, if any, shall be used for the economic development of the district surrounding the common.[22]

The administration of the state-owned commons is divided: the ownership aspects are dealt with by a special state body, the State Forest Administration. Regarding the questions actually concerning the forest, there is cooperation between this body and the common's council. The council is elected by the farmers with rights to the forest.[23] For the other types of use there is a so-called mountain council, elected by the municipal council.[24]

Rules of a Nonprivate Law Nature Controlling the Use of Range Lands

In modern society there has been an increasing tendency on the part of the state to regulate or control the use of real property.[25] Accordingly, one does not get a full picture by focusing on the traditional private law rules as has been done above. However, there is such a variety of state control that it is impossible even to give an outline. It is necessary to restrict the presentation to some remarks related to protection of the all men's rights or in a somewhat wider perspective the protection of outdoor recreational activities, which to a large extent are based upon the rules of all men's rights. This shall be done with the following examples:

1. There are rather rigid rules for building, be it houses for permanent residence or cabins for vacation purposes.[26]
2. Building roads[27] and developing waterfalls[28] are subject to state approval.
3. The forest legislation takes into account the recreational values connected with the forests, with the objective that forests may serve both commercial and noncommercial (ideal) interests.[29]
4. There is, it seems, an ever-growing legislation—and stricter enforcement—with respect to pollution prevention and environmental protection.[30] Generally, this is beneficial for recreational activities, but sometimes restrictive. In order to preserve nature, the freedom is curbed.[31]

Conclusion

The range lands in Norway are to a considerable degree used by more than one person. This is true regarding privately owned property (in particular when it is jointly owned, which is frequently the case). But even if there is one owner only, the land will in most cases be used by others as well, due to specially created rights over the property (encumbrances, seen from the owner's point of view), and due to the all men's rights founded on general law.

The pattern is notably difficult in the case of state-owned commons: there are a great number of farmers as users, as well as the general public by virtue of the all men's right. Otherwise put, there are possible conflicts between the owner and three groups of users: the farmers, the public, and those to whom the state has conferred rights in its capacity as owner.

The rights and obligations of the users—being owners or others—are to some extent defined by contract, but in important respects directly by written or customary law. Nowadays there is in addition a framework, consisting of a number of administrative law rules. Thus, the final solution to conflicts between the many users of range lands will depend upon the combined effects of traditional private law and modern administrative law. This synthesis may create problems, because we are at the interface of two different regimes of law. The traditional private law regime is court focused: if the parties are not able to solve the problems themselves, then the issues are decided by the ordinary courts (with the traditional possibility of appeal to a higher court).

In administrative law the conflict-solving mechanism is different. An administrative body, being responsible for a particular act, does not negotiate with the citizens to a comparable extent; administrative law is characterized by the issuance of decrees and granting permissions. And furthermore, the person not satisfied with a decision will make a complaint to the administrative body one step higher in the hierarchy. But above all, administrative decision making has (usually) a wider objective than that of the courts. The primary task of the courts is to solve conflicts between individuals; the administrative bodies in most cases have decided to take wider social aspects into consideration. But administrative decision making is in the end subject to court control, with the important reservation that a number of discretionary decisions (*skjønnsmessige avgjørelser*) cannot be challenged by the courts.

NOTES

1. The term *utmark* is defined in the Act on Outdoor Life of 28th June 1957, no. 16, sect. 1.

2. This term was used in the title of the conference from which the present volume grew (editor's addition).

3. For a more detailed explanation of the concept of ownership, as well as the manner in which the courts decide on who is the owner, see Falkanger (1990).

4. Here it must suffice to mention the "classical" conflict: "A" utilizes the grazing resources and "B" the forest resources. Is "A" or "B" to be considered as the owner with the right to have the benefits from other use of the land (e.g., the rent from leases for cabins or the right to develop waterfalls)? Or is the property jointly owned by "A" and "B", with the understanding that parts of the benefits are to be exclusively utilized by "A" (grazing) and "B" (forest yield)?

5. Act on Joint Ownership of 18th June 1965, no. 6. Further on this act, see Falkanger (1990, 71–124).

6. See section 9, compared with section 8 of the Act on Joint Ownership.

7. See Sections 4 and 5 of the Act on Joint Ownership.

8. See sections 6 and 7 of the Act on Joint Ownership.

9. See in particular the Farming Act of 18th March 1955, no. 2, chapter IX; and further the Game Act of 29th May 1981, no. 38, section 29.

10. See the Joint Ownership Act, section 15.

11. See the Joint Ownership Act, sections 13 (2) and 15 (5).

12. Act on Redistribution of Land of 21st December 1979, no. 77; see Austenå and Øvstedal (1984).

13. See the Act on Mineral Resources of 30th of June 1972, no. 70. Oil resources in Norway itself (the existence of which are highly unlikely) belong to the state (Act on Exploration for, and Exploration of, Petroleum below Norwegian Lands; 4th May 1973 no. 21), as does oil and other natural resources off shore (Act on Petroleum Exploitation, 22nd March 1985, no. 11 and Act on Scientific Research and Exploration for, and Exploitation of, Natural Resources other than Petroleum under the Sea Surface, 21st June 1963, no. 12).

14. See in particular the Act on Recreational Activities of 28th June 1957.

15. For an overview, see Falkanger (1986).

16. See, for example Supreme Court decision in Norsk Retstidende 1985, 247.

17. Act on Mortgages and Other Charges of 8th February 1980, no. 2. See as further examples the Act on Easements of 29th November 1968; Act on Leases of 30th May 1975, no. 20; Game Act (see note 9); and Act on Salmon and Inland Fishery of 15th May 1992, no. 47.

18. Act on Salmon and Inland Fishery (note 16), sect. 19.

19. The rules on commons have gradually been codified. On state-owned commons, see the Mountain Act of 6th June 1975, no. 31 and the Act on Forestry in State-Owned Commons of 19th June 1992, no. 60; for district-owned commons, see the Act on *Bygd* Commons of 19th June 1992, no. 59. These acts, together with the preparatory documents (*travaux preparatoires*), give a good overview of the present situation.

20. See, however, the Mountain Act of 6th June 1975, no. 31 (note 19), sect. 12 (3): Half the income from leases for cabins and hotels goes to the "mountain chest," from which expenses relating to the common are covered; see also section 11.

21. See in particular the Supreme Court decision in Norsk Retstidende 1963, 1263.

22. See the Mountain Act (note 20), chapters 11 and 12.

23. See the Act on Forestry in State-Owned Commons (note 19), sect. 1 on cooperation and chapter 3 on election of the council.

24. See the Mountain Act (note 20), chapter 3.

25. See Falkanger (1986, 20).

26. See the Planning and Building Act of 14th June 1985, no. 77, in particular sections 20–24.

27. Primarily regulated by the Planning and Building Act (see previous note).

28. See the Act on Regulation of Rivers of 14th December 1917, no. 17.

29. See the Forestry Act of 21st May 1965, section 1, and in particular sections 17a and 17b.

30. See the Pollution Act of 13th March 1981, no. 6.

31. See, for example, the Nature Protection Act of 19th June 1970, no. 63

and the Salmon and Inland Fishery Act (note 17), section 1 compared with chapter 3.

REFERENCES

Austenå, Torgeir and Sverre Øvstedal. 1984. *Jordskifteloven med kommentarar (Land Reallocation Act with commentaries)*. Oslo: Univer-sitetsforlaget.
Falkanger, Thor. 1986. *Eierrådighet og samfunnskontroll (Ownership and State Control)*, 3rd ed. Oslo: Universitetsforlaget.
Falkanger, Thor. 1990. *Tingsrettslige arbeider (Real Property Studies)*, 3rd ed. Oslo: Universitetsforlaget.

Hans Sevatdal   Chapter 7

Common Property in Norway's Rural Areas

Introduction

The phenomenon of land which is "common property" or in some sense collectively controlled, owned, or used in rural areas in Norway, is closely linked to the historical evolution of settlement and tenure patterns.[1] We call these lands by different names, for instance, *allmenning* and *sameige* in Norwegian. For reasons of convenience I will use the word "commons" for the whole group. Some of the laws governing use and management of the commons go right back to customary law in the early Middle Ages, which in Scandinavia means the tenth century. Statutory law dates mainly from the thirteenth century. I find it necessary to stress this point, because the phenomenon must be studied and understood in its proper historical context. As the commons are of very ancient origins, and such aspects as topography, climate, settlement patterns, and economy in Norway are varied, it should be expected that the commons are equally diverse. Classification in seemingly homogeneous groups thus becomes rather dubious. It is said that each individual common must be studied separately to get a true and precise understanding of its legal situation. This should be kept in mind as I present my classification scheme and the various features attached to each group of commons. Classification in itself always violates the realities to some extent.

There is a mutual relationship between land use, on the one hand, and ownership and tenure patterns on the other. Certain uses of land lead to the establishment of certain ownership and tenure patterns, which then influence further development of land use or vice versa;

certain types of ownership promote certain types of land use. (What came first, the bird or the egg?) These relationships could be extremely complicated and diverse, but for our purpose it is important to note that the ownership patterns often tend to lag behind. This means that certain ownership and tenure patterns can endure for a long time after the land use that created them in the first place has vanished. This general statement could easily apply to some aspects of the commons or Common Property Regimes. It should also be noted that the Nordic countries in general have a remarkable continuity in their legal systems concerning land tenure and ownership. This is mainly so because there has been very little migration into the area by alien groups of people in historical times, or other events causing sudden or revolutionary changes in land tenure systems.

Norway has a small population compared to the size of the land area. The actual cultivated agriculture area (arable land), however, is very small relative to the population (at present, about 0.2 hectares per person). And the rather marginal conditions for agriculture make this figure even smaller, so to speak, as compared with southern countries. This does not mean that the rural societies were proportionally "poor," it just means that the people had to utilize other resources. (By and large this meant maritime resources and the so-called "outfields." The outfields (woodlands and mountains) were of great importance for grazing, gathering of fodder (grass, moss, leaves) to keep the livestock during the winter, wood and timber, hunting, and fishing, just to mention a few important uses).

The predominant "original" settlement pattern was—we assume—composed of single farmsteads. Each farm could be very large in terms of area; most of the land was not cultivated, but was composed of woods, pasture, mountains, rivers, and lakes. The actual cultivated area was quite small in comparison. However, with the successive subdivisions of farms, clustered village-like rural communities developed from the seventeenth century onwards, particularly in coast and fjord areas. To a large extent, these villages have greatly changed due to the process of land consolidation, which occurred during the second half of the last century and the first decades of this century. This process included, among other things, consolidation of scattered plots and strips into single blocks of land for each farmer, rearrangement of management and use practices in land held in common, and also in many cases relocation of farm houses from the old clustered village to the new block of land. In other areas such as the southeast, single farmsteads have always been the dominant settlement pattern. New farmsteads were established at a certain distance from the old farmhouses.

Most types of commons are related to some sort of a "local community." In many countries this will not cause a definition problem; the village constitutes the obvious local community. The term "local community" will be used here in much the same sense as it would have been if we had real villages. But we have to keep in mind that in Norway the local community is seldom a village in this physical sense, meaning a clustered rural settlement (small rural town). We must imagine a combined "agroforest" landscape with scattered farms and single houses or small clusters of houses and farms in between. Small local urban centers have emerged all over the countryside in this century, but we do not call them "villages" mainly out of tradition, but also because they do not, as a rule, contain agricultural activities. We call them *tettsted*, or urban centers.

There are three important terms that describe settlement units at local levels: a smaller settlement is called *grend*, while a larger one is called *bygd*. Often each *bygd* has an urban center and contains several *grender*. A municipality (or commune) is normally composed of more than one *bygd* and always has an urban center. The three levels, *grend*, *bygd*, and commune, are relevant to our problem because they are units for groups of holders of different rights in the different types of commons.

The Commons

Definitions

The terminology in this field is problematic. What should be understood by the term "common property"? First of all, rights of some sort must exist that lay well within the concept (regime) of ownership in the legal system. This means that resources with access by everybody are not common property—they are common, but they are not property, and should rightfully be defined as "open access resources." The right of everybody (foreign tourists included) to roam everywhere in the so-called "outfields" (such as mountains and forests) in Norway is in this category. Secondly, more than one person (physical or judicial) must exercise rights, and these rights must have the nature of being property rights in the resource. From a legal point of view the rights of the different rightholding persons may not necessarily be of exactly the same type.

There are a lot of problems here. An interesting, but difficult one, arises when the resource belongs to a large group of persons somewhat vaguely defined, where each member has an equal and unrestricted access to the resource. Should this be called "open access" or "common property"? This and a lot of similar problems are discussed by Stevenson (1991). In my discussion here I will stress the nature of the right in the Norwegian legal system. If the holder or an organization representing the holders collectively is entitled to compensation in the case of violation of the rights, I will call it "common property." Normally this would involve a case of eminent domain (expropriation).

Even if the group is very large (several thousand families), the membership is vaguely defined and there are no regulations of the use. The members in this case undoubtedly have a kind of open access. For Stevenson the main criteria seems to be whether the use is regulated or not, and the example above would be classified "open access" by him. (The crucial point might be what should be understood by "regulation.") Protection against violation may take other forms than compensation, such as prohibition, or very heavy restrictions regulating use of the land and access to rights for those outside the group.

There is another difficult point which can be clarified by a simple example. Three neighboring farmers buy, inherit, or otherwise attain ownership of a piece of land, with equal or unequal shares. The land is thus held in common ownership by them, but it is not "common property" in our terminology, it is rather that the ownership (the title) is shared by them. The land is one estate (property unit) held in common that could at any time be sold to one person and thus pass into individual ownership again. In the classification in the cadastral system it would not be labelled "common property." Instead the register would show one estate (unit) with three individual owners holding the title in common ownership with equal or unequal rights. The reality is that the land is owned by three different persons. We call this "personal joint ownership," even if the "persons" are companies or other legal bodies, public or private.

Let us then assume that these three farms (not the farmers) own a piece of land jointly, in such a way that each share is legally (and in the cadastral sense) a part of each farm unit. This means that the share in the jointly owned land is so tightly connected to the farm that the two can not be separated in two property units without a legal (or in this case a cadastral) subdivision procedure. Such subdivisions are subject to very strict customary and statutory regulations, and permission to separate a farm from its share is very hard to acquire. Additionally, the other shareholders would have the right of first refusal. Land that is jointly held in this way, is by my definition, "common property." The actual wording in Norwegian is that the land is "jointly owned by farms," meaning that the shares are appendant to the farm of the holders of the right. It is a special category or type of ownership.

This example illustrates how difficult it is to define the term "common property" using legal categories. In this case we have two seemingly similar situations, both with several farmers as holders of joint ownership rights—perhaps even the lands in question are of the same type. One is by definition "common property." The other is not. On the other hand, to make the concept operational in everyday life it must be linked to legal classifications. In this case the legal conditions of the two categories are very different. For purely scientific purposes this base in the legal system may be of less importance.

The example also highlights another aspect: persons do not "own" a share in a commons in the same direct way, and in the same sense, that they own personal belongings or land held in fee simple. Most often the rights or shares are connected to or are contingent upon some sort of quality that the persons in question must possess. Most typically such a quality would be permanent residence in the local community, but it could also be status as a farmer, owner of a farm (as in the example above), tenant to a farm, or close kinship to a person who possesses such qualities.

One may also ask if cooperation, or collectiveness in actual use (operation), should be emphasized and added to the definition. Properties legally held in common may be, and often are, used independently (individually) by the shareholders. I have come to the conclusion that decisive importance must be placed upon whether the legal rights are common or not, and let the degree of collectiveness in use be part of the description. Consequently, I will classify some properties as "commons," even if the actual use is carried out mainly individually. In some cases the rights of management of the land is tied to more or less autonomous local bodies, based on the cooperation of the families living in the local community. But the land should not be municipal property; in that case it is owned by a public institution, which in itself does not qualify the land as "common property."

It once was quite common, and in some areas it still is, that two or more persons, or farms, would hold different rights to different resources in the same piece of land in the outfields. For example, one would own the trees, another the pasture. This situation in itself does not qualify as "common property," even if the usages are heavily interdependent and have to be coordinated. However, such arrangements are often found together with other conditions that would make it "common."

In Norway the two basic qualities that individuals must possess to have rights in commons are residency in the local community; and/or ownership of a farm, or at least ownership of a piece of agricultural land which once was a farm. Most typically, both residency and being a farmer used to be the standard norm, which in practice meant to own a farm and live on it as a farmer. It should be noted that for some types of commons residency and/or farming is not required, instead ownership to farmland is the decisive factor.

The demographic and occupational patterns in rural areas have undergone great changes in the last decades. For the purposes of this chapter the changes that concern the farms are most significant. There has been a steady decline in the number of farms *in actual use as farming units*. In fact the number of such units (with more than 0.5 hectares of cultivated farmland) has dropped by more than 50 percent, from over 200,000 in 1949 to well under 100,000 in 1994. (The rate of reduction may dramatically increase in the near future.) It is, however, most important to note that most of these farm units have by no means disappeared. They still exist as physical units in the landscape, permanent places to live for rural households, areas for recreational use, *and as ownership units with most of their rights intact.* The agricultural activity might have been abandoned totally, or may be kept at a very low level. Whenever possible, the agricultural land is leased to active farmers in the neighborhood. An extremely low amount (compared to other countries) has so far been sold and amalgamated with other farms in terms of ownership. There are undoubtedly numerous reasons behind this unexpected development. I will briefly note that this low rate quite possibly has to do with our peculiar *odal* (allodial) law, the taxation system with low tax on such properties, the relative attractiveness of a rural way of life, and the relatively abundant access to employment in some rural areas. My guess is that the situation is rather unstable.

There are, of course, great variations in the overall picture. In some areas that are in some sense marginal, most of the farms may be without permanent settlement. In any case, this means that a large and growing proportion of farm units, and consequently also the rights to the commons (in the cases where the rights follow the farm's ownership or residency) pass from the farming population to others. The rights follow the persons. And these other persons might or might not live in the local community. The often large areas of outfields that are held individually by the farms undergo the same process. This scenario gives an overall picture of rural resources passing out of the ownership control of the farming population, and in many areas out of the ownership of the local population. Ironically one may say that our *odal* law and kinship values and traditions, which are supposed to keep the ownership of farms in the hands of the farming population, produces exactly the opposite result because it is a right for the landowning families instead of farming families.

With this background I will briefly discuss the following main categories of common property:

1. State common land
2. *Bygd* common land
3. Common property owned jointly by farms (*sameiger mellom bruk*)
4. Common property in the counties of Nordland and Troms
5. Common property in the county of Finnmark

I will not include coastal waters, riparian rights, groundwater, and other special rights and types of ownership concerning salt or fresh water, even if these categories could definitely be of a common nature and related to a local community. The very special common property rights of the reindeer herders will not be discussed here either. The first three categories above are settled and can be described without much uncertainty. The situation in the three northernmost counties, categories four and five, is both controversial and partly in transition.

State Common Land

Area. These commons amount to an area of 26,622 square kilometers, which is close to 8.2 percent of the total land area of the country. Most of these commons are mountainous—only 7 percent is comprised of productive forests. These forests are distributed unevenly in the mountainous parts of southern Norway. The main land uses today are:

- forest (timber, fuelwood)
- pasture (sheep grazing)
- secondary summer farms with cattle grazing (*seter*)
- grassland for hay production (cultivated)
- fishing
- hunting
- tourism and recreational use of various sorts
- hydroelectric power

In addition, conservation must be mentioned, as several national parks and other protected areas are found in state common land due to provisions in the nature conservation legislation.

State common land once covered much larger areas. Over the centuries, parts of these areas passed into private hands in different ways. Commons could be sold by the king, to be subsequently subdivided between the buyers and those who possessed rights to the commons. The products of such a procedure could be held in fee simple as well as held in common by the local community, or this part could also be subdivided among the shareholders. When I say "sold by the king" I should be more precise: the king could of course only sell what rightfully belonged to him, or his share of the commons, not what belonged to the local community.

Owners of adjacent properties could acquire title to commons or parts of commons through long and exclusive use, in the sense that they (illegally) kept others out. By these and similar developments, state common lands were taken over by private owners on a large scale, especially in the eighteenth century (even if it was against the law, either as individual holdings or holdings held in common by the new owners). This process was more damaging for the local community at large than the king's sales, as the "new" owners acquired title to all the resources in the commons. The sale of commons is now prohibited by statutory law and has been so for more than a hundred years, except in cases where it is to be used for cultivation (reclamation). The final determinations of the boundaries for the state commons in southern Norway were made during court procedures by a special commission (*Høyfjellskommisjonen*) in the period from 1909 to 1954.

Legal situation. The basic principle of ownership and rights of use is as follows (Sevatdal 1985): The rights to traditional utilization of the resources in a specific commons belong to a specific local community, most often a *bygd*. However, the right of each rightholder in the local community is restricted by the concept of "household needs." The reality of this is that nobody is entitled to take anything away from the common and sell it. But this is not without some sort of exceptions; game and fish from hunting and fishing can be sold, but not the access to the hunting and fishing activity itself. This is similar to the selling of milk and meat, without the right to take in "foreign" cattle for pasture.

What may remain of resources when local needs are satisfied belongs to the state. Within this broad framework there are many refinements, the most important being the concept of "ownership to the ground." This is a direct, and most likely linguistically meaningless, translation of the Norwegian *eiendomsrett til grunnen*. Without going into legal refinements, the practical implication is that when all traditional, customary, and positively stated rightholders have received what rightfully belongs to them, there might still be something left. And this "something" is said to go with the ground as such, hence the need to appoint an owner of the ground or remainder. Often this rightholder is referred to as "landowner." There are many aspects of this principle, but by far the most important is that completely new types of exploitation that may come up belong to the one who "owns the ground." The same principle is applied outside the commons too. Development of hydroelectric power is a good example. It is an extremely valuable resource, and is plentiful in the Norwegian high mountain commons with their abundant precipitation and high elevations. It has been established—after dispute and court cases—that this resource belongs to the state as "ground owner" in the state commons.

Another example is the "selling" (or rather, long-term leasing) of building sites for recreational cabins, which is very common in the Norwegian mountains. This right also belongs to the state as "owner of the ground," but the annual income from this is shared with the commune in which the commons is situated instead of the *bygd* that has the rights in the common.

Rights to uses connected with farming, such as to pasture, firewood, and timber for building purposes, are reserved for the farming population in the local community. This means that (with some exceptions) all farming households in the local community have such rights. Absentee owners of farms, or resident owners possessing farms that are not actively farmed, make little use of such rights. But if the farm is "reactivated," or when new farms are established, the rights will come into being again. Whenever the farm is run by a tenant, he is in the possession of the right. At present, around 20,000 farms actually exercise such rights.

Everyone living in the municipality—which is in general a larger area than both *grend* and *bygd*—has equal rights to some sort of hunting and fishing. The public—that is, everyone living in Norway—also has access to certain limited types of fishing and hunting.

The legislation is derived from medieval times, but the actual laws have of course been modernized many times. A new codification of the laws concerning both state common land and parish common land has just been enacted; the "new" laws are dated June 19, 1992. A basic principle throughout the history of legislation in this field can be summarized in the following statement: the rights and legal conditions in each commons should be as they have been of old. This means, for example, that rights possessed by the actual local community should not change with shifting administrative units and boundaries. At present, the boundary of a local community relevant to a common remains the same, even if the boundary of a municipality has changed.

Management. The management and decision-making powers are divided between two, or in some cases three, bodies. These bodies correspond roughly, but not precisely, to the "interested" parties: the state, the local community, and the municipality.

An official in the governmental forest service takes care of the ownership interests of the state. The forest service also supervises most other activities that go on in the commons. The interests of both the local community *bygd* and the public within the municipality are taken care of by two bodies. It is important to note the distinction between commons with and commons without productive forests—that is, between forest commons and high mountain commons. In forest commons there is a board, elected by those who have rights to the wood, that makes all decisions concerning the collective use of the resources. In mountainous commons (as in areas over the timber line), a municipal board is responsible for organizing the use. This board is elected by the municipal council, but the majority of the members of the board should always, by law, be persons living in the local community (*bygd*).

Bygd Common Land

As mentioned above, *bygd* in Norwegian is a local community smaller than a municipality. This type of commons differs from state common land in the actual ownership (title) to the land itself. While the ownership of the land (ground) in state common land rests with the state, the *bygd* common land belongs to those farms which possess rights to the wood in the common. This is a rather formal legal definition; in practice we may say that the commons belongs completely to the local farming population. Or put in another way, the commons belongs to the farms in the local community. So we have two local groups—the owners of the farms that own the ground itself, in much the same way as the state owns the ground in state commons, and the other right holders in the local community (*bygd*). Neither the state nor the municipality council have significant power, as all decisions are made by a board which was elected by those farmers who have use rights in the commons.

Parish common lands cover an area of 5,500 square kilometers, and of this area 1,700 square kilometers are productive forests. Most of them are found in two counties in southeastern Norway (Hedmark and Oppland), and around 17,000 farms have rights in these commons. A significant fact is that all forest use in these commons is now organized on a collective basis. Each common is managed as one unit as far as utilization of the forest goes. This includes commercial sale of timber and often, wood products. As most of the commons also have sawmills, the right holders get the wood products they need from the sawmill, instead of logging themselves. Quite often the sawmills have developed into wood-based industries owned by the common.

Land Owned in Common by "Farms" (Sameige Mellom Bruk)

The translation of the name of this type of commons is difficult, and quite probably impossible. In Norwegian we call them *sameige mellom bruk*, which literally means "land jointly owned by farm units." These lands are completely "private," in the sense that no public institution as such, at state or local levels, has rights or any other kind of power within the regime of ownership. These commons constitute a very heterogeneous group, with various types of ownership. What combines and characterizes them is that they are owned in common—not directly by persons or organizations, but by other properties (usually farms). This type of commons is very common in outfields in general (lakes, pastures, mountains), but it is not so extensive in productive forests as it used to be, because in such areas it has been largely subdivided (individualized into plots) among the farms, by public or private land rearrangement procedures. But in many cases of such subdivision, especially in the eighteenth and nineteenth centuries, most other resources except timber remained common property. The subdivisions included the wood producing capacity of the land only, excluding the land itself and pasture and other resources. *In many mountain areas in southern Norway this type of commons is the dominating type of ownership.*

Laws and regulations for management are very different from those that govern state common land and *bygd* common land. The origin and historical backgrounds of the two arrangements are also different, but are still interwoven in rather complicated ways.

There are mainly two ways in which such commons originated. One is by way of subsequent subdivision of farms into new farms, in such a way that the outfields of the original farm were kept in various types of joint ownership according to the feasibility for the use of the resources. Each new farm established by the subdivision process got a share in the outfields, usually according to the assessed value of the farms in the tax rolls. The various uses (not the ground) could, however, be treated like an "estate" or become an object of ownership in itself and be subdivided in various ways, as need arose. For example, one farm might possess the right to the trees (or even to certain types of trees), another the hay-harvesting rights, while the grazing, hunting, fishing, and the ground itself could be held in joint ownership by the farms, all in the same piece of land. The basic principle was to subdivide, or individualize, each type of resource when the need arose, and keep the rest in common. So the extensive outfield areas of the original farm units, historically and presently, have to be conceptualized as dynamic systems in which economy, practical feasibility, legal considerations, and tenure systems are the main ingredients.

Some of these commons have, however, originated another way through collective action by a local community such as a group of farms. In organizing, the group would obtain collective ownership, and over time the share in the land would amalgamate with the farms. The collective action could be as simple as buying, but it could also mean acquiring state common land by excluding others from its use for a long period of time, and performing transactions that rightfully should belong to an owner. In the past the state has been in a relatively weak position to protect public interests against such actions. At some stage in history the acquirement would get public recognition, most often by a court procedure. In the late eighteenth century there were a lot of such court cases in the central part of southern Norway, but probably the bulk of such recognition took place from 1909 to 1954 with the work of the so-called Mountain Commission. Some of the really large commons and individual estates of this type in the high mountains most probably originated this way.

Since 1860 it has been the responsibility of the Land Consolidation Service to clarify legal matters and settle legal disputes, and individualize and/or lay down rules for the collective use of such commons (in short, to readjust the ownership and tenure patterns in the individual common to the changing needs in the use of the land). This service works within the framework of the judicial system, and its decisions have the power of court rulings. Any case will be activated after a formal request from one of the rightholders. (Even if there are hundreds of shareholders in that particular common, a request from one of them is enough to start a land consolidation court procedure, if the court itself rules that it is necessary and beneficial to all shareholders.) This has many important implications for the management of these commons, probably the most important is that each member knows that if they do not reach an agreement by themselves, it is always possible for one of the rightholders to bring in an independent decision-making body. Although the process could take a lot of time and money, quite often the shareholders prefer to ask these independent "outsiders" to make decisions on tense matters instead of reaching an agreement collectively by consent or by majority rule.

It is, unfortunately, impossible to present comprehensive and reliable statistics for this type of commons. The reason for this strange fact is mainly that each individual unit of these commons does not show up as a separate unit in the cadastral records and tax rolls. It is the shareholding farms that show up. This is now being changed, but it has not yet been completed and statistics are still not available. What can be said is that these commons certainly are more extensive than both state common land and parish common land put together in southern Norway. (In northern Norway it might be otherwise, as will be discussed later in this chapter.) While approximately 20,000 farms have access to (rights in) state common land and 17,000 in parish common land, more than 50,000 farms hold shares in commons held jointly by farms. (These figures cover the entire country.) This is the predominant type of ownership in the mountainous areas in southern Norway. Modified forms such as the timber rights individualized in plots—while pasturing, hunting, fishing, etc. are still common—are very common in forest areas in the western part of southern Norway.

Legislation in this field has recently been modernized; the most important laws are the Act on Joint Ownership of 1965 and the Land Consolidation Act that took effect in 1979. The majority of owners have the power to decide upon the management of the whole property, within the limits, roughly speaking, of suitable land use. They can elect a board to take care of the management and the day-to-day decisions. Neither the majority nor the board can decide to sell any part of the commons without everyone's consent. There are also other regulations that protect the minority. Most commonly, the shareholders elect a board to manage mutual interests, but the actual use is mostly individual rather than collective.

As stated above, it is always possible for one or several shareholders to apply for assistance from the land consolidation service. In that case, all disputed questions will be decided upon by the Land Consolidation Court, which include both judicial matters (disputes concerning ownership and rights) and rules and procedures for the management. For example, the court can decide whether the use should be collective or individual. In case the court decides on collective use, it will organize proper institutions for implementation. When individual use is decided upon, the court will lay down direct regulations regarding each shareholder's conduct.

Discussion

There are certain trends concerning the management and use of the above-mentioned three types of common property:

1. Traditional agricultural uses of the commons have declined. For example, hay harvesting in the uncultivated outfields—once very important—has hardly been practiced for several decades. Pastures in the outfields for milk-cows without any kind of cultivation are of little importance. Grazing areas for sheep are important, and some types of cattle (such as heifers and those used for meat-production) are of some importance. In some areas the practice of using secondary summer farms in the mountains is still in use, in modernized forms.

2. The commons have always been reservoirs of arable farmland used for the establishment of new farms, and recently have been important in the enlargement of the cultivated land of existing farms. The land thus cultivated is either sold from the common or is rented out on long-term contract.

3. Forestry is of great importance and value. In this field there is a tendency towards collective forms of use. For *bygd* common land this is the dominant form, for the two other types of commons both the legislation and actual policy try to encourage collective forms of use. "Collective" may, however, be a somewhat misleading term. It often means that the property is managed businesslike as one single unit, and not necessarily that the shareholders actually work together.

4. Recreational use, such as the building of recreational cabins, hunting, and fishing in the commons as well as in the outfields, has been rapidly increasing in importance and value. Selling the access to such commodities is rather important, and requires collective action. The hunting of big game (elk, hart, roe-deer, and to a lesser extent wild reindeer) is of special relevance to our discussion here for several reasons. First of all, the stocks of these animals have *increased* dramatically in the last three decades, and have also spread to new regions. Secondly, the market for recreational hunting has grown, thus creating the potential for a rapid rise of prices. And thirdly,

utilization of this resource requires collective action on the part of the rightholders. This is, of course, the way the commons operate, but most other property units are so small, or of such configuration and situation, that cooperation among several owners is necessary for them too. The bulk of the holdings are in themselves simply not suitable units for the management of big game, and quite often the municipal wildlife management board that decides which animals will be felled in the commune and establishes each hunting unit simply will not give permission if the owners do not amalgamate their properties into suitable units. In most local rural communities big game hunting traditionally has been an extremely valued activity by the landowners and by rightholders in common land, as well as among the rural population in general. But it is an increasingly expensive activity for the rightholders, in the sense that the alternative sales value is high and increasing. One may therefore expect growing tension and stress in the relations among the rightholders, between those who want to hunt and those who want to make maximum profit from the resource by selling the access to hunt. Until now, the local culture and values have had the upper hand. This certainly creates favorable conditions for the study of collectiveness in rural Norway. In addition, there is a great need for alternative employment, and tourism combined with hunting and fishing is certainly one option. Hence a certain political drive towards the selling of such resources to outsiders follows, combined with service activities that render employment.

5. It should also be noted that in Norway, as in Sweden, Finland, and many other countries, there has always been a common right of way for the general public. That right is valid for everybody and for all kind of outfields, regardless of type of ownership. During the winter the right applies to the frozen infields as well. Besides walking on foot or skiing, the right includes picking of wild berries and mushrooms as well as camping for a few days. This right is of course of great importance for recreation and tourism.

In relation to general regulation and planning of land use by the various authorities on municipality, county, and national levels, the commons are mostly regarded just as other types of properties. There is, however, one important exception for state common land concerning nature conservation, especially the so-called national parks. The Act on Nature Conservation (June 19, 1970) states that to protect special values, state land and adjacent private land could be given the status of national parks. The use of adjacent private land (instead of state land) for this purpose has created much debate and the subject has garnered a ruling in the Supreme Court. There is, however, another controversial aspect which in general has been neglected and should be taken into account. The type of state land that is relevant for national parks is, from a nature conservation point of view, often state commons. The legal position of the state in the state commons has been interpreted by the conservation authority so that national parks preferably and easily could be created on such land. This may be legally sound, but it makes people tend to forget the psychology and use rights of the locals. In some cases this attitude has challenged and provoked a powerful and most able local opposition among the rightholders in the commons. This is especially the case with the largest, and certainly one of the most important national parks, Hardangervidda National Park in the central mountain plateau in southern Norway. Using various combined judicial and political actions and mass media campaigns and professional use of strategic alliances with political parties and organizations (and certain institutions within the administration hostile to this type of conservation), the opposition has managed to severely violate the protection practices. The most harmful was the building of a road for cars right the heart of the supposed wilderness of the national park, years after the park was established. The example is a good lesson showing what strong and able collective action from united local groups of this kind can do, and how disastrous it can be for central governmental bodies and the interests they promote to ignore collective rights.

The last point to be discussed has to do with local institutions for management of the three types of commons. It is an astonishing fact that remarkably few standardized local institutions seemed to have existed until quite recently (essentially this means our own century). How can this be? The resources in the commons were important—at times there was even a scarcity of resources and strong competition. In the other Nordic countries—Denmark, Sweden, and Finland—we find such institutions as the *Byalag*, which have developed in large districts from the late Middle Ages to our own time. They are well visible in documents and the legal system, and were integrated into the administrative system (Erixon 1978, Meyer 1949). We would expect to find such institutions in Norway too, especially since Norway and Denmark were in very close union from the fifteenth to the nineteenth centuries. A closer look at the sources reveals institutions, or at least traces and remnants of institutions, at multiple farm and *grend* levels (Frimanndslund 1956, Løne Vinje 1991). They are, however, very heterogeneous, and seldom appear in legal documents, public administration practices, or statutory law. They have even to a large extent escaped the attention of local historians. They are characterized by local needs and conditions in the sense that they address problems and do not seem structured as general management institutions for mutual affairs at a local level. Still, these institutions can have multiple functions, one of which seems to have been to carry out the minimum organized actions required for the management of common property owned jointly by the farms. I have so far found no case where customary institutions of this kind have been active in the management of local interests in state and parish common land.

The research base is at present far too weak to draw conclusions, but a qualified guess is that problems related to the collective governance in all three types of commons were mainly solved within two regimes—that of customs and tradition, including standards for decent behavior (customary law), and statutory law. Formalized institutions at local levels seem to have been of minor importance after the old Norse institution the *Thing* disappeared. We have many cases where the parties' discussions enter into an agreement in the form of a written or oral contract, obviously based upon statutory law modified for local conditions. If they did not reach an agreement the case would be taken to court. Today, management of state common land and parish common land is very much institutionalized, both at local and national levels.

Commons in Northern Norway

Northern Norway covers the three counties of Nordland, Troms, and Finnmark. Various questions concerning common property in this vast area are discussed in other chapters of this book, as are the special rights for the Saami reindeer herders. I will confine my remarks to a minimum to round off my discussions.

There is (at present) no *bygd* common land in northern Norway. Common property owned jointly by farms is essentially the same for southern and northern Norway, and what has been said in this chapter about this category covers northern Norway too. In Finnmark there is very little, if any, common land of this kind. That leaves us with the category "state common land." In the three northern counties of Norway, most of the outfields in some sense belong to the state. These are vast areas, covering approximately 68,000 square kilometers, which is 21 percent of the area of Norway. Several important aspects should be kept in mind:

1. We must distinguish very clearly between Nordland and Troms on one hand, and Finnmark on the other.
2. The legal conditions of the outfields are at present in a process of transition.
3. Historically, the state has played a crucial role, and state ownership has traditionally been more extensive and important than in the South.

From a legal point of view these areas used to be classified as a category of their own. They were not regarded as commons, but they were not held in fee simple by the state, either. For classification purposes state common land is defined in our national atlas (NGO 1985) as "[s]tate land in Nordland, Troms, and Finnmark, which in historical time has not been subject to private ownership." These areas, or at least a huge proportion of them, were once state commons, but were supposed to have lost that status. The state has had a stronger legal position than in state common land in the South, but the local population had a variety of different rights—partly by law, partly by tradition, and partly by acquiring rights and property from the state. These rights were not supposed to be "proper common rights." The rights to use may belong to individuals, to farms, to the local community, to inhabitants of the municipality or the county, to the Saami who keep reindeer, and to the public in general. So they were indeed "common" in one sense or another.

Briefly, there are three aspects that should be mentioned concerning today's changes:
1. A special commission (*Utmarkskommisjonen*) for Nordland and Troms
2. A popular action launched by the farmer organizations in Nordland and Troms
3. A special commission, *Samerettsutvalget*, that among other functions investigates the present state of land rights in Finnmark.

As mentioned above there was a judicial commission at work in southern Norway from 1909 to 1954 to determine the boundaries for state common land. A similar commission has been at work in Nordland and Troms since 1985 to work in disputed areas. In addition, to clarify and determine the boundaries between state and private land, the commission decided upon the status of state land, as well as the nature and extent of existing use rights in this land. The commission has finished several cases and also made some rulings of general validity for the two counties. The most important one is that the land in question is not a special category of its own, but is state common land. The rights to use must be decided for each common, and might differ very much. This ruling has led to political action by the farmer organizations, especially in Nordland. They claim that the same local institutions and powers as in the South should be introduced for management and decision making. In practice this means a board, *fjellstyre*, at municipal level, but the outcome remains to be seen.

The situation in Finnmark is very special. A group of specialists within the framework of the commission, *Samerettsutvalget*, have published their results concerning the present state of the rights to land and water in Finnmark (NOU 1993, 34). It should be noted that this commission has no power to decide. The commission shall investigate, express its results and opinion, and give advice to the government for further action. The members of the commission are specialists and representatives for the various group interests, and its conclusions are therefore seldom unanimous. The group that has been working on the land rights since 1985 consists of independent specialists only. Their conclusions concerning the nature of the state land are formulated with all the proper reservations real specialists should take, but are essentially as follows: The state land in Finnmark can not be classified as state common land in the same meaning this concept has in southern Norway (NOU 1993, 34). There are, however, many similarities—for instance, the position of the state as owner of the "ground" and the collective nature of rights to use many of the resources. It is a "common" of its own kind. In 1997 the commission proposed a new act for these lands (NOU 1997, 4). Their suggestion is to make them into a special type of commons (see chapter 13).

Summary Statistics

All numbers are approximate.

State Common Land in Southern Norway

Type of land: Productive forest amounts to 7 percent; the rest is comprised of mountain areas above the timber line
Area: 26,600 square kilometers
Number of commons: 195
Number of rightholding farm units: 20,000
Owner of the "ground" and other "residuals": The state
Access to resources:

1. Pasture, secondary summer farms, cultivation: local farming population, according to need
2. Wood: Local farming population according to household need, residuals to the state
3. Hydroelectric power: The state
4. Hunting and fishing: Everybody in the municipality and the general public

Decision-making bodies:

1. The State Forest Company (*Statskog*)
2. An elected municipal board (*Fjellstyret*)

3. An elected board at *bygd*-level (*Allmenningstyret*)

Collectiveness in use: Individual use dominates
Alienation: The commons as such can not be sold or subdivided, neither can rights to use, which are inseparable from the farms. With subdivisions of farms the new farms get individual rights to use. Land can be sold or leased for cultivation (reclamation).

Bygd Common Land

Type of land: Productive forest amounts to 31 percent; the rest is comprised of mountain areas over the timber line
Area: 5,500 square kilometers
Number of commons: 51
Number of rightholding farm units: 17,000
Owner of the "ground" and other "residuals": Local farms
Access to resources:

1. Pasture, secondary summer farms, cultivation: local farming population, according to need
2. Wood: Local farming population according to household needs, residuals shared by the group of farms that "owns" the common
3. Hydroelectric power: The local farms that "own" the common
4. Hunting and fishing: Everybody in the local community (*bygd*)

Decision-making body: Elected local board
Collectiveness in use: Collective use of forest dominates, otherwise individual
Alienation: The commons as such can not be sold or subdivided, neither can rights to use, which are inseparable from the farms. With subdivisions of farms the new ones get individual rights to use. Land can be sold or leased for cultivation (reclamation).

Land Owned in Common by Farms (Sameige Mellom Bruk):

Type of land: Predominantly comprised of mountainous areas, but also other outfields and coastal areas (*fjøra*, small islands, skerrings)
Area: No statistics available
Number of commons: No statistics available
Number of rightholding farm units: More than 50,000, but no better statistics available
Owner of the "ground" and other "residuals": Specific local farms units— the shareholders
Access to resources:

1. Pasture, secondary summer farms, cultivation: the shareholders only, according to their share

2. Wood: The shareholders only, according to their share
3. Hydroelectric power: The shareholders only, according to their share
4. Hunting and fishing: The shareholders only, according to their share

Decision-making body: The majority of the shareholders according to their share, or an elected board. The Land Consolidation Court may be brought in.
Collectiveness in use: Both collective and individual
Alienation: Shares can only be sold together with the farm, or a part of the farm. Individualization (subdivision) of the common land into plots can be done by the Land Consolidation Court. To sell any part of the land requires consent of all shareholders.

State Common Land in Nordland and Troms

Type of land: All types of outfields—some productive forests, mountains dominate
Area: 20,000 square kilometers
Owner of the ground: The state
Access to resources:

1. Pasture, secondary summer farms, cultivation: Local farming population according to traditions
2. Wood: Local farming population according to traditions
3. Hydroelectric power: The state
4. Hunting and fishing: Everybody in the municipality and the general public

Decision-making bodies: The State Forest Company (*Statskog*), no local board so far
Collectiveness in use: Individual use dominates
Alienation: The commons as such can *probably* not be sold or subdivided, neither can rights to use, which are inseparable from the farms. Land can be sold or leased for cultivation (reclamation) and other purposes.

State (Common) Land in Finnmark

Type of land: All type of outfields—forests, mountains, water, coastal
Area: Over 90 percent of the land in the county (close to 48,000 square kilometers)
Owner of the "ground" and other "residuals": The state

Access to resources:

1. In general : Local population according to traditions and special laws and regulations
2. Hydroelectric power: The state
3. Hunting and fishing: Everybody in the municipality and the general public.

Decision-making bodies: A special board with an attached administrative service (*Jordsalgsstyret* and *Jordsalgskontoret*); both at county level
Collectiveness in use: Individual use dominates
Alienation: Land can be sold or leased out by the state according to laws and regulations, "especially the Act on State Land in Finnmark county, dated March 12, 1965."

NOTE

1. An earlier version of this analysis was presented to the International Symposium on "Village Communities and Common Property in Italy and in Europe," Pieve di Cadore, September 1986.

REFERENCES

Erixon, S., 1978. *Byalag och byaliv*. Stockholm: Nordiska museet.
Frimannslund, R. 1956. "Farm Community and Neighbourhood Community." *Scandinavian Economic History Review* 4, no. 1. Uppsala.
Løne Vinje, T. M. 1991. *Gardsstyre. Ein analyse av organisert samarbeid i klyngetun, Hovudoppgave ved Norges landbrukshøgskole*. Inst. for Planfag og Rettslære. Agricultural University of Norway.
Meyer, P., 1949. *Danske Bylag*. Nyt Nordisk Forlag Arnold Busck. København.
NOU (Official Norwegian Report) 1993: 34. *Rett til og forvaltning av land og vann i Finnmark*. Bakgrunnsmateriale for Samerettsutvalget. Oslo: Statens Forvaltningstjeneste.
NOU 1997: 4. "*Naturgrunnlaget for samisk kultur*" (*The natural resources foundation of the Saami culture*). Oslo: Statens Forvaltningstjeneste.
Sevatdal, H. 1985. "Offentlig grunn og bygdeallmenninger" ("Public land and parish common land") in *National Atlas of Norway*. Norges Geografiske Oppmåling (National Survey of Norway).
Stevenson, G. G. 1991. *Common Property Economics*. Massachusetts: Cambridge University Press.

PART THREE

Saami Reindeer Herding in Northern Norway, Sweden, and Finland

Introduction

"Picture a pasture open to all. It is to be expected that each herdsman will try to keep as many cattle as possible on the commons. Such an arrangement may work reasonably satisfactorily for centuries because tribal wars, poaching, and disease keep the numbers of both man and beast well below carrying capacity of the land. Finally, however, comes the day of reckoning, that is, the day when the long-desired goal of social stability becomes a reality. At that point, the inherent logic of the commons remorselessly generates a tragedy," wrote Garret Hardin (1968), starting a generation-long controversy about the nature of utilization of a pasture open to all. The argument shall not here be thrashed out again. The gist of the debate was that the empirical relevance of the powerful model was rather small (but was perhaps somewhat more appropriate with regard to the sea [see Feeny, Hanna, and McEvoy 1996]). All the interesting empirical cases lay somewhere between the open access producing the tragedy of Hardin's model and the complete control of access needed by the single omniscient and omnipotent decision maker to avoid the tragedy. And the rather mundane discovery is that we know all too little of the variety of pastoral praxis, of how and to whom pastoral resources are allocated in the different parts of the world. To increase our knowledge, we here turn to the rights and duties involved in the pastoral praxis of the Saami on the plains of northern Fenno-Scandia.

Kaisa Korpijaakko-Labba traces the "history of rights to resources in Swedish and Finnish Lapland" in chapter 8. The paper, based on her doctoral thesis (Korpijaakko-Labba 1989, 1994), maintains that in the Lapp areas of Torne and Kemi in the northernmost parts of Sweden and Finland and in the southern parts of the Finnmark county in Norway, the Saami were treated by courts and other authorities as owners of the land (or, more precisely, possessors of taxpayers' rights that later developed into ownership) at least until about 1740. This holds true of Saami villages as well as individual Saami.

Between the middle of the eighteenth and the twentieth centuries the Saamis' status as landowners was forgotten and replaced by institutions based on the premise that nomads could not acquire title to land. Hence the land belonged to the state. In the mid-twentieth century it has come time to ask the question that might reveal how this came about.

The historical perspective on what we today consider just can be supplemented by considerations of utility. Irrespective of historical justice, are our current institutions able to guide various actors towards an economically sensible and ecologically sustainable resource mangement? Thráinn Eggertsson (See chapter 2) implies that the action of the state has been a slow transfer of the range lands of the reindeer-herding Saami from a regime of common property to a regime more resembling open access with clear signs of crowding and overgrazing. In Norway much of this process must have taken place during the last thirty years, judging from the situation described by Robert Paine.

In "Kautokeino 1960: Pastoral Praxis," Robert Paine presents a status report on the Norwegian reindeer-herding Saami from around 1960. He opens with a brief overview of the social organization and inherent constraints of Saami reindeer praxis. The condition of the herds at each season and the differences in ecological conditions among the Kautokeino pastoral groups are described. Paine then turns his attention to the significance of these differences for pastoral knowledge and their implications for the adoption of separate production strategies. In view of this, he is critical to the simplistic use of concepts like "carrying capacity" and "sustainability."

The heritage of the reindeer-herding Saami as reported by Korpijaakko-Labba and their recent historical situation as described by Paine must be kept in mind when turning to the long-term issues of the rights to the resources on the plains of northern Fenno-Scandia. The Saami of Finland, Norway, and Sweden claim property rights to the resources their culture and livelihood have been tied to. In this claim they have the support not only of new investigations of the history of resource control, but also of some developments in international law (for example, the ILO Convention no. 169 on indigenous and tribal peoples).

Even though some general conditions (such as the state ownership of the land the reindeer herders use and a history of nation-state building hostile and detrimental to Saami culture), are similar in the three countries, the situations for the Saami of Norway, Sweden, and Finland are not quite comparable. The differences are important to take into account in any institutional matrix designed to manage the resources. For example, even though the majority of all Saami live in Norway, they represent a smaller fraction of the total population in their main area (Finnmark) than comparable areas in Finland and Sweden. And furthermore, only some 10 to 15 percent of the Saami population in Finnmark depends on reindeer herding for their livelihood. This will obviously cause problems in defining rights for the Saami in general: how should one treat those who are not reindeer herders? Property rights to the Saami as an indigenous people must address the thorny issue of ethnicity. In the proposal of a new act for the land in Finnmark (NOU 1997), the Saami Rights Commission has consciously tried to avoid defining rights to land in terms of ethnicity. The proposal is briefly summarized by Torgeir Austenå at the end of this section. The proposal combines elements from the two types of commons (state commons and *bygd* commons) with established traditions in the south of Norway.

This proposal, as any legal development in a rule-of-law system, stems from an understanding of current law and the context where it is applied. An introduction to these questions is provided in this section. Chapters 8 through 10 provide an overview of the current legal status of resources in northern Fenno-Scandia.

"In the Legal Status of Rights to the Resources of Finnmark with Reference to Previous Regulations of the Use of Nonprivate Resources" by Torgeir Austenå and Gudmund Sandvik, details about the legal history of resource control in Finnmark are presented. The parcelling out of lots for private ownership began in 1775 and has continued since then without amounting to large areas. Of Finnmark's 40,000 square kilometers, 96 percent is still owned by the state. On the state land, as well as some of the private land, the reindeer-herding Saami enjoy some specific usufruct rights, such as pasturing along the coast in the summer and on the Finmarksvidda plateau during the winter. They have special rights to the use of wood and to fishing the rivers of Alta, Tana, and Neiden. They enjoy other rights such as hunting or fishing in the sea on the same basis as other inhabitants in the county. The right which has caused the most conflict between the reindeer herders and others throughout history is the right to pasture. The state has introduced different clauses in sales contracts of land and there are now at least five different classes of private land regarding this issue. With modern herding technology and the current institutional matrix, however, the problem is cropping up in new forms. In addition, new encroachments from tourists have complicated the situation. A governmental task force, "The Saami Commission," has been working since 1982 to establish current status and suggest new solutions.

In "The Legal Status of Rights to Resources in Swedish Lapland," Bertil Bengtsson reviews the current status of rights to resources for the Swedish Saami reindeer herders. Their rights are regulated by the Reindeer Farming Act (RFA) and to some extent by the Forestry Act. The complicated system of the RFA implies that the right of the Saami in areas delineated as reindeer herding areas (approximately one-third the area of Sweden) is a kind of usufructuary right comprising reindeer grazing and other rights connected with this, such as the construction of certain facilities, using lumber for household needs, using migration paths, hunting, and fishing. Certain of these rights can be exercised by the reindeer-herding corporation (called a Saami village), certain others by the individual Saami; further, in some respects the party entitled is the Saami (trust) fund, which benefits the whole Saami population. However, the legal status of the Saami is changing. In a case decided by the Supreme Court in 1982, a Saami corporation claimed ownership rights to their grazing areas. The litigating corporation lost, as their claim for ownership rights was not well founded. But the judgement opened the possibility for a different ruling in other cases. This ruling, taken together with demonstrations of ownership in the middle of the seventeenth century, seems to indicate that the rights of the reindeer herders are stronger than their position in current legislation indicates and may provide a counterweight against infringements from industry and tourism.

"The Legal Status of Rights to Resources in Finnish Lapland," by Heikki Hyvärinen, outlines the current legal status of the Saami rights to resources in contemporary Finland. The state of Finland considers itself the owner of over 90 percent of the land used by the Saami on the premise that the land has never been owned by anyone (*terra nullius*) and should therefore revert to state ownership. Recent research findings (see chapter 8) show that state officials at one time treated the Saami as owners of what is now considered state land. This right of the Saami has never lapsed legally to anyone's knowledge. In 1990, the ILO Convention no. 169 could not be ratified because of its implications for Saami right to the land they traditionally have depended on. Work on a special Saami bill confirming their right to land has been going on for a long time, but has not progressed to parliament for a decision as the government has been delaying the matter. Recently, Finland's joining the European Union has further complicated the matter. The rights to land, water, all renewable resources, and to engage in reindeer herding remain unclear in relation to the Saami as an indigenous people.

REFERENCES

Feeny, David, Susan Hanna, and A. F. McEvoy. 1996. "Questioning the Assumptions of the Tragedy of the Commons Model of Fisheries." *Land Economics*, May 72 (2),: 187–205.

Hardin, Garret. 1968. "The Tragedy of the Commons." *Science* 162, 1243–1248.

Korpijaakko-Labba, Kaisa. 1989. *Samelaisten oikeusasemasta Ruotsi-Suomessa*. Helsinki.

Korpijaakko-Labba, Kaisa. 1994. *Om samarnas rättsliga ställning i Sverige-Finland*. Helsinki: Juristförbundets Förlag.
NOU 1997:4. "Naturgrunnlaget for samisk kultur" ("The natural resources foundation of the Saami culture"). Oslo: Statens Forvaltningstjeneste.

KAISA KORPIJAAKKO-LABBA   CHAPTER 8

The History of Rights to Resources in Swedish and Finnish Lapland

The problems concerning the existence of indigenous peoples are similar all over the world. The key conflict involves the right to utilize natural resources: the situation varies from a total lack of such rights to situations where the rights are legally and economically in danger of extinction. Indigenous cultures are tightly integrated, and breaking bonds with traditional livelihoods poses a threat to entire cultures. One such group are the Saami in northern Fenno-scandia. Their habitat stretches approximately from Røros in Norway through the northern parts of Sweden and Finland to the Kola Peninsula in Russia. Conservative estimates put the total number of Saami at seventy thousand.

Life in strictly local terms has required extreme adaptability from the northern indigenous peoples. Nature is barren and unproductive, a circumstance readily apparent when examining the livelihoods of the people. Traditionally, throughout the region the most important livelihoods have been hunting land and sea mammals, fishing, and, particularly in northern Eurasia, reindeer herding. Subsistence has necessitated great mobility, which is reflected in the social systems developed among the peoples—permanent habitation in villages has not existed. Usufructuary areas, whatever the basis for their use may have been, have been extensive throughout the region.

Globally, the issue of rights for indigenous people is often referred to as "aboriginal rights." This term is an effort to recognize that indigenous peoples have a right to land and water through some principle of natural law; namely, that they are the original inhabitants of their land and they have utilized the land to this day. In Scandinavia, particularly in Sweden and Finland, the situation is decidedly different. The Saami, while never actually having been conquered, became true judicial subjects of the state very early on, and the prevailing legal system recognized their various rights to land and water use. For this reason the question of Saami land rights in Sweden and Finland is not only a question involving history and international law, but a question concerning the legal history in the whole country and Lapland specifically.

The Origins of Saami Culture

It is difficult to talk about Saami culture or analyze its past without at least some discussion on who the Saami are. The origin of the Saami is, of course, a matter independent of official government definition and goes back far into prehistory. Although it is not possible to present a completely accurate account on the origin of Saami culture, a useful factor seems to involve the so-called asbestos ceramics in northern Fenno-scandia. Asbestos ceramics spread to the North around 1500 B.C. and came to prevail over all Finnish Lapland, northern Sweden, Finnmark in Norway, and the Kola Peninsula—the very areas that have been traditionally populated by the Saami. In other words, from the geographical range of asbestos ceramics we are able to extract an historical cultural range for the Saami that is internally homogenous and clearly distinguishable from the surrounding cultures. The second factor common is—naturally—language. The division of language into a Saami "family of languages" is seen by linguists to have taken place simultaneously with the above material process. Thousands of years ago, the Saami culture, as well as many other cultures, was something other than the state culture. The process which has resulted in distinction between the Swedish, Finnish, Norwegian, and Russian Saamis was a lengthy one. The institutional methods employed by the later states in their rivalry for dominance over the northern areas involved a combination of taxation, the building of churches, trade, and, above all, setting up a juridical and public service administration in the area. The success of Sweden in this rivalry was considerable, resulting in Sweden's control over a vast area that comprises much of today's Sweden and Finland: the Lapland region of West Bothnia (Västerbottens Lappmarker). It is the juridical-historic past of this area that I have examined in my dissertation (Korpijaako 1989) and will shortly sum up in this chapter.

The Nature of the Swedish-Finnish Legal System

Until the beginning of the thirteenth century, law in the area of what is today Sweden and Finland could mostly be termed as customary law. This applied to Lapland as well. Gradually, province by province, written law came to be developed. Initially this resulted in many different provincial laws. The first laws pertaining to the entire state date from the mid-1300s and from the year 1442.[1] In 1809, when Finland and Sweden separated, the general law in force in the country was the Code of 1734.

Due to these facts, Sweden-Finland could clearly be considered, in its legal development, to be among the rank of countries with statutory law. In reality this has meant that legal guidelines in a dispute must first be sought in written law (that is, not case law). Additionally, in our legal system disputes concerning land and water areas have traditionally been so-called nonmandatory disputes. They could not even have come up for action without the initiative of the private plaintiff. These facts prompt an exploration of the nature of Swedish-Finnish ownership according to written law.

Title to Land in Sweden-Finland

A closer analysis shows that land ownership can be understood as the sum of two bodies of statutes dealing with the same matter from two different perspectives. On the one hand, the law defined the basis by which a person in general could be considered a landowner. In this context the law exhaustively listed various forms of recognized legal title: inheritance, contract of sale, gift. The idea was, and still is, that in case of dispute, such title had to be proved in order to verify the existence of a right.[2] On the other hand, the law provided for protection under law that a landowner enjoyed the concrete use of his land. In this case the law listed all actions that, from the point of the landowner, constituted illegal interference with his property.[3] Legal acquisition and protection under the law formed the core of Swedish-Finnish land ownership rights.

These rights of ownership had a strong background in customary law, which also became evident in earlier legislation on land ownership. From the 1500s, changes in political and economic life began to affect the legal development to an increasing extent, which spurred a significant breakdown of the traditional concept of land ownership. Sweden-Finland did not enter a wholly feudal period in its development as did most countries in Europe, but for the peasant, beginning in the mid-1500s the title to land became subject to a large number of previously unknown restrictions.

In spite of everything, if a peasant succeeded in remaining independent under increasing burdens of taxation and other pressures, he came to be officially called a freehold peasant.[4] The privileged freehold, controlled by the nobility, formed a second category of real estate in addition to taxable land. Both of these titles can nowadays be deemed to have included the most elementary criteria of land ownership rights. The rest of the land, the third category, belonged to the crown. A tenant on crown land did not, of course, have ownership rights to the land he used, but used it against a rent.

Rights of the Saami in Sweden-Finland

Saami Land Use

The tax rolls for Lapland, which represent the earliest primary sources of the past from the mid-1500s, indicate that the Saami (or better the Lapps) inhabited and used lands within certain defined units of land called Lapp villages. A Lapp village used to cover a vast area of land and should, in modern perspective, be compared to municipalities rather than typical villages. The Lapp villages were, however, not the only units of land partitioning in Saami society. Within a Lapp village, each family controlled and used a clearly defined area of land, which the documentary sources have termed as "hereditary" or "tax lands." In Sweden-Finland, the life of each Lapp family and its means of livelihood was decisively focused on one unit of such land.

The sparse population within the Lapp villages can be seen as a direct relation to both the natural conditions of the area and to the means of livelihoods available within this area. The principal means of livelihood of the Saami were hunting, fishing, and reindeer herding; hunting and fishing were indigenous to Lapland while reindeer herding was developed at a later date. Since the Arctic environment is barren and unproductive, a great deal of territory was (and is) needed to make these traditional pursuits into a viable means of livelihood. The problem of scarcity has to be seen from an entirely different perspective in the North than in the South: the stage at which scarcity of land and water areas emerged in the Saami livelihoods occurred much sooner than in agrarian livelihoods.

Since the mid-1600s cases between Saami have been decided in annual rural court sessions, of which exact records were kept on how justice was administered among the Saami. The local court sessions were held once a year in each Lapp village within the vast area of Lapland. Besides the Swedish judge, the jury or panel served an integral part of the court. The twelve-man jury was elected from the Saami in each village, and eventually new settlers were included in the pool. Litigation among Saami was both common and diverse, involving all conceivable aspects of human culture that might come up in such communities. A prominent concern in Saami litigation was a need to protect sources of income—the right to use land and water. Although usufructuary areas—the so-called tax lands belonging to individual families—were large, the individuals involved never seemed to have an excess of land, or too little. In the light of the principle of sustainability, this approach is wholly consistent: In the Saami land use system the people lived on what nature could give them annually, but on the other hand, only certain places—a good fishing place, a rich pasture for the reindeer, and so on—yielded enough nourishment and other necessities. That is why these were important and had to be protected.

Legal Title to Land

Even a brief look at the documentary sources, especially the court records, shows that the families within the Lapp villages considered the tax lands they used as their "own": "To own" is actually the very concept they used when describing their relation to the land. The privileges connected therein were strongly defended against other families and outsiders. In other words, the nature of Saami livelihoods has been no obstacle for very strong ambitions constituting a kind of private right to the lands.

The key problem of my research has been to clarify, based on documentary sources, the precise nature of the legal status of the Lapp villages and the private tax lands within the judicial system of the time. The problem can best be summarized by the following question: Were the hereditary lands of the Saami or the Lapps[5] regarded as their "own" only according to local practice and custom, or was this title legally recognized in the same way as were, for instance, the farms of peasant proprietors during the same period of time?

The right of a freehold peasant can be seen as a collection of various elements—rights as well as obligations—described in detail in legislation. If a peasant met the criteria, he could be considered in a later examination on the matter to have owned his land by dint of so-called taxpayer's right. If the criteria were not met, the answer is naturally the opposite. As far as the Lapps are concerned, the matter can be—and I think it must be—considered within the same framework and the same principles. If we consider the extensive range of documents related to the legal position of the Lapps on the whole, particular criteria based on specific points in legislation can, indeed, be distinguished. The overall picture of Saami land rights as found in historical documents is summarized below.

The courts in Lapland did apply Swedish law, but the laws and legal practice did not, in my view, endeavor to dismantle or change the unique system of land use among the Saami or the legal principles connected with it. On the contrary, Swedish law in a way expanded and stretched, as it were, to cover and protect the Saami system of land use. One of the most important rights was that the property was transferable (i.e., it could be inherited or sold). A closer analysis of the Lappish court records shows that this was the case in Lapland, too. The tax lands belonging to private Saami families were transferable property to the same extent as the land of a freehold peasant. Under the conditions of Saami society, the most common and important form of land transfer was inheritance: this guaranteed undisturbed land use from generation to generation. Succession of land proceeded in accordance with the principles set out in the inheritance code.

A Lapp could also resort to other measures concerning his or her land, although this was not as common. The source material does, however, indicate some actual sales of land and especially of certain lakes. In other words, it was legally possible. As in the case of peasants' land, the land was subject to redemption on the basis of rights of inheritance, a characteristic which distinguished the land clearly from crown land. A tenant on crown land could not transfer it to his heirs any more than he could to anyone else.

The Lapps certainly enjoyed so-called protection of possession with regard to their taxable land. Where violations of rights occurred, provisions on unlawful use were applied. In practice this has meant hundreds of cases concerning fishing, hunting, or reindeer herding without permission on another Lapp's tax land. The court documentation also contains a great deal of information on the boundaries of Lapp villages and also private tax lands. Court decisions on such boundary disputes indicate precisely, place name by place name, where the boundary between the villages or the lands lay. It is also clear that the tax which the Lapps paid on their lands was a land tax in nature. Land registers detailing taxes and taxable lands were drawn up from time to time, and they were the same as those drawn up for peasants´ property. Given the size and difficult conditions of Lapland, preparation of such registers was an accomplishment in itself.

Neither did the colonization of Lapland take place with the conscious intention to disregard or override the rights of the indigenous population. New farms were to be established primarily in unused areas. If these did not exist, the matter had to be negotiated with the Lapp concerned. Mines and the like could not be set up just anywhere on a tax land. The procedure followed was the same as that specified in a law protecting a freehold peasants' property in connection with such measures.

Put briefly, if these same criteria prove that a peasant once had ownership rights to his land, the corresponding conclusion ought to be—or, in my opinion—*must be* admissible in the case of the Saami/Lapps. If all criteria are met, and this is what the documents show, the Lapps must be seen as having owned their land in the same way as the freehold peasants of the time did. This right is the direct predecessor of today´s right of ownership.

The Reduced Legal Status of the Saami

In the present-day situation, the circumstances described above may seem almost unbelievable. However, one must remember that the perspective is crucial when evaluating the past. From the point of view of Sweden-Finland the Saami living in the periphery of the country were in fact politically very important. At that time it was thought that only the Saami were capable of permanently inhabiting these regions and thereby ensuring the sovereignty of the state in the area. Their loyalty had to be secured through favorable treatment on the part of the state.

It has become customary to point out that the change in Finland's status in 1809 did not entail the loss of previously acquired rights. In 1809 the bond between Sweden and Finland broke and Finland became a Grand Duchy within the Imperial Russian Empire. In his sovereign pledge Alexander I promised to strengthen and solidify both the previous constitutional laws and the liberties which each social class had obtained. Generally speaking, this has also been the case: the change in status did not affect the citizens' property rights. In the North the assurance made by the czar did not, however, have a complete application. Over the years, the land title of the Saami/Lapps gradually disappeared from most of the official records and was forgotten.

The most crucial and devastating circumstance was obviously the fact that the northernmost regions of Finland existed for long periods without the local court institution that had earlier been so important. Local court sessions were only occasionally held in the North and, even then, these were presided over by judges unfamiliar with previous praxes and local customs. The people therefore lost not only the forum which had been so instrumental in assisting them in maintaining a necessary order concerning land use, but also the possibility to get proper written documents of their rights—documents that could be used as proof in later disputes concerning the same land.

There were naturally many other reasons for the development, but a common denominator is seen in the goals and ideals connected with the creation of nation-states. Doctrines of race and so-called racial hygiene were becoming virtual scientific disciplines at the time. The brachycephalic feature of the Saami was invoked to dub them an inferior race compared to the Swedes, Finns, and Norwegians. Their entire culture was termed barbaric and a manifestation of culture at a lower level of development than civilization proper.

In addition to losing the previous important court institution, the land title of the Lapps gradually came to be omitted from public land registers as well. The Finnish Lapps (or Lapp villages) did pay land tax on the land and water within the village as recently as 1924, but this very year all the prevailing land taxes in the country were abolished by law. The result was that no information on the previous land title of the Lapps existed when the general parceling of land was carried out in northern Finland by law in 1925. The end result was that the hereditary lands of the Lapps came to be categorized as some kind of undefined "government lands." In Finland this has meant the transfer of some three million hectares of land to state ownership (i.e., 10 percent of the whole area of Finland). This land is currently administered by the National Board of Forestry.

Conclusions

As it is clear that the state authorities of Finland, Sweden, and Norway take it for granted that the state owns the lands and waters in the Saami area, it is interesting to examine which legal grounds are the foundation for these claims. After having familiarized myself with all possible literature on the question and after having followed the official decision-making process concerning the northern areas in general, I think it is fair to say that the explanation is, by nature, very simple. The claims of the states have been essentially based on the premise that in the course of history, the Saami never acquired a real land title to the lands they used. In other words, the areas inhabited by the Saami have been considered as ownerless areas, *terra nullius*, where, according to law, the state acquired title in the absence of another owner. This explanation has also enjoyed the support of several disciplines. According to historians, lawyers, economists, and the like, the development of all peoples progresses from a hunting and fishing stage to nomadism and after this to the highest possible level—to the "civilization," of the agricultural stage. As hunters, fishermen, and nomads it was not possible for the Saami to acquire a title to the land according to these theories.

In my view, the extensive documentation on the legal rights of the Saami/Lapps proves without doubt that the typical relations and elements constituting land ownership rights developed irrespective of whether the land in a particular case was used for farming or not. The rules of behavior associated with land ownership rights seem to have no logical connection with the nature of the livelihood engaged in; the institution has been governed by the need for legal regulation stemming from scarcity of resources. Hunters, fishermen, and nomads may have needed—and the Saami certainly did need—ownership rights to land as a means of regulating their mutual relations. Contrary to earlier interpretations, the Saami/Lapps recognized the basic problem of ownership centuries ago, and have sought to apply standards associated with ownership to their legal relationships. This goal did not go unrealized, for the legal system of previous Sweden-

Finland did protect the rights of the Saami in the same way it protected the rights of the peasants.

NOTES

1. The country and city codes of King Magnus Eriksson and King Kristofer.
2. This was done by presenting the original document concerning the legal act.
3. These lists were in older legislation very detailed: using another's fields without permission, fishing and hunting without permission, etc.
4. In Swedish, "*skattebonde*; in Finnish, *verotalonpoika*.
5. The term "Saami" is based primarily on ethnic criteria and is not used consistently in old documents. The people who lived in Lapland and used land and water for fishing, hunting, and reindeer herding (and payed a land tax for these livelihoods) were called Lapps.

REFERENCE

Korpijaako, Kaisa. 1989. "Saamelaisten oikeusa semata Ruotsi-Suomessa." Helsinki. Swedish translation in 1994.

ROBERT PAINE   CHAPTER 9

Kautokeino 1960: Pastoral Praxis

In 1961–1962, field research among Saami reindeer pastoralists of Kautokeino afforded some understanding of the principles of Saami pastoral praxis—practitioner praxis, that is—in action. At the present time, when government-regulated pastoralism and pastoral problems appear to run together, there may be things worth learning—or relearning—from the 1960 picture. It is with this in mind that I attempt, in this short essay,[1] a description of the pastoral year from that time of one Kautokeino group. Before venturing into the description, the ecologic and social constraints under which the pastoralists operate are delineated; on completion of the description, some conclusions are drawn.

Constraints

Crucial to reindeer pastoralism as a practical enterprise is practitioner knowledge of animals, terrains, and fellow pastoralists. Much is dependent, then, on the individual and the group working within a feasible ecologic and social scale. For example, the thousands of square kilometers over which Kautokeino pastoralists work is far too large (and variegated) for any one herding group to master (see Figure 1), it has to be scaled down or divided up into pastoral ranges over which each group possesses the requisite level of knowledge. This also becomes an arrangement of social knowledge between herders, for reindeer pastoralism throws herders into working partnerships—often in contingent fashion, that is, as the needs of the situation dictate; and within each of the pastoral ranges it should be feasible for a family, even an individual, to possess a practical social lexicon concerning his/her fellows. Such is simply not possible across the total Kautokeino

FIGURE 1
The Scale of Reindeer Pastoralism in Finnmark 1961-62
(adapted from Ø.Vorren 1962)

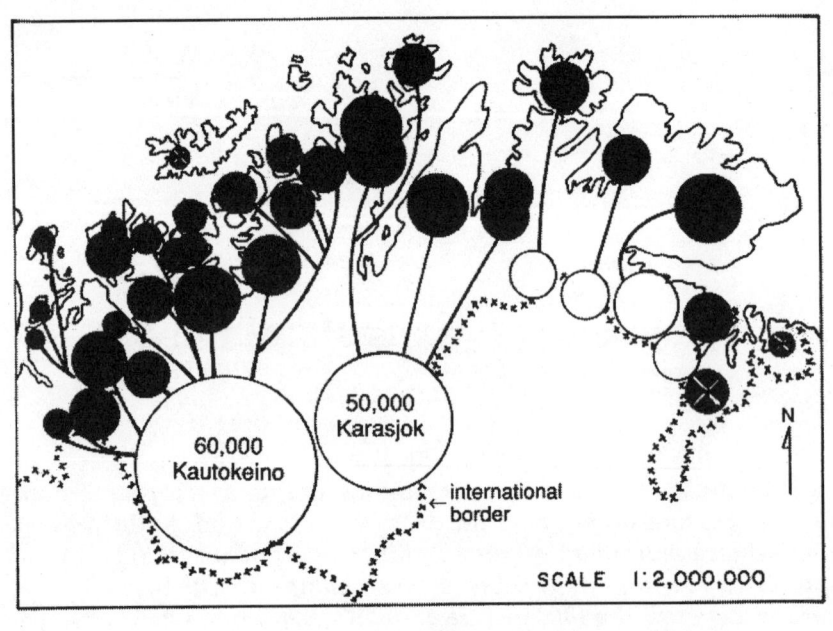

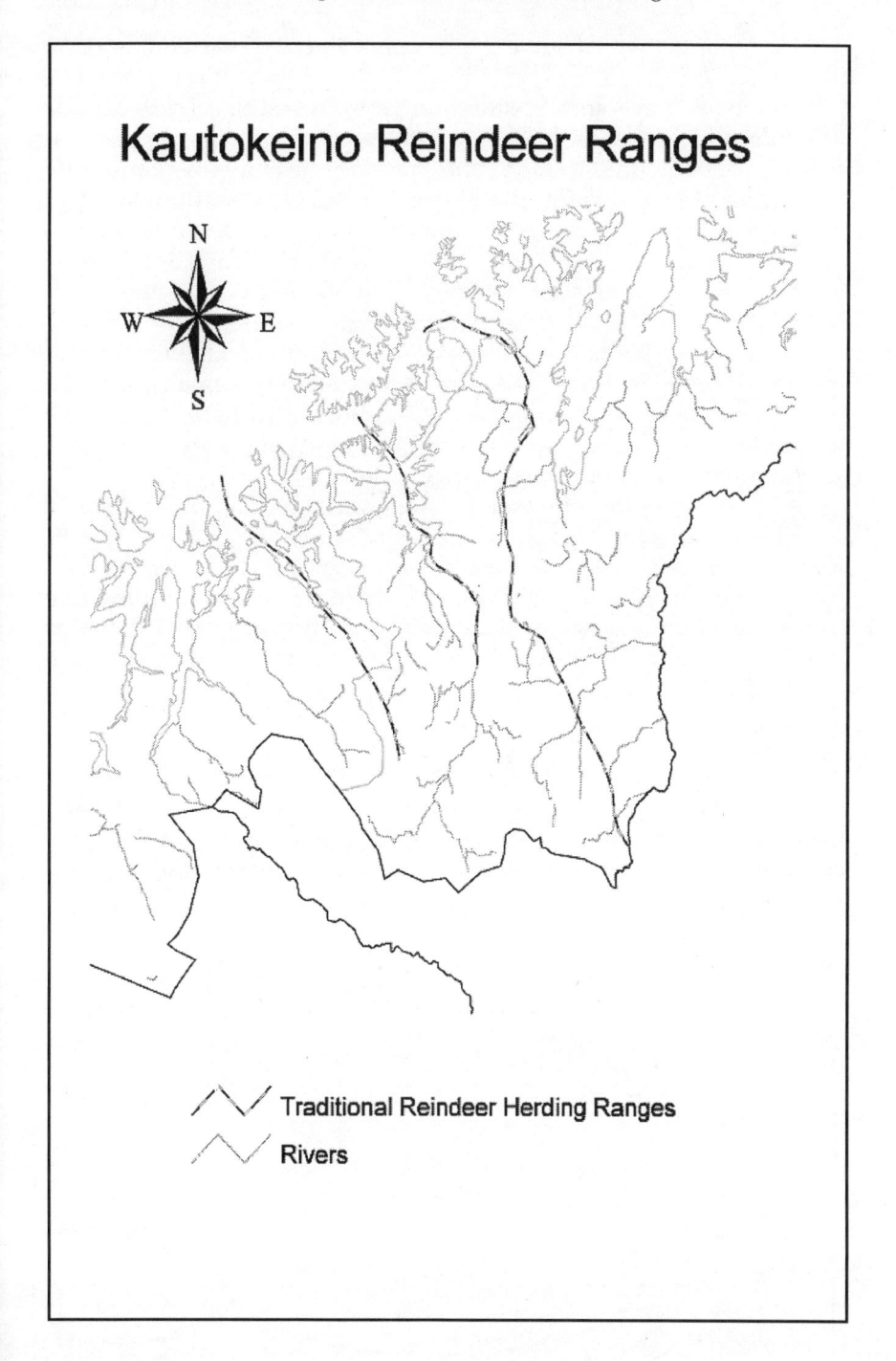

FIGURE 2
Kautokeino Reindeer Ranges: West, Middle, and East Ranges

pastoral population of some 800 persons and 170 households (1958 figures).

There are three such pastoral ranges—West (*Oar'jebelli*), Middle (*Gow'dojottelit*), and East (*Nuor'tabelli*) (see Figure 2). Most pastoral work is done within a range: thus the scale of things is reduced by (let us say for sake of argument) one-third and pastoral work is embedded in a social network and the ever-present uncertainty of this work is made manageable. The emergent point, then, is the social density of pastoral relations within a pastoral work area—the range. Consider these figures: 84 percent of married men and 65 percent of married women remain on their natal range; in 59 percent of marriages both spouses remain on their natal range; close to 10 percent of the marriages are between first cousins; additionally, there is a notable occurrence of separate groups of siblings marrying each other.[2]

The basic unit inside the range is the *sii'da*, composed of both people and animals, and exercising usufruct over different areas of pasture at different seasons of the year (see Figure 3). Thus a *sii'da* has ecologic and economic connotations, and as a work (and hence social) unit its precise composition changes from time to time (see Figure 4). There are two analytic points of primacy here. The first is that the three factors of production—herd, pasture, and partners— are brought together in a *sii'da*. Ideally, these should be present in commensurate proportions and the commensurability retained even as the size of a *sii'da* changes.

And second, work relations within a *sii'da* fall into two separate domains: herding, which is a collective or joint responsibility, and husbandry, which—pertaining to the individual ownership of animals—is not. Herding is the day-to-day work with a herd; it concerns the herd/pasture relationship as directed to the welfare of

FIGURE 3
Sii'da as Reindeer Management Unit (with four family herds)

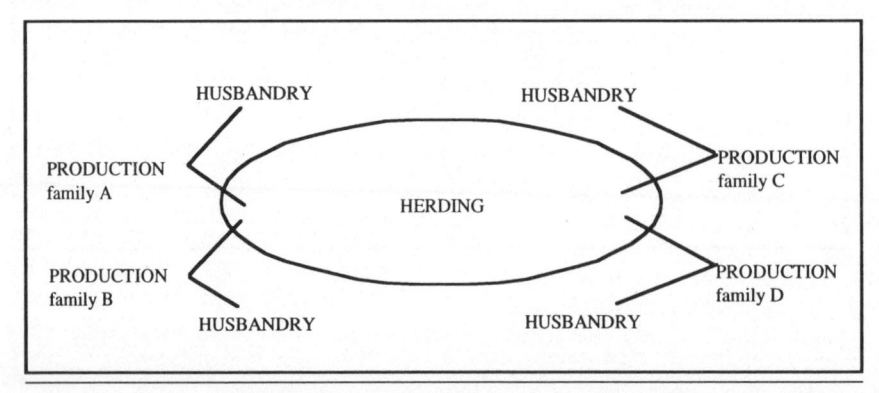

the animals and, if necessary, to the exclusion of the comfort of the herders themselves. Husbandry, on the other hand, has to do with the herd as the harvestable resource of its owners. While the tasks of herding, then, are those of the control and nurture of animals in the terrain, husbandry is the efforts of the owners in connection with the growth of capital and the formation of profit. The problems of herding are those of economy of labor and they may usually be solved by owners in conjunction with each other; those of husbandry concern the allocation of capital and here each family is wholly responsible unto itself.

In concluding this briefest of accounts, notice must be taken of the means by which pastoral knowledge of both the herd and individual animals is generated and maintained. As seen with a husbander's eye, there are six basic components to a herd: the calves and yearlings (two designations), junior and senior cows (three designations), junior and senior bulls (six designations), and castrates (two designations). Important for the identification of individual animals (within the same age and sex class, for instance) on which good husbandry depends, are antler structure, body color, and the earmarks of ownership. Behavioral clues, of course, are noted as are social groupings of animals within the herd: thus bells (hung around the neck) placed on selected animals impart much information, at night as well as by day.

The description that follows draws on my field trips with a Middle Range *sii'da*.

FIGURE 4

The Sii'da and Herd Management of Commensurate Proportions

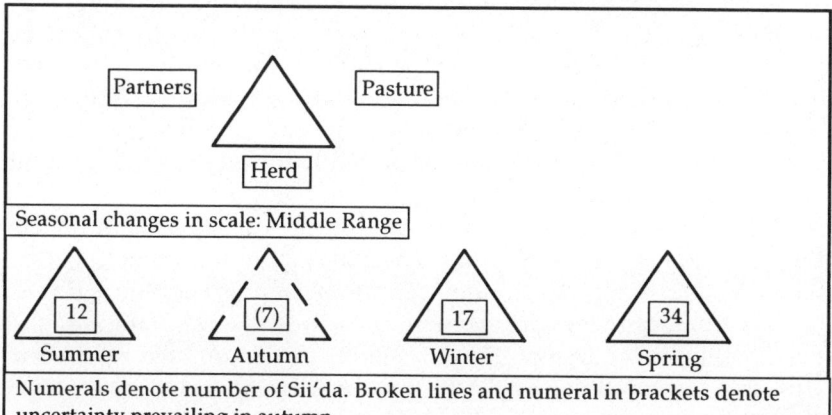

Partners — Herd — Pasture

Seasonal changes in scale: Middle Range

Summer 12 | Autumn (7) | Winter 17 | Spring 34

Numerals denote number of Sii'da. Broken lines and numeral in brackets denote uncertainty prevailing in autumn.

Description

Summer

Most summer pastures of the Kautokeino herds are out on peninsulas—the *njar'ga* herds. A few are on off-shore islands to which the animals swim—the *suolo* herds; some others are inland, out of sight of the coast—the *nanne* herds; and some of these are on the high terrain around the tree line—the *or'da* herds (see Figure 5). The *njar'ga* type of herd management I knew from Middle Range reaches the summer pastures in June, after calving, and already by the beginning of September the pastoral summer has passed by and herders are preparing to move or are already on the move to autumn pastures.

Yet it is the brief northern summer of constant daylight that is the important season of growth and body building for the animals. The protein-rich diet of grasses and fiberless foliage is easily digestible and its mineral content gives quick nutrition. The summers are particularly crucial for the calves and yearlings. The most rapid growth in the life of a reindeer takes place in the first sixteen months of its life, and that growth is almost wholly confined to the two summers within this period; in the intervening winter the young animal often has a hard time maintaining its own body weight. For the sexually mature animals, the nutritional value of summer pastures is a determinant of their virility or fecundity, at the rut in October. The pastoralists are acutely aware of these predetermining consequences of the summer season. A poor winter, I was taught, can be quickly compensated by a good summer, but animals may well have trouble surviving through the winter if the summer pastures have not been optimal.

For all that, the summer is a slack season for the *njar'ga* pastoralist. Until towards the close of the summer, herding is minimal or absent. (Quite different conditions exist for the *nanne* pastoralist.) Instead, the animals are left to themselves to best explore the varied diet of the summer ranges. The pastoralist reasons that beyond bringing his animals to a "good" grazing range, there is little that he can do to forward the nurturance of his animals at this time of year. Indeed, optimal nurturance in the summer correlates closely with free-range movement, and—in contrast to what is the case at other times of the year—the pastoralist does not lose much knowledge of value about his herd by this arrangement.

Let us now consider the herds and pastures of the *sii'da* (numbers I through VII in Figure 6). The male herds arrive some weeks before the cows and their calves, the separation having been undertaken before calving, and they are taken to the farthest reaches of the summer pastures; in the course of the summer they will (in most cases) mix with the cows and calves. Each of these pastures contains ecologic alternatives necessary for a herd during the summer: access to the

FIGURE 5
Summer Herds: Njar'ga, Suolo, and Nanne

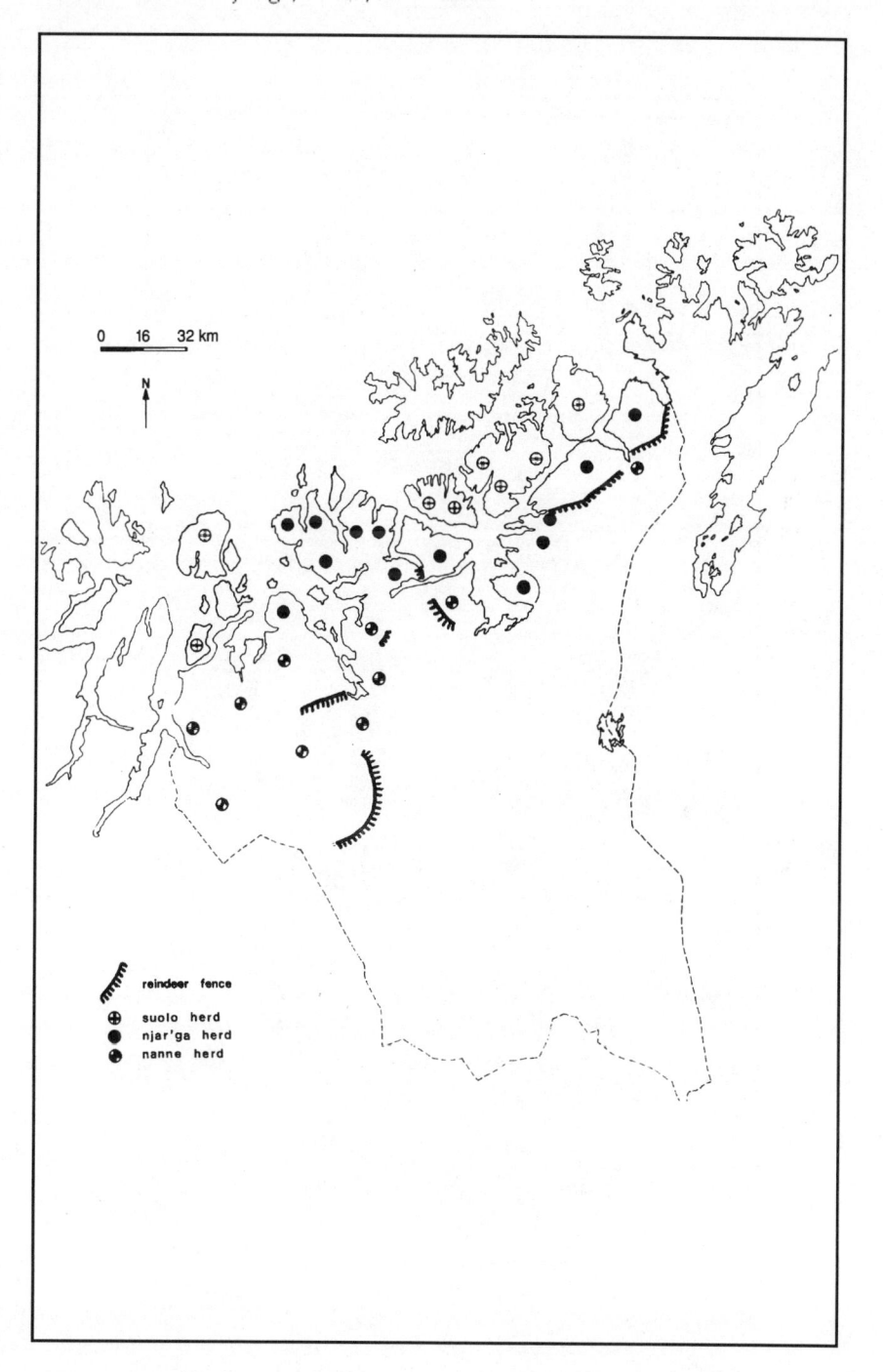

FIGURE 6
Njar'ga Pastures of Gow'dojottelit

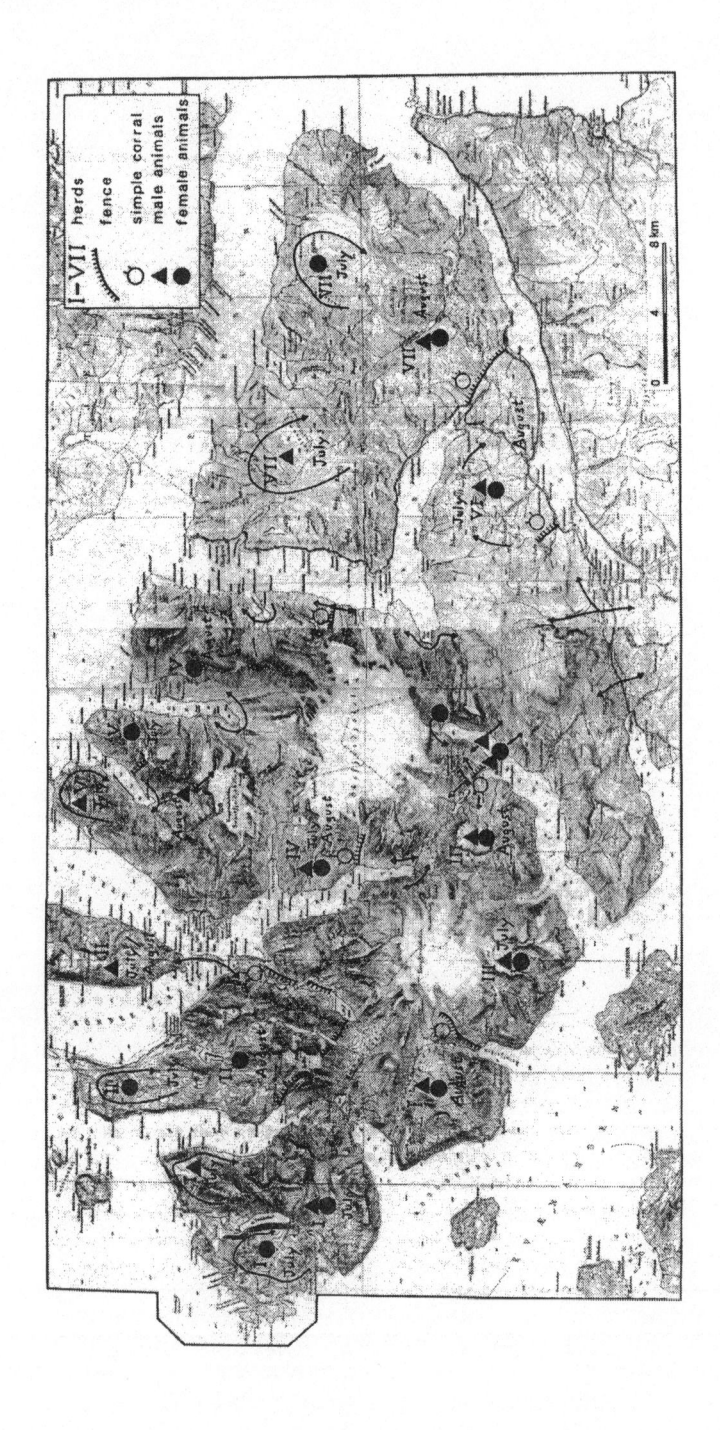

shoreline for salt, to low-lying pastures for early vegetation, and to mountain pastures for cool in the heat of the summer and relief from insects (respite may even be sought on the glaciers). This means there are likely to be patterns—both by the month and by the day—in the wanderings of animals between locations within a specific pasture. Clearly, movement onto the peninsulas in July and off them in September has to be synchronized, although it is done informally. As shown in Figure 6, there are strategically placed short stretches of fences that help control the movement of animals in the late summer, and in the vicinity of each fence is a small corral used for husbandry tasks (i.e., earmarking, castrating, and slaughtering).

Autumn

As summer closes, the days become shorter, the weather harsh; animals move up an age-class (except calves), the rut approaches, and movement of all herds and camps is in over the tundra. Not for nothing is this season also known as *hil'bad ai'ge*—the time when animals are least tame. The bases of grasses and sedge plants that remain green longest, and fungi—especially mushrooms—are searched out as the herds move across the autumnal landscape. If not hindered, animals will range widely, especially the mature males (bulls and castrates). Herders have codified the autumnal landscape: *varre* or hilly, open terrain (in the cool temperatures of autumn) was associated with *logjes* or "tame" animals; *vuobme*, or low-lying with lush undergrowth, was associated with *hil'bad*. Indeed, the cows were kept on the higher ground because if they first get a taste of the rich vegetation in a *vuobme*, it is difficult to keep them moving as a herd. The mature males, on the other hand, were allowed to enjoy the *vuobme* both because it would overtax the herders to prevent them and it is there that the animals—many of whom were to be the studs in the rut—add weight and strength. Some of the castrates might be left behind in this way, but one can depend on the bulls moving inland for the rut. There is a hazard, though: some bulls may well wander off with those from other herds that have been attracted to the same *vuobme*. It is only if there are no other animals in the vicinity that the herd will be brought for the night to the richer vegetation of a *vuobme*. More usually on autumn migration one collects the herd as best as one can in the afternoon as dusk falls early, and releases it on a long, dry hillside with a wind blowing down its slope.

Calves are becoming fairly independent, and herders are well aware of the chances of calves becoming lost at this time of herd mobility. It can happen when the deer scatter in search of mushrooms, especially under cover of darkness or in mist and rain, and towards the end of the rut when older bulls, through fatigue, lose control of their harems and the younger males begin to pursue the cows. It happens especially in the confusion around the large separation corrals (near the perimeter of the winter pastures) when animals are passed through the corrals not once but several times.

The introduction of these separation corrals with multiple pens has to be considered in conjunction with the crowding of animals behind summer fences followed by a rush of thousands onto the autumnal lands—all within a few weeks. Inevitably, there is some loss of control over herds and loss of knowledge as to the whereabouts of animals. The purpose of the corrals is to restore control and reconstitute the herds. But the costs are high in wear-and-tear on animals and herders. So in strongest contrast to summer, autumn is a season of exertion; more significantly, it is the one time of the year when these pastoralists sometimes end up not working with the herd, thereby adding to toil and strife. Dispersal of animals at this time of year makes good ecologic sense and benefits herds and herders alike; it is the crowding of animals and the mixing of herds of separate owners, followed by their forcible separation, that runs contrary both to the welfare of the animals and to the interests of their owners.

Before the fence-and-corral complex stamped its character on autumn herding, the rut was that season, above all others, when herders reacted in response to the behavior patterns within the herd (competition among the bulls, harems "herded" by a senior bull, etc.). This rut happened in specific places and it was a notable time of close observation of one's animals—hence of important husbandry knowledge, too. Now, though, animals are in rut on the way to the corrals, while they are passed through them (which can take a couple of days), as well as after the corrals. The trauma can be considerable for a thousand or so animals milling around within an enclosed space.

Buyers come to the corrals. It is convenient for owners to sell some animals here, and it is a time of year when cash is needed for domestic reprovisioning. But the drawbacks are also prominent in the minds of the owners: rupture in their herd knowledge at this time and temporary loss of condition of the animals around the corrals. Based on his observations of his animals through the seasons, an owner will likely have individual animals in mind, but he may not find them at this time.

Often, then, the owner of a *njar'ga* herd can expect to leave the separation corrals with a good fraction of his animals missing (temporarily, he hopes), and it may well be December before he has them all together. By the same token, the herd which he takes into the winter pastures will include a number of animals that are not his. Nevertheless, the autumn slaughter takes place then. Yet already in these first years of their use, pastoralists have different perceptions of the corrals. I think, for all concerned, the corrals are a new meeting place that they value—watching others' animals as well as one's own, catching up on news, certainly listening and perhaps telling. I also noticed a generational difference. For the young men, unlike their elders, work at the corrals has an ambience of tournament: beyond the simple opportunities that corral work offers to demonstrate physical prowess, there is the competitiveness centered on the acquisition of unmarked calves that are without their mothers.

Winter

As October passes into November, autumn changes into winter, but there is no sudden metamorphosis; rather, both seasons are present for a while. Stretching beyond this intermediate period and into January is the period of winter that the Saami know as the time of darkness (*skabma-ai'ge*). Not only are the days now the shortest in the year—from around the end of November to the middle of January the sun is below the horizon—but snowfall is heaviest.

The short daylight hours notwithstanding, much activity is pushed into this first part of winter. The principal business is reconstituting herds from the autumn and then, during the last days of December, separating into the smaller camps and herds of the later and longer period of winter. So there will be men, and some women, away on a round of visits to other herds; and every camp receives its visitors. There will be many deer separations, all handled without recourse to the big corrals. Draught animals no longer carry packs but now pull sleds. Dominance within the herd passes from the postrut, and now antlerless, bulls to the cows who retain their antlers until after calving.

Then there are the shifts in the herders' technical conversations. Early in October, talk was peppered with topographical references incorporating the term for "bare of snow," but as November approaches it is the extensive vocabulary for snow that one hears. Snowcover—and its changing texture—has multifarious effects. First, with regard to herd behavior, animals will leave the herd one behind the other, in Indian file, whereas so long as the terrain was bare of snow the animals were more likely to leave bunched together or in a broad phalanx. The texture of a snow surface becomes a factor controlling the mobility of animals and men and—particularly significant—their relative mobility. Snow cover, too, offers the herder many clues: it carries evidence of recent and prevailing wind and weather conditions; it is "read" for what it tells about the movement of a herd or the whereabouts of animals that have strayed. Perhaps most important of all, though, reindeer have to dig through snow to reach the lichen beds, so the condition and depth of snow affects, quite directly and at all times, the physical well-being of the animals. Of course, the precise significance of snow crucially changes, as the landscape changes topographically, climatically, and calendrically.

In this early period of winter, then, the pace of things has not slackened much from autumn. But by January, if not before, that changes, too. The landscape is under a thick blanket of snow and the temperature stays well below freezing: winter literally envelops the pastoralists and their animals. The area over which a herd pastures shrinks as the animals spend more time, and burn more energy, digging through to their food supply—they stay closer together. Likewise with the herders: now in their smaller winter camps with, at last, their own animals, the Saami know the months of deep winter (January to March) as a time of peace (*rafes-ai'ge*).

Whereas summer herds are large aggregations on physically separated pastures, on this winter terrain herds are smaller (so there are more of them) and pastures constitute an overlapping quilt (See Figure 7). The absence of physical obstacles means that animals could wander of their own accord from one end to another of the Kautokeino tundra. That herds, by and large, stay put on their pastures is accountable, first and foremost, to the natural attraction each pasture holds for the animals. The total pattern amounts to a fairly equitable spatial distribution of herds across the total area, with separated herds pasturing separately without commotion, often in near proximity to each other. Of course this would not happen, let alone be sustained, without the intervention of herders—and that would be worth little without up-to-date knowledge of local changes in climatic conditions. Temperatures? Wind force? Depths of snow? To complete this necessary knowledge, the herder needs a mental case history of how the snow fell during the preceding months or weeks: all may augur well, or, the indications may be such that he devises possible alternative pasturing strategies with his fellows. So as is true of all seasons, one is safest where pasture is ecologically varied, even on a micro scale. Thus, herders stress the importance of being able to move locally between *vuobme* and open tundra: in the *vuobme*, snow is less likely to become tightly packed than it is on the windswept tundra; however, it often becomes too deep, especially for the younger deer— it is then that one might take the herd to pasture on the open tundra. But the viability of that move will depend on several natural factors, one of which is that the sun of late winter does not "bake" a snow crust on the exposed slopes. Perhaps by that time, though, the depth of snow in the *vuobme* will be reduced anyway. In short, one looks to trace a viable path between changing alternatives.

The peace of this season may be threatened, however. There is always the possibility of a diminishing food supply or one which the animals cannot reach at all on account of ice over the lichen beds. Then the animals will want to wander, and it is left to the herder to compensate for the constraint he has imposed (through herding) on the animals' freedom of movement. Using his ski staff he will test the depth of snow and the strength of a crust; in worsening conditions, he may dig some "craters" to help the animals reach the lichen. When that access is impossible, he will cut foliage and pull down hanging

FIGURE 7
Winter Herds

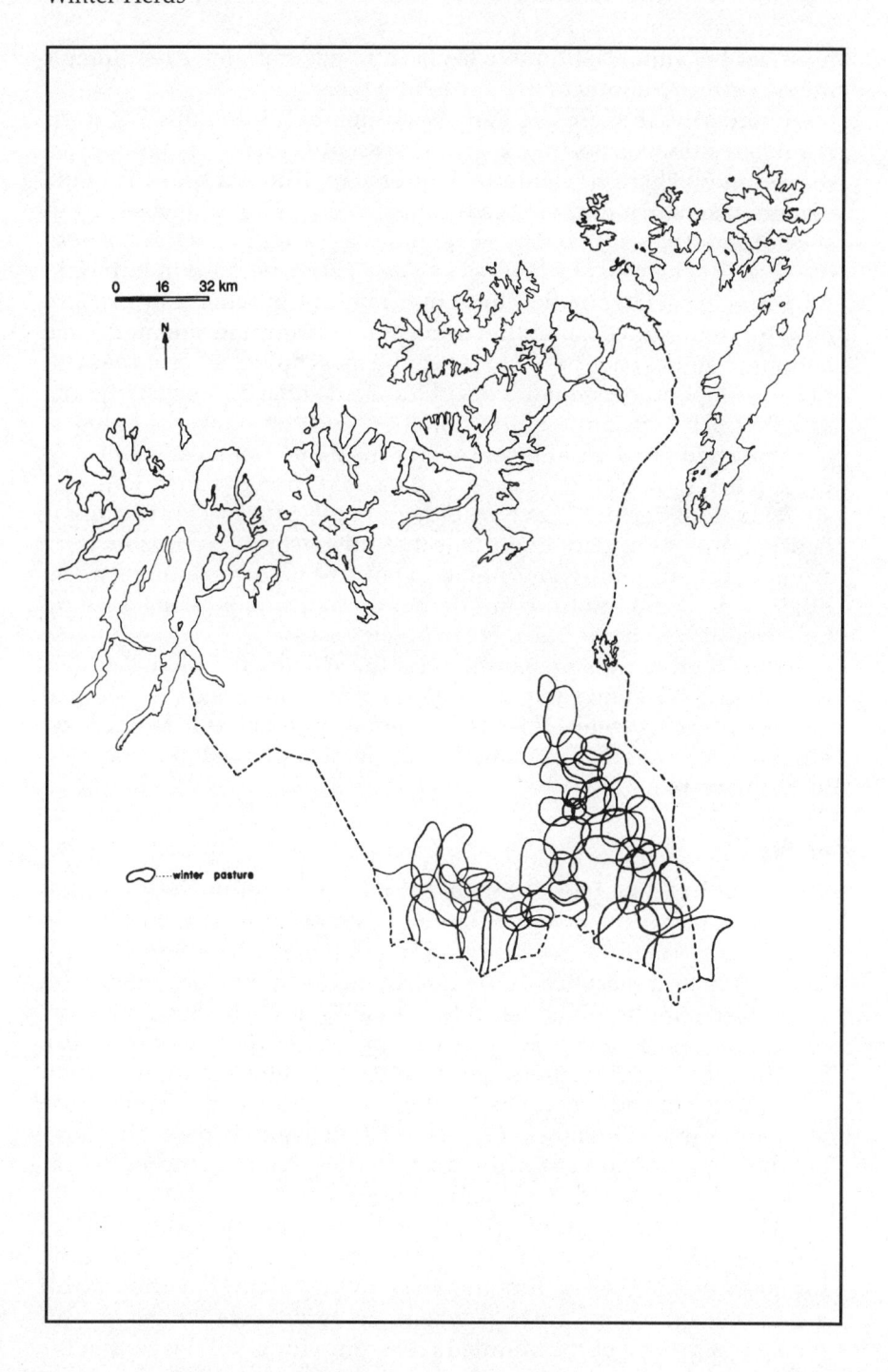

moss for his animals, or move his herd to another area even though there is already another herd (or herds) there.

Ordinarily, though, the daily problems of herding during deep winter are minimal. So, time is given to essential undertakings beyond the routines of herding, undertakings of very different kinds. For one, owners take stock of their herds. Small numbers of animals—up to several hundred, say, among several owners—will be herded across the tundra to be sold on-the-hoof in Kautokeino, perhaps in Karasjok. And it is especially at this time that families butcher and prepare meat for domestic consumption through the spring and summer. Since autumn, families have been mostly eating meat boiled fresh or smoked; those with poorer economies may have sold some of the better joints and delicacies (marrow bones, tongues, and the like); the blood is never wasted but cooked in a gruel for the dogs. The meat now being prepared for the spring, however, is salted, hung, and dried—the staple that herders will have with them in their rucksacks. Because it is dried, it is especially important that this meat is taken from a fat animal. Indeed, some owners make a point of taking a young female; others who could afford to but do not do so, perhaps taking a young bull instead, will be ridiculed behind their backs.

Then there are the preparations for the spring migration. Families who shared herd and camp through the winter may each be going to their own spring camp. If so, their animals must be separated. Soon afterwards, in the case of many family herds, bulls will be separated from the cows.

Spring

There is a confluence of factors in the movement of the herds off the winter pastures. The new vegetation draws animals: grasses take the place of the winter fare of lichens, and for *njar'ga* herds at least, there is probably the associated desire for salt in their diet which the coastal vegetation supplies. If these factors weigh most with the males, for the cows it is the return to their calving places (or places of birth), and for the herders, the movement is also part of the essential rotation of feeding grounds: it saves the delicate lichen beds—soon to be without a protective snowcover—from being overtrampled. The move off the tundra also spares the animals the worst of the mosquito plague.

The final destination of the spring migration is the summer pastures. Some herds reach these pastures before calving. However, for many others there is first the move to the calving grounds which are in much the same areas as where the herds were for the rut. All the *njar'ga* groups of the Middle Range camp for several weeks in the vicinity of their calving grounds and only after that do they begin on

FIGURE 8
Spring Herds (1961)

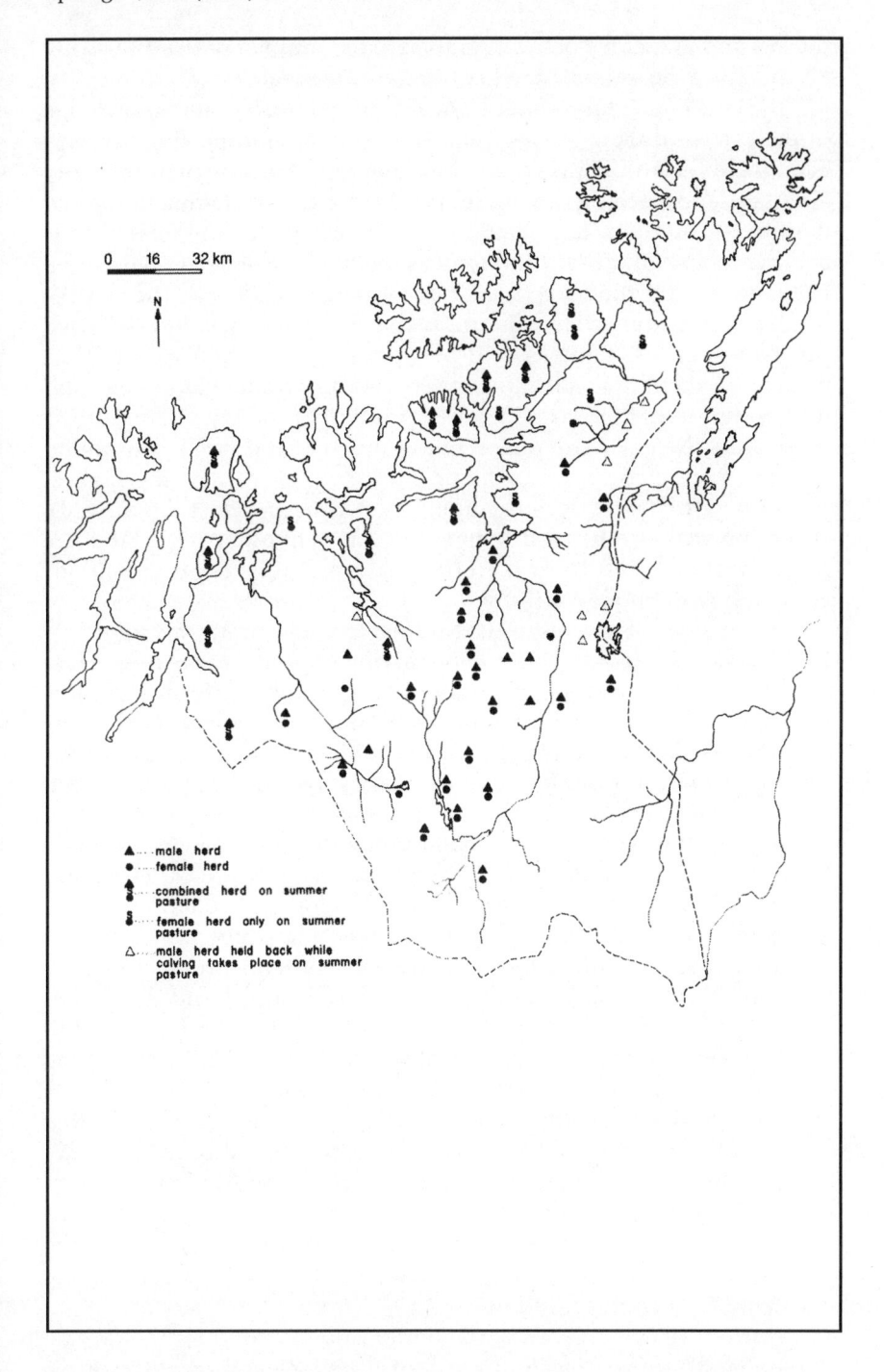

the last and longer leg of the migration to the summer pastures. During all this time the animals are kept under close watch.

Even though owners have been able to acquire a sound knowledge of their herds during the last months of winter and the animals have been relatively undisturbed, there is a pervasive feeling of uncertainty about the approaching calving season. In talking about what the spring may bring, herders look back over the seasons of the year that is coming to an end. "Last summer was too hot," or "The animals stood too long behind the fence," are typical reflections on which rest the herders' forebodings. How these adverse conditions may have affected the two-year-old females will be uppermost in their minds. The number of them that calf can vary appreciably from year to year, and it is widely regarded as a significant index of the well-being of the herd. Nonetheless, herders don't have much confidence in the ability of these young cows to nurse and nurture their calves (the highest percentage of lost calves is among first-time mothers, I am told). Attention will also be paid to how the calves of older cows fare: Are their mothers able to graze efficiently enough to provide their offspring with enough milk?

All the *njar'ga* groups of the Middle Range separate male animals from cows before setting out for the calving grounds. There are several reasons. For one, the bulls and castrates (especially the former), unless held back, will range widely in search of fresh vegetation that the spring thaw is uncovering. It is quite usual for the departure of the male herd from the winter pasture to be delayed until calving is well advanced at the spring camp. The concern of the cows, on the other hand, is to find sheltered places in which to drop their calves, and they have to be herded carefully to prevent them from scattering and "hiding" in the landscape. Another reason for the separation is that although a cow which is pregnant or has a newborn calf tolerates other females, she will be nervous and restless in the presence of male animals. The date on which the first calf is dropped and the period that elapses until the last calf is born are likely to vary from herd to herd, and even from one year to another within the same herd. Among several factors, the most important is the duration of the rut in the previous autumn and the conditions prevailing in the herd at that time. While it is usual for calving to be concentrated in the first half of May, some calves may well be born before the herd of pregnant cows sets out for the spring camp and even while still in the same herd as the males. More calves are likely to be born on the way to the spring camp, perhaps a journey of two days; these will be taken along on sleds, their mothers following.

Herders usually have a particular location in mind for the calving ground and "nursery" (*aldo manus*), such as a long and gentle southern slope with optimal exposure to the sun and good drainage. Since the newborn calves sleep much of the time, with their mothers grazing or dozing nearby, it is important that the ground be relatively dry. On wet ground reindeer become restless, and the openness of such a site (over which constant watch is kept) helps to give protection from predators. It is not uncommon for a *sii'da* to return to the same location each year; whether they occupy the site in any particular year, however, depends on a couple of factors. First, the convention of usufruct with respect to calving grounds is left broadly interpreted and a principle of "first there" is also accepted. Second, the number of calves already born while en route may cause the herders to abandon their original plans. So they may have to settle for a calving ground which is, in their opinion, less than ideal: its selection may be forced upon them in a situation of decreasing options as calving progresses. Yet the consequences of this are usually not too serious; the terrain is, by and large, suitable and herders have an intimate knowledge of it.

By being in attendance at spring camps, herders gain valuable knowledge of their animals at this critical phase in their life cycle. Most valuable of all, perhaps, one is able to distinguish between the different circumstances attached to cows without calves: cows that may be sterile (e.g., a three-year-old or more that still has to calve); cows that failed to calve this time but have done so in earlier years; and in the case of cows that gave birth but lost their calf, one wishes to know how they lost it. By the time a cow has calved for the second time, it has a "biography" on the basis of which its owner is able to predict her behavior in various situations. With this kind of knowledge, decisions regarding which cows to slaughter will be better informed.

The migration to the coast cannot begin in earnest until a few weeks after calving. There may be several short moves to new pastures, but the calves must be allowed time to gain strength before the long journey. This period is known as the spring of summer, and it is the time of the spring camp proper. The few herders who watched over the calving are now joined by their families, who bring with them the male herd.

Herding routines now encompass the two herds. Although they are still kept separated, herdsmen are able to move from the one herd to the other. The male herd is taken on relatively wide pasturing circuits and brought back each day to a position that is "in front of" the cows and their calves. This way the cows are left undisturbed (for it is unlikely that any of the males would wander back towards the winter pastures). Moreover, any cows that manage to wander (in the general direction of the spring migration) are likely to be observed by the herders who are with the male herd. Were the arrangement the other way around and the cows pastured in front of the males, the encroachment of the males into the cow herd would always be a likelihood; should any of the cows wander, there would be less chance of finding them.

The landscape steadily changes character at the spring camp. After the twenty-first of May, the sun does not dip below the horizon. Despite brief snowstorms, and even in those years when overcast skies withhold the sun for many hours, the snow retreats almost daily and the spring vegetation begins to grow apace. These changes mean that the decision to move out to the coastal summer pastures must soon be made. But in deciding *when* to begin the move to the summer pastures, opposing considerations have somehow to be balanced. The longer one waits, the stronger the calves will be. But as time goes by, the more difficult the journey for the calves on account of the thaw and spring floods (for rivers must be negotiated).

Usually, in the first days of June, preparations will be made to move before it is decided exactly when to move. Rain or cold winds from the interior can delay departure (even on migration, herds tend to veer into the wind); another cause of delay can be the movements of other herds in the vicinity. But the prospect of an exhaustion of good pasture in the spring camp area brings urgency to the move. Typically, a period of warm winds from the coast, winds that will draw the animals forward (and which may defy all efforts made to keep the male herd pastured in the vicinity of the spring camp), will end such a period of indecision.

The male herds reach the coast in a matter of a few days, following an alpine route (not manageable for the nursery) on account of its snow cover, and traveling by night (which is now light) for the sake of lower temperatures. Along with this herd goes the baggage train—fully loaded sleds pulled by draught animals—together with most members from the spring camps (certainly any children and old people).

It is left to a few herders (men and women) to undertake the longer and more difficult journey with the cows and calves. Meager supplies are packed on the backs of draught animals as the route renders the use of sleds impractical at this time of year. Whatever route is chosen, the calves will need much rest and physical help from the herders, especially when transversing rivers and ravines. All the while, they (and their mothers) need to graze and if for no other reason than that the high altitude routes along which the (fast-moving) male herds are taken are not practicable. There would always be the risk of not enough pasture easily available (the terrain may be stony; where there is pasture it may be buried under ice crusts). Herders speak loosely of expecting to reach summer pastures near Midsummer's Night, the twenty-third of June.

Whether it is more advantageous to be behind or in front of another herd is a particularly pressing question when traveling with the nursery. However, there is no uniform answer. In general, those who are behind have to take care to hold their animals back, and those in front can be reasonably sure that most of their stragglers will be herded by those behind and thus still reach a summer pasture—if not the owners'. On the other hand, animals that are behind can sometimes draw advantage from following the already trampled snow or the smell of the herd in front, and herders may draw advantage from learning about problems those in front of them are experiencing as they traverse the terrain. But much depends on the local, variable natural conditions and, ultimately, on who is in front or behind.

Conclusions

This portrait of the pastoral year as an ecologic system, with space and time components changing in tandem, raises several analytic issues worthy of brief comment.

Knowledge

The annual cycle of herd knowledge on Middle Range ("A" in Fig. 9) differs significantly from that on the other two ranges. The difference springs from the following alternatives shown below in Table 1.

Only the few *nanne* herds in "A/B" have unbroken herd knowledge of quality: calving takes place on spring pastures, and the move to summer pastures is a relatively short one. But unbroken herd knowledge means an unbroken work cycle: some watch will be kept over the herd even through the summer to avoid undue dispersion, and the herding watch continues through the autumn—this time to hinder mixing with other herds, principally *njar'ga* herds as they pass by.

Now, differences as to when in the pastoral year herd knowledge is optimal have a strong—if not determining—influence on pastoral production profiles. Let me demonstrate this by comparing the production of two owners, one from Middle Range and the other from West Range, whose herds are approximately the same size (one thousand animals before calving). In the case of the former, the high quality of herd knowledge he amasses at the spring camps directs his attention to the multi-variable permutations of his herd as a reproductive unit; the West Range owner lacks that kind of herd knowledge, but what he has in appreciably greater measure is sound knowledge and herd control through the autumn. He applies this knowledge to production—and for him herd composition maximizes one value.

TABLE 1
The Annual Cycle of Herd Knowledge

A	A/B	B
Calving grounds on spring pastures with herders in attendance	Same as A	Calving grounds on summer pastures without herders in attendance
Critical crowding behind fences and around autumn at fences and corrals	Same as B	Avoidance of critical crowding corrals
The distribution of the alternatives are:		
Most *njar'ga* herds	A few *nanne* herds	All *suolo* herds, most *nanne* herds, a few *nar'ga* herds
Characteristic of West & East Ranges		Characteristic of Middle Range

To illustrate that distinction, consider the following: Both owners have more bulls than castrates, but for quite different reasons. For the Middle Range owner, Iskun Biera, the bulls are studs for the increase of the herd; for the other, the bulk of them are to be marketed—nor does he bother much about castration as a means of increasing body weight for he concentrates his production on the sale of young males. There is a clear economic rationale in play here: the greatest growth of a reindeer is in its first two years; thereafter there is slow incremental growth, and each year animals are lost. So the sensible time to sell an animal is at the end of the two years. But Iskun Biera does not do that, his rationale is of another kind: just as he keeps biographies of his cows, so he likes to see his bulls grow and pass through the different age-classes. If for the West Range owner optimal herd size is constantly reviewed as an economic matter, for Iskun Biera the aesthetics of herd composition are no less a concern.

A glance at some slaughter and sales percentages of the two owners adds a further dimension to this difference between them. Slaughtering three times as much as the other, the West Range owner sells them all on-the-hoof (less those kept for home consumption) which speaks to an economy of decisions; Iskun Biera, on the other hand, sells some on-the-hoof and others he himself slaughters to sell as joints of meat. The overall slaughter percentages for the two owners

FIGURE 9
Annual Cycles of Herd Knowledge

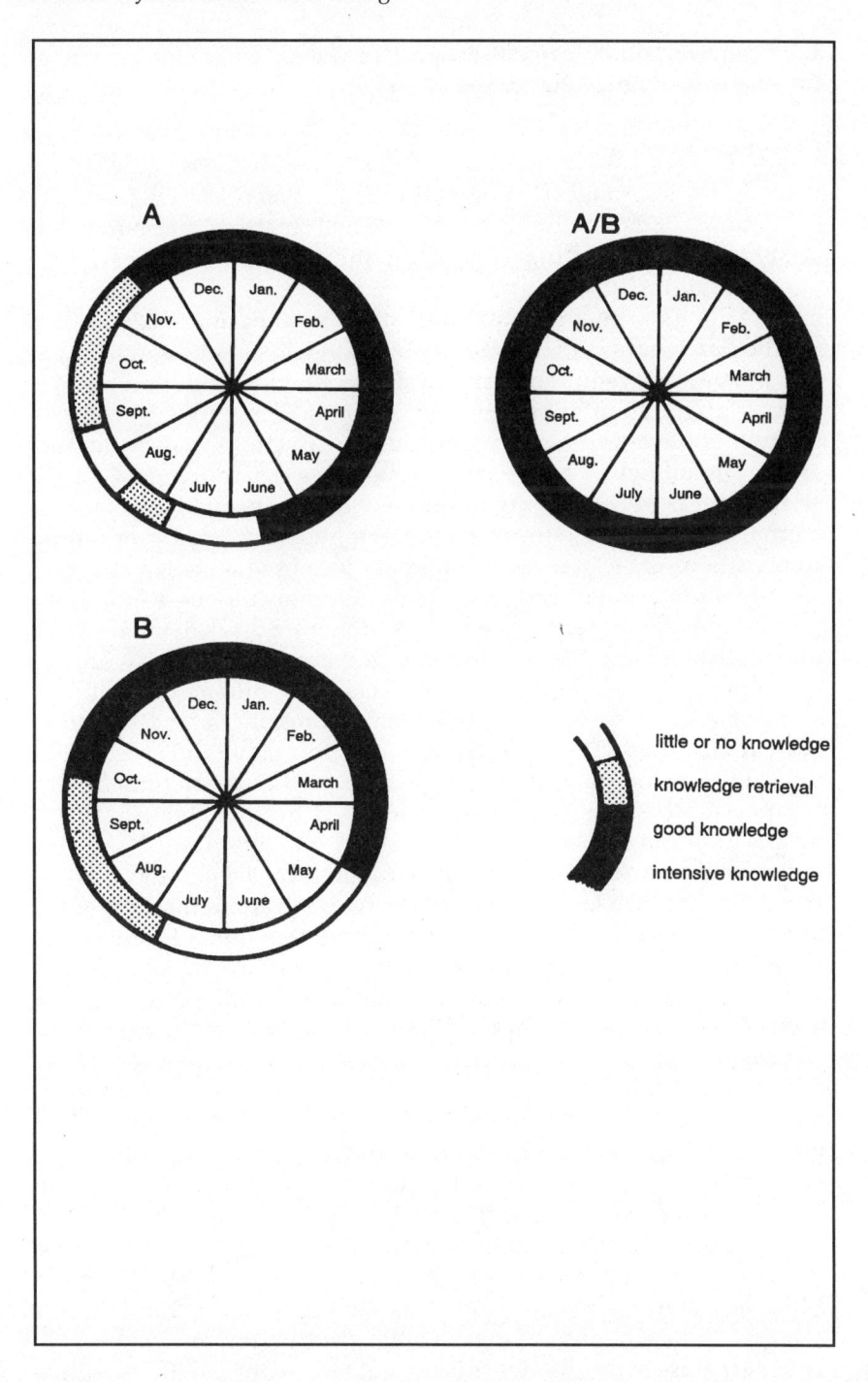

are 7 percent and 20 percent, respectively, and the kroner income of the one is four times that of the other.

Carrying Capacity

It also follows to suggest that the prevalent notion of carrying capacity as "natural" and therefore determinable by "objective" measurement[3] is seriously misleading regarding the nature of this reindeer pastoralism, for several reasons. First, in no two years do pastures necessarily have the same "natural" carrying capacity. Second, there will be differences in the suitability of seasonal pastures in relation to specific pastoral requirements. Third, there is the need for a pasture of each season to offer both ecologic combination and, especially in the winter, access to alternatives. Such distinctions and desiderata are not quantifiable. And fourth, pastoralists determine not just the size of their herds but the composition and, in particular, how long an animal shall live; again, these determinations rest upon subtle combinations of factors—with different outcomes (as shown above)—among which "natural" carrying capacity is simply the one of last resort.

Carrying capacity, then, has much to do with what these pastoralists desire. This means moving carrying capacity, in our analyses, into the active voice with expectation of different practitioner strategies according to their particular circumstances, and also their individual values. These values may be appreciably independent of the kind of circumstances we have had under discussion thus far—as comparison between Iskun Biera and his brother, Iskun Mikko, demonstrates. Along with other close kin, the two have shared the same summer pastures (VII in Figure 6) for years. Iskun Mikko is two years younger than Iskun Biera, but the two men are similarly situated in terms of family development cycle (if anything, it is Iskun Mikko who is the more favorably placed regarding a domestic labor force). However, he has less than half the animals his brother has. I believe this is largely accounted for by differences in the personalities of the two men—their desires and their abilities and hence their respective self-images. Iskun Biera is "energetic" where Iskun Mikko is relaxed, but it is he—rather than Iskun Biera—who has reindeer "talent"; Iskun Biera, for all his wealth, is not "proud" but definitely "cautious," even "miserly," whereas there is a touch of extravagance about Iskun Mikko.[4] Their production profiles offer corroborative testimony: Iskun Mikko actually slaughters rather more animals than his brother (his percentage of slaughter is on a par with those of the West Range)—including many more cows. For these two men (and many others, I wager) "optimal herd size" means quite different things, and the clue as to the nature of the difference is in (what we might call) the "optimal life fulfillment" of each. In short, carrying capacity should not be taken as analytically "given" and based on generalized energy ratios and—

an even more grievous sin—predetermined and unproblematized values such as "profitability."

Sustainability

So we are led to the question: *whose* standards of sustainability? To neglect the question exposes the very notion of "sustainability"—a current shibboleth—to the risk of being used as a science alibi for a political warrant to reorder practitioners' ecology and economy according to the values of the state. I was alarmed, therefore, to read in the 1992 Preliminary Programme for this MAB conference about the definitive conclusion that "overgrazing" is the problem with current Saami reindeer pastoralism (p.2). Further, regarding primary resource livelihoods in general, the Programme states that to change the regimes of utilization in a direction approaching a more sustainable resource use pattern means changing the structure of property rights to the resources (p. 3–4).

This sounds to me very much like giving "models for" analytic primacy over "models of." Perhaps the difference between the two may be justly put thus: In the "model for," the analyst constructs a scheme that is as close as is possible to certainty. However, from the practitioner's point of view, this likely means forcing certainty onto a world full of uncertainties. It also means the analyst uses abstract logic to gain control over the interaction between practitioners and the environment in which they operate, and consistency of action is seen as a virtue. In the "model of," uncertainty is recognized and incorporated, hence ambiguity and contradiction are also recognized as inevitable constituents of reality. This leads to a praxis in which contextual knowledge is central, and contextual knowledge is closely allied to practitioner experience, and thus, praxis has a strong pragmatic character (Wynne 1992). While in the field recording a pastoral year, I was often reminded that "this is how it is this year, but next year may be different"—*uncertainty*, in other words, was a pervasive element in the pastoralists' understanding their occupation. And responding to it, they drew upon contextual knowledge as a guide to action. Care must be taken, then, that in our search for the holy grail of "sustainability" we don't erode practitioner responsibility for what they do, thus risking, as I argue elsewhere (Paine 1993), the creation of the conditions under which the Hardinian "tragedy of the commons" emerges.

NOTES

1. For a full account see my *Herds of the Tundra* (1994).

2. From a sample—gathered in the field—of 170 marriages for the thirty-year period of 1924–1953.

3. Of the so-many hectares of pasture containing so-many tons of nutrient for so-many animals consuming so-many kilograms per so-many units of time kind.

4. The operative Saami words here (as spoken by informants) are: *saerra, fitmat, 'c´æu'lai, i duost,* and *hanes.*

REFERENCES

Paine, Robert. 1993. "Social Construction of the 'Tragedy of the Commons' and Saami Reindeer Pastoralism." *Acta Borealia* 9:2.
Paine, Robert. 1994. *Herds of the Tundra.* Smithsonian Institute.
Vorren, Ørnulv. 1962. "Finnmarksamenes Nomadisme" (The nomadism of the Saami of Finnmark), 2 vols. Oslo, Universitetsforlaget.
Wynne, Brian. 1992. "Misunderstood Misunderstanding: Social Identities and Public Uptake of Science." *Public Understanding of Science* 1:3.

Torgeir Austenå and Gudmund Sandvik CHAPTER 10

The Legal Status of Rights to the Resources of Finnmark with Reference to Previous Regulations of the Use of Nonprivate Resources

Introduction

In the Norwegian language, *Finnmark* refers to an entire county in the northern-most part of Norway, while *Finnmarksvidda* denotes only the plateau east and south of the fjords. In this chapter, "Finnmarksvidda" is used when we mean the plateau, but it is necessary to have the whole county, and not only Finnmarksvidda, in mind when discussing the rights to common resources in this part of Norway. The county of Finnmark measures 48,000 square kilometers (by comparison Denmark measures 43,000 square kilometers). The legal situation for Finnmark as a whole is different from the rest of the country because of historical reasons and the fact that only 4 percent of the county has been transferred to private ownership.

Furthermore, it is necessary to say a few words about the term "Saami people" as used in this chapter. "Saami people" can often be used to mean Saami reindeer herders. Here we use the term in a much wider sense to describe a particular ethnic group. Although Saamis often speak of themselves as "a people in four countries," most of the Saamis live in Norway, especially in the northern-most counties of Finnmark and Troms. The reason is evident: the coastal region of what is now northern Norway has always offered fish in the fjords, rivers, and lakes; sea animals, game, and birds along the shores and in the mountains; grass and moss for reindeer and cattle; and wood for many human needs. One consequence is that today only between 10 and 15 percent of the Saamis in Norway are reindeer herders, while others are fishermen and farmers, or have gone into secondary and tertiary occupations like many of the other inhabitants. In comparison, in Sweden and Finland only Saamis who were reindeer herders, hunters, and salmon catchers had the status of *lappar*. Saamis who established themselves as colonists or farmers were no longer *lappar*, but Swedes, Finns, or Kvens. This seems to be the reason why Saamis in Sweden and Finland today are generally reindeer herders. A few Saamis in Sweden and Finland are hunters and fishermen along rivers and lakes. Other ethnic Saamis have been assimilated.

In Norway, reindeer husbandry is by law generally reserved for Saamis from Saami husbandry families (this is not the case in Finland). So, for the sake of clarity, "Saami reindeer herders" will be our terminology when that is meant. Another problem to be considered is the phrase "rights to the resources." Throughout the centuries the inhabitants of Finnmark have worked as fishermen, farmers, herdsmen, and reindeer herders, without particular reasons to try to find out what kind of rights they had to the resources they used. During the last few decades, however, the question of ownership of land, water, and other resources in Finnmark has emerged. The traditional view has been that the state has the rights of ownership. In contrast to this view some representatives for the Saami people have asserted that the Saami are the rightful owners of areas they have been using to make their living for the past several hundred years.

The arguments for the two different views on this matter can be summarized as follows: Under the main legal theory of the past it was not possible to obtain ownership to land by nomadic use. According to this theory, one could only acquire ownership right to unowned land by living on it and using it in some specific way, for instance cultivating it (the specification theory). Both the Danish-Norwegian and the Swedish kings declared that land not taken into specific use was unowned land. The specification theory was stressed by the Norwegian government in its proposal to Norway's parliament in 1848 (Ot. prp. 21, 1848). There is reason to believe that the theory is a part of the basis for the opinion that the state was the owner of land and water in Finnmark. The administration was built on that view.

According to an order in council from May 27, 1775, the state began to parcel out lots for private ownership. It was said that this order introduced private ownership to land in Finnmark. This point of view has more or less clearly been laid down by the administration and in legislation since then, and also earlier than 1775. It can also be found in some court rulings. The state will now assert that the people in Finnmark have adopted this view in practice and that the state's ownership to land and water is based on what Norwegian courts call *festnet bruk*, or publicly accepted use and opinion for such a long period that it can not be altered.

The Saami people were the first to use these areas. They used the resources for their living without any interference from the state or anyone else. They had no reason to think about the term "right of ownership." The resources were shared between separate reindeer herding groups (*si'idaer*). Inside the group several resources were used in common, but some were separated for individual use. The Saamis will assert that none of them have transferred their rights to the state in any form of contract and the area has not been expropriated from them. The use of land and water was a necessity for their living and culture, and thus in their opinion the resources were theirs.

The question may be raised why in 1775 and afterwards the Norwegian institution *almenning* (commons) was not applied in a legal system in Finnmark. In fact, civil servants used this term from the 1690s to around 1850, probably by analogy of the general articles on commons in the royal lawbook of 1687, Christian V's *Norwegian Law*. The articles 3-12-1 to 8 date back to chapters VII 61 to 64 in King Magnus the Law-Mender's national lawbook 1274–1276. Article 3-12-1 begins by asserting that the "common shall remain as it has been from ancient times," thus ensuring the people traditional ways to use resources on the "king's common" (3-12-2). But south of Finnmark these articles applied to established farms in definite rural districts, not to anybody outside. Established farms were new in Finnmark. Qualified legal work would have been necessary to elaborate precise rules that suited Finnmark. Even for southern Norway, such work was not undertaken until the end of the 1850s and then in order to protect the pine forests in the Southeast. In Finnmark the old collective use remained outside the established farms, with some regulations by the county administration (especially on the use of forests). A formal transformation of the different uses of the "common" resources into legal rights was not undertaken. The Saami Rights Commission has been asked to give its opinion on this question (see chapter 13).

One will encounter another complicating factor when considering the legal situation of the rights of ownership to land in Finnmark. It could be said that the state of law is different for the southern or interior part of Finnmark (Finnmarksvidda) and the rest of the county. Beginning with a peace treaty of 1613, Finnmarksvidda had the administrative name of *fellesdistriktene* (the common districts) until 1751. That meant that it stood under Swedish ecclesiastical and legal jurisdiction and that both Sweden and Norway had a right to taxation. This fact could have an impact on the legal status today. In this area the impact of Swedish rights in similar areas in Sweden could be of interest. The Saamis could point to views in the judgement of the

Swedish Supreme Court in 1981 (The Taxed Mountain Case) and to the historical investigations of Kaisa Korpijaakko-Labba about Finnish rights concerning Lapp areas (*lappmarkerna*) of Torne and Kemi in the northernmost parts of Sweden and Finland (see chapters 9 and 11). In this part of Finnmark the Saami people were the predominant group using the resources. The Danish-Norwegian administration had full jurisdiction over the rest of Finnmark (the *norske privative sone*)—this fact could have impact on the legal status. In this area the people were more mixed with Norwegians, Finns (*kvener*), and Saamis. The variety of resources utilized was larger.

Also, according to international law, Norway has some obligations towards the Saami people. These obligations are founded on the United Nations 1966 Human Rights Covenant on Civil and Political Rights for Minorities and the ILO Convention 169. Norway has, unlike Sweden and Finland, ratified the ILO Convention 169.

In 1980 the government set up a Saami Rights Commission. One part of the Commission's task is to give an opinion on who has the right of ownership to land and water in Finnmark. The Commission has not yet given their conclusion concerning the ownership question. Thus the following discussion about rights to the resources of Finnmark will be based on the traditional opinion of the state as owner of land and water in Finnmark.

The Local Administration of the State's Right to Land and Water in Finnmark

The rights of ownership are delegated to the local office for administration and selling of land in Finnmark (*Jordsalgskontoret*). This office is located in Vadsø. The legal basis for the administration is the act of March 12, 1965, concerning unregistered land belonging to the state in Finnmark. According to this act, land can be sold or rented to private persons, municipalities, and companies. Land which—in the opinion of the authorities—is needed as grazing land for reindeer can not be sold. The same applies to land used for moving reindeer on traditional migration paths between the coast and the inland.

The parliament has also stated in the act that land which, according to the view of the administration, ought to be in ownership of the state for the sake of forestry, mining, fishing, outdoor life, and nature conservation should not be sold.

The act of 1965 is a continuation of the two previous acts of June 22, 1863, and of May 22, 1902. The act of 1863 intended to finance the administration of forest in Finnmark by sale of land through auction. That proved a failure because the incomes were small. By the act of 1863 all previous *amtssedler* (county deeds) for homesteads were transformed free of charge into titles of ownership. In pursuance to the act of March 12, 1965, the king was given the authority to lay down more detailed regulations concerning the use of resources in Finnmark. Such regulations were laid down on July 15, 1966. We will later return to some of these regulations.

The Right to Reindeer Herding in Finnmark

The right to reindeer herding in Norway is regulated in the Reindeer Farming Act (RFA) of June 9, 1978. The act was amended in February 23, 1996. Norway is, according to section 2, divided into six reindeer herding areas (*reinbeiteområder*). These areas are situated in the counties of Finnmark, Troms, Nordland, Nord-Trøndelag, and Sør-Trøndelag/Hedmark. Reindeer herding inside these areas cannot legally be done by anyone other than Norwegian citizens of Saami descent, according to section 3 in the RFA. Reindeer herding outside these areas can be done by Norwegian citizens, but only through special concession from the king. Finnmark is divided in two reindeer herding areas which cover practically 100 percent of the county.

On a national scale, reindeer herding is a small industry. The total industry comprises less than 700 management units. Only approximately 2,500 persons have reindeer herding as a main or subsidiary trade. Domesticated reindeer numbered 220,000 in April of 1991. Even though reindeer herding is a small industry, it is of great importance for the Saami people both economically and culturally. The industry has always been looked upon as a Saami industry. Finnmark is the main area of this industry.

The rights of the reindeer herding Saamis to utilize outlying fields have been recognized since ancient times and were supported by the Lapp codicil, which is an appendix to the border treaty in 1751 between Sweden-Finland and Denmark-Norway. Along with the progress of the Norwegianization policy at the end of the 19th century into the 20th century, the theory emerged that the use for reindeer herding purpose was only a tolerated use.

According to this understanding, the legal basis for reindeer herding was the law, as it was at any time, given by the parliament. This view was stated by the Department of Agriculture in 1976 (see Ot. prp. nr. 9 for 1976–77, p. 42 and p. 47). The Norwegian Reindeer Herding Saami Association was of another opinion. They pointed out that reindeer herding has been exercised in the area for centuries and had established a legal basis for reindeer herding that was independent of the law given by the parliament. The agricultural committee in the parliament would not decide which of these two views was right or wrong (see Innst. O. nr. 98 for 1976–77, p. 3 and 5). The majority of the Saami Rights Commission proposed a new section in the Reindeer Farming Act stating that the legal basis for reindeer herding is independent of the law given by the parliament. If the final conclusion should be that the right to reindeer herding had been established before the border treaty of 1751, the existing law of 1978 concerning reindeer herding is regulating the extent and contents of the right and is not establishing the basis for the right.

The organization and management of the reindeer herding industry is regulated in the act of 1978. Specific for this industry is that the reindeer owners themselves are given great influence and responsibility over the administration. The Ministry of Agriculture appoints a *Reindriftsstyre*, the central administration of reindeer herding. Persons who are active in reindeer herding must be appointed to this board. Finnmark is divided into two reindeer herding areas (*reindriftsområder*), and a board must be appointed to administer each area. Some of the members of the board must be active reindeer herding persons.

Each of the reindeer herding areas is divided into several reindeer herding districts with their own boards consisting of reindeer owners. The district board can represent the owners and the district as legal persons in lawsuits. This organization gives the reindeer herding industry administrative boards on national, regional, and local levels. The industry comprises the majority on all of these boards, even if the election procedures of members to the boards vary. The right to reindeer herding inside a *reinbeiteområde*, as mentioned above, depends on one's descent. That his or her parents or grandparents have had reindeer herding as their main trade is a condition.

The basic unit for reindeer herding is called a management unit (*driftsenhet*). This term was introduced in the act of 1978 concerning reindeer farming. In principle a management unit would be owned and managed by one responsible owner. The management unit could include reindeer belonging to the owner´s spouse, their descendants, brothers and sisters, and their children. The condition that the manager be of Saami descent and not an owner of another management unit applies. Reindeer belonging to several persons or families exercising reindeer herding in common could be accepted as a management unit on condition of consent from the board of the reindeer herding area. A reindeer owner is not allowed to move his unit from one district to another without consent from the board of the reindeer herding area. An existing management unit can normally not be divided. Sale and inheritance of the unit is regulated by the law. An undivided unit can be transferred to the spouse of the owner without approval. This is also the situation concerning transfer or inheritance of the undivided unit from parents or grandparents. Any other form of transferring or establishing management units cannot be done without approval. The right to herd reindeer includes the right to use available resources and to put up necessary erections for the trade (for example, the right to take materials from the forest for firewood in order to erect necessary structures like fences and turf-huts and make equipment). The right to reindeer herding also includes the right to hunt and fish in the lakes and rivers of the district. In addition, the reindeer herding industry is legally protected against encroachments. If range land is taken for other purposes such as roads, power lines, mining, and so on, the reindeer owners, often represented by the board of their district, can demand compensation for lost range land and for disadvantages for the industry.

The reindeer farming industry in Finnmark has large problems today. The situation is particularly difficult in the core areas of Finnmarksvidda. Too many reindeer have exploited the grazing resources, which has caused low productivity and unsatisfactory economic development. The regeneration of the lichen fields in the worst regions could take up to twenty to thirty years. These factors cause conflicts and social unrest, and it seems to be difficult for the reindeer herding industry to solve the problems. This is one of the reasons why the parliament amended the Reindeer Farming Act in February of 1996.

Rights to Resources in the Underground and to Water Power in Finnmark

Minerals with a specific weight lower than 5 belong to the owner of the ground. In regulations laid down by the Ministry of Agriculture the rights to underground resources in Finnmark do not follow the ground when the state sells it to private owners. Minerals with specific weight higher than 5 belong to the state everywhere in Norway (as a *regale* since the 16th century), but they are free for everyone to search for. If something is found, the finder needs official permission to start exploration (or exploitation) of the resource. The owner of the ground has the right to a sort of fee maximized to U.S. $3,000 per year in the period from 1994 to 2003. Even if the state has sold the ground in Finnmark, the owner's right to this fee according to the act of June 30, 1972, is reserved for the state.

The right to water power belongs as a main rule to the owner of the ground, according to the Watercourse Act of March 15, 1940. If the state has sold the ground in Finnmark, the right to water power is reserved according to the regulations of July 15, 1966.

The Right to Materials from the Forest

In the inland areas of Finnmark, birch and willow (*vier* and *salix*) are fairly common in valleys and wet areas. Pine is found in Alta and in some areas in Porsanger and Karasjok, as well as in the Pasvik area. Wood was a vital resource for all people in Finnmark until the age of coal, petrol, electricity, and modern communications began. Wood had been necessary for heating and cooking and as raw material for many sorts of equipment. The coast being rather bare, turf, heather, willow, and driftwood from Siberia filled some of the needs of the local population.

The very first known regulation concerns the use of the pine forest in Alta. In 1693, the *fogd* (sheriff and judge) of Finnmark declared that the "commons of his majesty in Alta" should only be used by the population of Alta and to a certain degree by the population of western Finnmark farther out. Fishermen from the counties of Nordland and Troms were forbidden to cut timber in Alta, and the merchant company had to ask the county administration for permission to cut timber for its buildings. This regulation was confirmed by prescripts in 1753 and royal resolution in 1775 and extended to all Finnmark, while the use of birch was reserved for the local populations. As a whole these regulations are still in force. An embargo on the export of pine timber from Finnmark since the end of the 17th century was extended to an embargo for the whole of northern Norway by an ordinance of 1752. This embargo was cancelled only in 1925. The regulation of timber cutting in Karasjok began in 1776. Here pine was cut by the local population (Saamis) in winter, transported down the river Tana to the Tana fjord, and used especially for houses and for boats in the western part of eastern Finnmark. The Varanger region seems to have got their timber from the (until 1826 Norwegian-Russian) Saami *si'idas* of Neiden and Pasvik.

The copper mine of Kåfjord in Alta from 1826 to the 1880s was the most important industrial enterprise in northern Norway during the 19th century, but the enterprise was not allowed to make charcoal from the Alta forest to melt the ore into copper. Coal had to be imported from England. In 1863 two laws included Finnmark in the wood protection policy in Norway. The income of state ground sold to private individuals should, as mentioned above, be put in a fund and used to pay for the forest administration of Finnmark.

What about the Saami reindeer herders? They were supposed to conform to the same rules as other inhabitants. But of course it was usually impossible for them to ask for permission to cut wood, so the general rule applying to their population entitled them to use dry wood freely and to cut fresh wood (pine and birch) for their specific needs for different equipment. In the 1860s reindeer herders were forbidden to pass through pine forests with their herds in winter, because pine sprouts tend to become brittle in frost.

The order in council of May 27, 1775, founded the legal basis for selling ground in Finnmark to private persons. The right to the conifer forest on the ground did not follow the ground to the new owner. Since 1820 it has been possible to add a clause to the deeds giving the new owner the priority to take wood necessary for his household on the property he had acquired. The new owner would receive first choice of any lumber that he needed for the maintenance of his household. According to the act of 1965 the forest will always follow the ground to the buyer and new owner.

Because of the different practices the state has followed during the history there are different categories of forest in Finnmark. The greatest part of the forest is, however, on ground belonging to the state. The question is then what usufruct rights the local inhabitants have to take materials from the woods. In the 1965 act on selling ground in Finnmark it is said that people from the county of Finnmark have the right to have singled out enough birch necessary for firewood for their households. It is a question under discussion whether the local inhabitants have a more extensive right to wood than stipulated by the act of 1965. The answer depends on what is to be found in court rulings, administrative practice, and the opinion of the inhabitants and the state. The Saami Rights Commission has proposed that the municipality should have the right to manage the forest resources in Finnmark. The use of birch should be reserved for the local populations, and pine timber should be reserved for farmers and reindeer herding Saami people.

The Right to Use Outfields as Grazing Land for Domestic Animals and as a Place to Gather Winter Fodder

As far as we know, the oldest regulation of this resource took place in 1590 between the Saami reindeer herders of Varanger on the one side and the inhabitants of Vadsø and two nearby communities on the other side. The two parties agreed that the reindeer herders should not pass over the pastures near the coastal communities later than the end of May. The agreement was confirmed by the county governor (*lensherre*) and it was referred to in 1622 in court proceedings (Sandvik and Winge 1987, 132). Such an agreement reflects the opposite interests of reindeer herders and cattle owners.

Since the end of the 17th century this situation has been common. There were also—though less frequent—the opposite interests of the separate groups of Saami reindeer herders from the inland and the coast about grass in summer and lichen in the winter. We have not seen any real regulations, but there have certainly been customs as to where the reindeer herds were supposed to be. The *ting* protocols contain admonitions from the *fogd* (sheriff and judge) that the inland reindeer herders should not infringe on the pastures and range lands of the coastal population. We have noticed one interesting proposal in the 1820s from the county governor (*amtmann*) that the reindeer should leave the coast region at the same time as cattle owners sent their sheep to the mountain pastures.

Since the 1760s and the royal resolution of 1775, the authorities have allotted homesteads practically free of charge and without ethnic discriminations. The homesteads were generally used for a combination of activities, especially fishing in the fjords and along the coast, as pasture for one or two cows and some sheep, hunting, salmon fishing in rivers, and picking cloud berries. Generally, the only "agriculture" in Finnmark was cattle breeding for the use of the family, not cultivation of the soil for cereals, potatoes, and hay. Because of the very long winter season, winter fodder for the cattle was a vital necessity. It had to be collected on the hillsides, bogs, and on the clearings of former dwelling places, and then be dried into hay and stacked, to be transported home when snow came. But the winter fodder was scarce, so in spring the cattle needed the first green grass along the shore as badly as the reindeer. And in the autumn, the reindeer much preferred the hay in the stacks to the brown and wilting grass on the ground. Fencing in was impossible, especially as the homesteads were composed of a site for the dwelling house and a great number of "hay fields" not seldom far away. It is easy to understand why interests clashed and conflicts increased as the numbers of both homesteads and of reindeers grew.

Problems became acute after 1852. Until 1852 the reindeer herders from Finnmark could have their herds on the extensive lichen fields in Finland in winter. But in 1852 Russia closed the frontier and denied access to the grand duchy of Finland to reindeer from Finnmark. Norway followed suit by denying access to the fjords to reindeer and fishermen from Finland in the summer season. The lichen fields in Finnmark now became the minimum factor in the reindeer husbandry. In order to protect the lichen, it was vital that no reindeer should be allowed to remain on the plateau during the summer season. The government (by its commissioners) and the reindeer herders agreed that every herd had to move to the coastal region in summer. Accordingly, the clash of interests between reindeer herders and cattle owners intensified.

When the *Storting* in 1863 decided that homesteads should no longer be allotted freely but sold, it put as the first paragraph of the law that "the ground of the state" should not be sold when it existed in different forms of collective use, notably as grazing land for reindeer and cattle and as moving routes for reindeer between the coast and the inland (and as ground for cod drying along the coastline). This was a "major clause," the parliamentary committee said, and it is still the leading principle in today's law. The *Storting* followed suit a few years later by ordering commissions to round out or regroup the numerous lots of "hay fields" belonging to the homesteads.

Nevertheless, it is safe to say that this problem was not really solved. Compared to the 19th century, however, the problem has taken new forms. Earlier the reindeer were more tame. Female reindeer were milked and castrated bucks were used for transport. Herds were smaller and had to be guarded against beasts of prey, at least in winter. With modern herding techniques there is now more distance between the herders and reindeer and apparently less control of the herd.

Nowadays, meat production for the market is the important aspect of reindeer husbandry. It is not quite a new aspect. Modern agriculture in Finnmark is still chiefly comprised of cattle breeding and milk production, but now with few units, each with many cows. The modern and highly productive cow is only able to pasture on flat and cultivated fields (so-called culture range lands, or *kulturbeiter*). Grass for ensilage and hay is grown on deep-cultivated fields. Pasturing on the hillsides and in the mountains is now only for sheep and perhaps for some young cattle.

Before the order in council of 1775 there was no regulation of using outfields as grazing land in summertime for domestic animals in Finnmark. The resolution altered nothing in this practice. In 1863 the parliament enacted a law on selling ground in Finnmark. According to this act and the following regulations, it was an assumption that the owner of domestic animals did not have the right to use outfields for pasture unless there was basis in the contract for it. In 1902 the state was given a stronger right to regulate rights to pasturing given in contract from that time on. And in 1955 a new clause was taken into new contracts. The state was given the right to terminate the grazing right giving notice one year in advance. In the resolution of 1965 the right of grazing in the outfield established in new contracts was weakened even more. According to this historical background there could be up to five groups of rights to use the outfields as grazing land for domestic animals (Tønnesen 1972, 1979; 264–267). There could, of course, be some owners of husbandry who had obtained more extensive grazing rights than the rights based in their contracts. Such a conclusion can only be reached by concrete judgement of use and court rulings. Whether there exists a general rule of grazing rights more extensive than that following from the contracts was discussed by the Saami Rights Commission. The Commission's proposal was to give the farmers a legal right to grazing land.

The Rights to the Resources in the Zone Next to the Seashore

The general rule in Norway is that an owner of ground bordering the sea is also the owner of a part of the ground out into the sea. The border for his property out in the sea will differ a bit depending on the deepness and the slope of the sea ground. The owner of the seashore has the right to build out into the sea. Another important right is the ability to move back and forth from his land undisturbed in a boat. The seashore owner's rights include several other possibilities for dispositions, but the right to shoot or hunt is afforded to everyone. The right to fish is also given to everyone, with the exclusion of salmon fishing with some types of fishing tackle.

These general rules are also laid down in Finnmark, but there are still two questions to be discussed. The first problem involves homesteads where the state reserved a small corridor between the new homestead and the sea. The conclusion as to access and rights out in the sea depends on what is written in the contract when the homestead was sold. The other question concerns the legal situation for seashore not sold by the state. The state claims to have the same rights here as owners of ground elsewhere in Norway. Who holds the rights to this type of property is uncertain. The inhabitants have used the resources on the shore with few or no restrictions. The Saami Rights Commission has proposed that the resources in the seashore zone should be managed by the municipality.

The Right to Fish, Hunt, and Pick Berries

Fishing; picking berries, mainly cloud berries (*molte, Rubus chamaemorus*); and hunting wild animals and ptarmigans (*rype, Lagopus*) was earlier an important part of the daily bread winning for a great part of the inhabitants in Finnmark. These resources were parts of the diet of the local population and to varying degrees were also sold to local merchants for "export" to towns south of Finnmark. In relation to the small population these resources were rich, at times abundant. The local communities had their traditional resource areas, and it was considered bad manners to trespass into neighboring areas. This was the situation for the reindeer herding Saami people and for the inhabitants living alongside the rivers and the fjords.

This situation has changed during the last decades. Change from a subsistence economy to money economy, more leisure time, and better communications have resulted in a use of the mentioned resources as an important part of the leisure time activity, both for the inhabitants of Finnmark and for tourists from other parts of the country and abroad. In the reindeer herding industry all these leisure time activities often mean disturbances and difficulties, and can cause confrontations concerning regulation and use of the outfields in Finnmark.

Fishing in the Sea

Since the 1830s the principle of "free fishing" has been gradually put into effect through laws. From the 1850s Finnmark has been the main "fishing county" in Norway. The fisheries in the fjords and along the coast attracted farmer-fishermen from southern Norway, who settled in the coastal region. The immigration from the south is probably the main reason why the population of Finnmark increased from 0.9 percent (7,700) of Norway's population in 1801 to 1.5 percent (32,800) in 1900 and to 2 percent (80,000) in 1975.

Whaling with modern technology (from southern Norway) started in the Varangerfjord in the 1870s. Modern whaling seems to have reduced or even extinguished most of the former whale populations within a generation. Commercial losses were probably the main reason why the whaling company then retired from Finnmark, rather than the uproar of the fishermen in 1902, or 1904 state regulations. The fishermen stuck to the very old belief that the whales sent the *lodde* (*mallotus villosus*) and the cod to the coast and to the fjords—indeed, fish catches around 1900 were poor. But the real reasons behind the declining fish catches were most likely due to over-fishing along the coast and in the fjords with huge nets and the new trawlers, seal invasions, and—very probably—periodic oceanographic changes.

From the 1680s to 1830 the inhabitants of the fishing stations (*fiskevær*) along the coast had the exclusive right to the nearby fishing grounds. Fishermen from southern regions had to fish elsewhere. The reason for this exclusivity was that merchant companies were under the obligation to supply the inhabitants of the station and the surrounding regions with necessary goods on long-term credit, mostly paid in dried cod (*stockfisch*), salted salmon, and cloud berries in barrels. Casual fishing in the fjords was free. The Saami reindeer herders themselves fished in the fjords during the summer season, but in preparation for winter they exchanged dried reindeer meat against dried cod fished in the fjords during the previous winter. Sea fish was a normal part of the diet of the Saami reindeer herders. From 1830 to the First World War, fishing in the fjords and on the coast was gradually liberalized. Since the 1930s (and especially since the end of the 1980s) regulations here have become common. This is a very complicated subject. An issue to be explored is if it might be possible for the Norwegian parliament to reserve fishing in the fjords and near the coast for the fishermen from the coast of northern Norway and to allow only "passive" fishing gear here, thus excluding use of "active" fishing gear like trawls. The goal would be to protect the "local" fish, and a "sustainable" use of such resources. While no definite answer can be given at the moment, the entire coastal population of northern Norway depends on fisheries for its existence, no less today than during the 18th century.

Fishing Rivers and Lakes

The salmon have been the main basis for Saami settlement along the rivers in Finnmark, especially along the great rivers of Tana and Alta. The county governor decided in 1763 that the salmon in the river of Alta should be reserved for the local community, a principle that was extended to the river of Tana by the royal resolution of 1775. This is still Norwegian law. For the other rivers the salmon fishing was free— the general principle in Norwegian law, that salmon fishing belongs to the individual owner of the riverside, was not introduced in Finnmark. Since around 1900 the state has usually rented the salmon fishing in the small rivers to local salmon fishing associations, who usually demand small fees from local fishers and greater fees from external salmon fishers. The income is used for supervision and development. Finnmark has a lot of lakes, rich in trout, char (*røyr, salvelinus alpinus*), and whitefish. Regulations here have only recently been introduced for southern Norwegians and foreigners. Saami reindeer herders take what they need of inland fish during the winter season.

The right to fish salmon in the rivers of Alta, Tana, and Neiden. The right to fish salmon in the rivers of Alta, Tana, and Neiden is different from the rest of the rivers and lakes in Finnmark. In the Alta River, owners, as well as ground users who are living on their property, have the right to fish salmon from June 24 through the rest of the season. Up to June 24 the right to fish is free for all the inhabitants of the municipality. In the Tana River the right to fish salmon belongs to owners of properties not farther away from the river than two kilometers. In addition, it is a condition that the owner grow at least two thousand kilos of hay. The right belonging to this group is limited to net fishing. A much debated question is whether all inhabitants of the Tana Valley have the right to fish salmon by rod or not. (The state has acknowledged owning this right.) The Saami Rights Commission has proposed a new act concerning fishing in the Alta, Tana, and Neiden rivers. In section 2-2 it is proposed to give people living in Tana, Karasjok, and a smaller part of Kautokeino the right to fish salmon by rod. Currently the income of licenses to fish salmon by rod is used locally for supervision and management of the river, and on the county level for the administration of the inland fisheries. The inhabitants of the Tana Valley (comprising the municipalities of Tana and Karasjok) pay a trifle for one year's license. In the river Neiden the right to fish both salmon and other sorts of fish belongs to the Neiden Fishing Community (*Neiden Fiskefellesskap*). The conditions for membership in this community are almost similar to the conditions for fishing salmon in the Alta and Tana Rivers. The conditions of agricultural activity are, however, not so clearly formulated. The Saami Rights Commission has proposed special rights for the Neiden Saami concerning their special salmon fishing in the Neiden waterfall.

The right to fish in the rest of the rivers and lakes in Finnmark. Fishing of salmon and other sorts of fish in the rest of the rivers and lakes in Finnmark is regulated in the act regarding salmon and inland fish of May 15, 1992. The main rule in Norway is that the owner of the ground bordering rivers and lakes is the owner of the right to fish. Many of the private properties bought from the state in Finnmark are not, however, bordering the rivers because the state has held back a small corridor of ground between the private property and the rivers (see "Innstilling om lov og forskrifter om statens umatrikulerte grunn i Finnmark fylke," 1962, 17). In these cases and everywhere else, the state is, as owner of the shores of the rivers and lakes, the owner of the fishing right. The right to fish with fishing rod in rivers and lakes belonging to the state is free for Norwegian residents. Fishing with all types of fishing tackle is free for residents of the county of Finnmark. All the same, there are some regulations about how many fishing nets each resident can use in some specific lakes. Foreigners are allowed to fish with fishing rods within a distance of main roads of five kilometers. The right to fish both salmon and other types of fish in rivers is often rented to organizations.

The Right to Hunt

The right to hunt is regulated in the Act of Small and Big Game Hunting of May 29, 1981. The general rule in Norway is that the owner of the ground has the sole right to hunt on his own ground (section 27). The right to hunt on ground belonging to the state is regulated in section 31. The right to small game hunting is allowed for all residents of Norway.

The right of reindeer herding Saamis to fish and hunt is regulated in the Reindeer Farming Act of June 9, 1978. In connection with legally executed reindeer herding the Saamis are allowed to fish and hunt under the same conditions as the residents in the district they are passing through with their deer.

The Right to Pick Berries

The right to pick berries is open to everyone except the picking of cloud berries. Picking this sort of berry is regulated in the act of March 12, 1965, section 5a. The right to pick cloud berries is reserved for residents of the county of Finnmark. (The only exception is berries an individual might eat on the spot.)

REFERENCES

"Innstilling om lov og forskrifter om statens umatrikulerte grunn i
  Finnmark fylke." September 1962. av. 29. Oslo:
  Landbruksdepartementet.
Sandvik, Hilde and Harald Winge. 1987. *Tingbok for Finnmark 1620-1633.*
  Oslo: Norsk lokal historisk institut.
Stortingsforhandlinger. 1848. 1. del., Propsisjon no. 21, p. 1 and 23-24.
Stortingsforhandlinger. 1976-1977. 4. del., Proposisjon nr. 9, p. 42 and 47.
Stortingsforhandlinger. 1976-1977. 6. del. B. Innstilling O. nr. 98, p. 3 and 5.
Tønnesen, Sverre. 1972, 1979. *Retten til jorden i Finnmark.* Bergen, Oslo,
  Tromsø: Universitetsforlaget.

BERTIL BENGTSSON   CHAPTER 11

The Legal Status of Rights
to Resources in Swedish Lapland

The Reindeer Farming Act

Most of the inhabited mountains of northern Sweden are owned by the state, although the ownership has not been registered in the land register; in certain parts, there are also private owners, above all forest companies. However, at the same time the Saami—or, more correctly, the reindeer herding Saami—have particular rights to land and water in those parts and the adjacent forest territories; in all, these rights cover about one-third of the area of Sweden. The rights are regulated in a detailed way in the Reindeer Farming Act of 1971 (RFA). As will appear from the following, this act does not render the whole truth about the legal status of the Saami, but by way of introduction it may be useful to describe the act's main principles. Some of these rules have recently been changed; a government bill concerning amendments in the act has was passed by the parliament in December of 1992. The new rules came into being on July 1, 1993. In this chapter, the earlier rules will also be dealt with to some extent.

The RFA can partly be regarded as a kind of monopoly legislation which grants an exclusive right to the Saami for reindeer grazing in Sweden. However, an essential part of the act deals with the rights of the Saami to land and water, called reindeer herding rights. These are described as the right of the Saami to use land and water for their own support and the maintenance of their reindeer. This description implies that it is a special sort of usufructuary right for which the act does not prescribe any time limit. The reindeer herding right includes reindeer grazing, hunting, fishing, and some felling of lumber. The right has until now belonged primarily to such persons of Saami heritage whose parents or grandparents (maternal or paternal) had reindeer herding as a permanent occupation. According to the new legislation, the reindeer herding right belongs to the Saami population and is founded upon the usage of time immemorial. However, the right can only be exercised by Saami villages, and the amendment will not mean any real change in this respect. (It should be noted that the term "village" here denotes a particular type of legal entity, as in a reindeer corporation rather than a village in the usual sense of the word.) Although it is not clearly stated in the act, it is evident that a reindeer herding right is not transferable, nor can it be mortgaged.

The reindeer herding may be carried out year-round in the mountain lands along the Norwegian border in regions defined in the act; in the counties of Norbotten and Västerbotten above what is called the Cultivation Border (an historic term denoting the line above which the land should be, in principle, reserved for the Saamis) and elsewhere on crown land where such herding has been carried on of old; and in the county of Jämtland and Kopparberg—moreover, in certain districts particularly designated for reindeer grazing. Further, during the period of October 1–April 30, reindeer herding is permitted in such areas outside the regions just mentioned where grazing has traditionally taken place during certain parts of the year (in most of the forest regions of Northern Sweden and in certain parts extending even to districts not far from the Guld of Bothnia). In the latter territories, the reindeer herding right may be compared to a kind of easement (*servitut*) connected with the more extensive right that the Saami can exercise in the mountain area, in about the same way as an ordinary easement in Swedish law is connected with the ownership of real estate profiting by the right. The two types of regions mentioned will here be called "whole-year herding area" and "winter herding area," respectively.

The reindeer herding right is carried out by Saami villages on separate areas assigned to each village. As mentioned above, the function of these villages is in the common interest of the members: to manage the reindeer farming in the grazing area of the village. They are not allowed to engage in any economic activity other than reindeer herding. When registered, the Saami village becomes a legal entity. In questions concerning reindeer herding rights the Saami village represents its members. In principle, solely certain persons who carry on or have carried on reindeer farming, as well as members of their families, may be members of a Saami village. If membership is refused to a Saami intending to exercise reindeer farming, the county government can grant him admission, provided that there are particular reasons. Thus, the Saami village cannot even decide on its own which persons should be permitted to utilize the village land for reindeer grazing.

The authority founded on this usufructuary right is divided between the Saami villages and its members in a rather complicated way. The Saami village has the right, for the common need of its members, to use the grazing area allotted to the village for reindeer grazing. Within this area, it may erect certain facilities required for reindeer husbandry, with the permission of the owner of the land. For this purpose, the necessary lumber may be felled in the reindeer grazing mountains and to a limited extent in other parts of the area, too. However, the felling of growing coniferous trees requires the permission of the owner and user of the land, and remuneration shall be paid for the root value of growing trees; exceptions are made for lumber felled on certain crown land and for deciduous trees otherwise felled in the reindeer grazing areas. Moreover, the Saami village has the right to move reindeer from one part of the village grazing area to another. The members, for their part, are permitted to construct small facilities required for reindeer husbandry and to fell the necessary lumber for this purpose as well as (above all) to fell lumber on crown land for the construction or renovation of their family homes. They may also hunt and fish in the outlying parts of their village's grazing area in the reindeer grazing mountains. According to the new legislation, other Saami may also be permitted to take wood intended for handicraft on the reindeer mountains and on certain other crown land.

A person holding reindeer herding rights cannot be deprived of this usufructuary right on the grounds that he has violated the rules of the act or otherwise neglected his duties in the exercise of this right. By decisions of certain agricultural authorities, however, his use of land may be limited insofar as concerns the size of the herd, reindeer grazing, and felling of lumber. Further, usufructuary rights may be terminated against the will of the holder in certain other cases. Until now, the government could decree that the use of a particular area should cease if it was required for some purpose which could motivate expropriation according to the Expropriation Act or else was of vital importance to the public interest; if the area had small significance for reindeer herding, the use could even be terminated in this way as soon as the area was required for the public benefit. However, according to the new legislation, the rules of the Expropriation Act shall be exclusively applicable in this situation. Damage and inconvenience to reindeer herding or hunting or fishing rights shall be compensated; if the damage or inconvenience does not affect any particular person, it is generally divided evenly between the Saami village in question and the Saami Fund—a public fund used to benefit reindeer herding, the Saami culture, and Saami organizations. Otherwise, the owner or user of whole-year herding land must not take any measure causing considerable inconvenience to the reindeer herding, unless the land shall be used according to a municipal plan or for other activities that can be authorized according to special rules. In these cases, the Saami are not entitled to any compensation.

Neither Saami villages nor members of such villages may grant rights which are part of the reindeer herding rights, except that ex-members may be allowed to hunt or fish for their household needs, free of charge, in the village area. Otherwise, the authorities of the state are in charge of all granting of rights in the reindeer grazing mountains. Usufructuary rights may be granted only if it is possible to do so without any considerable inconvenience to reindeer herding; as for hunting and fishing, the right will be granted on condition that it is compatible with good game management or fishing conservation and does not encroach to any appreciable extent upon legislated hunting and fishing rights of the village members. Except when exclusive fishing rights to a given body of water are granted, no permission from the Saami village is required. If the right granted involves exploitation of natural resources, the state is to make compensation for the damage or inconvenience caused to reindeer herding, otherwise a fee shall be charged (except when the right implies lumber felling), unless there are special reasons for granting the right free from charge. The compensation or the fee is divided between the Saami Fund and the Saami village in question. It appears from the above that the RFA builds on the assumption that the areas covered by the act are not owned by reindeer herding Saami. However, the text does not expressly deal with the ownership of the land.

To sum up, the complicated system of the RFA implies that the right of the Saami in these areas is a kind of usufructuary right primarily comprising reindeer grazing, but also involving other rights connected with this, such as constructing certain facilities, using lumber for household needs, and using migration paths (hunting and fishing are included). Certain of these rights can be exercised by the Saami village, certain others by the individual Saami. Further, in some respects the party entitled is the Saami Fund, representing the whole Saami population.

If a nonmember of the village interferes with the enjoyment of such a right, the village or person entitled can claim a remedy in an ordinary court of law, in the same way as an owner whose right has been infringed. In this way, damages may be claimed, as well as an injunction to cease an illegal activity (although at least the latter expedient does not seem very practical). In case of a legal exploitation of the resources, the Saami village can claim compensation as an owner according to a similar procedure. An intentional violation of the Saami rights to natural resources can involve criminal liability; here, too, the village or the individual Saami has the same legal position in the trial as other injured parties. However, the power of the Saami to use and exploit the resources in question is limited in several ways: in civil law, the reindeer herding right enjoys less protection against measures taken by the authorities and the landowner than other similar rights to use land and water (for instance, easements created by the land authorities). The weak position of the Saami in this respect appears clearly when the rules concerning the abolishment of reindeer herding rights and the granting of usufructuary rights are compared to the ordinary principles of Swedish law. Here, attention should be called to the imperfect protection against encroachment, and, above all, the ability of the landowner to change the use of the land or take any other measures affecting the reindeer farming without any compensation to the Saami, provided that the inconvenience cannot be regarded as considerable. Further, there are reasons to emphasize the lack of influence of the Saami when fishing and hunting rights are granted on the whole-year herding land. Also, in other respects, there is no way of transferring a reindeer herding right except by accepting a Saami as member of the Saami village.

An essential idea behind these rules—most of which originate from the first Reindeer Grazing Act of 1886, is that the limitations of the Saami rights are justified by the monopoly exercised by the Saami concerning the reindeer grazing business; the RFA is regarded as part of the public law rather than the land law, and consequently it will seem natural that the act regulates the reindeer herding right according to what seems appropriate essentially from an economic point of view. In general, this approach seems to have been predominant among Swedish jurists for more than one hundred years, at least among those who have not studied the historical background of the rules. Even the bill of 1992 expresses this attitude to some degree as is discussed below.

The rules of the RFA concerning the protection of the reindeer herding right are supplemented by some important regulations in the Forestry Act. Evidently, the possibility of reindeer farming depends to a considerable degree upon the state of the forests used for grazing. According to the Forestry Act, lumber felling on whole-year herding areas may not take place without prior consultation with the Saami village affected. Further, the County Forestry Board (*skogsvårdsstyrelsen*) shall decide whether timber felling in certain slow-growing forests will have such detrimental effects upon the reindeer farming that it should not be permitted. The grazing areas would then be reduced to such a degree that the possibility to keep the number of reindeer permitted is affected or the ordinary gathering and moving of the herd is impossible. If felling is permitted, the board can lay down conditions that certain measures shall be taken that are obviously required to protect the reindeer farming.

The *Taxed Mountains* Case

The predominant outlook in the RFA was challenged by the Saami in the much-discussed *Taxed Mountains* Case (1981). In the first instance, the case concerned the ownership of certain areas in the province of Jämtland known as taxed mountains (after an admini-strative proceeding in the 1840s involving taxation and land partitioning). A number of Saami villages claimed, on historical grounds, to be rightful owners of these areas or, secondly, to have several types of limited rights to the same areas, among others rights of reindeer grazing, hunting, fishing, felling of lumber, harvesting, cultivation, gravel mining, other mining, minerals, landowners' share in mines, and hydraulic power. The Saami claimed a declaration that all these rights existed on the basis of civil law, irrespective of the RFA. The state maintained that it owned the properties in dispute and that only the rights specified in the RFA belonged to the Saami.

The process finally reached its conclusion by the judgement of the Supreme Court in 1981, where several issues of essential importance for the Saami rights were dealt with in a thorough way. The Saami parties lost on all points; as most jurists did not have time and patience enough to read the whole report, running to 253 pages, a common opinion was given that the Saami claims were altogether groundless. However, the matter is more complicated than the decision may seem to imply.

First, it should be emphasized that the judgement does not definitely solve the problems concerning Saami rights in other parts of Sweden. The Court underlined that it was not possible to form an opinion regarding the legal status of the northern mountains, which were not a subject of the dispute in question. However, in the discussion of the material presented by the parties, the Court made some statements concerning the earlier rights of the Saami that have considerable interest in this context. Thus the Court found it necessary to examine the rights that the Saami would have had to the taxed mountains irrespective of the RFA, which implied an investigation of the historical background not only in Jämtland but also in the rest of the Swedish mountain areas. As a link in this analysis, the Court declared that it was possible, at least during the seventeenth century, to acquire land by using it for reindeer grazing, hunting, and fishing, without cultivation of the land or even permanent residence in the area. In doing so, the Court disclaimed the common supposition that "nomads cannot acquire ownership rights." The statement has no counterpart in previous Scandinavian precedents and should be of great significance for future standpoints on the rights of the Saami in Sweden (and in Finland, too), provided that legislators and courts will pay due attention to the position taken by the Court; as will appear from the following, most jurists are apt to disregard the statement, probably in view of the possible implications for the ownership of the state.

However, this pronouncement by the Court had no decisive influence upon the outcome of the case. According to the court, the requisites for this type of land acquisition by the Saami would be in cases where they were not permanently domiciled in an area where their use of the land had to be intensive, long standing, and basically undisturbed by outsiders; further, somewhat fixed boundaries for the area in use should be required. None of these prerequisites were regarded to have existed in Jämtland at the critical time—that is, in the middle of the 17th century. According to the Court, it was at this time that the Saami had the chance to be considered owners of the mountains, since the state eventually caused the unowned land in northern Sweden to come under state ownership through later legislation (the forest regulations of 1683).

It appears from the decision that the possibilities for the Saami to acquire ownership rights might have been better further north in Sweden, where the use of mountain land was more intense and undisturbed, and where there were also forms of village organizations which could be considered as owners of the land. In fact, the only valuable evidence pointing to Saami ownership concerned the northernmost parts of Sweden and the present Finnish Lapland. The Court had no reason to deal more thoroughly with the Saami rights in these areas, as the litigation did not apply to them. But the decision did state that the state would not have been able to refer to the regulations just mentioned with regard to land owned by the Saami at that time, as the regulations only applied to unowned land.

As for the limited rights that the Saami claimed in the second place, it should be noticed that they demanded that the Court should establish that these rights were still in existence, irrespective of the legislation; the limitations prescribed by the RFA should, in consequence, be deemed invalid. Such a claim would only have been approved if the legislation had been considered clearly unconstitutional; this was denied by the Court. (Although one member dissented concerning the regulation of hunting and fishing rights, which was regarded as discriminatory insofar as the Saami lacked any influence upon the granting of such rights.) The Court had no reason to discuss whether the Saami could claim compensation because they had been deprived of certain rights to natural resources through the reindeer farming legislation. However, the Court (as a kind of obiter dictum) concluded that their right of use was constitutionally protected in the same way as ownership rights; this did not mean that it was protected against expropriation and similar measures, but that the rights could not be taken from the Saami without compensation being made for the loss, according to the Swedish Instrument of Government.

Later Development

As mentioned before, the Swedish authorities and jurists in general mostly regarded the judgement in the *Taxed Mountain* Case as a confirmation of the traditional view that the rights of the Saami did not amount to ownership rights in any part of Sweden, and that the reindeer herding right essentially was based upon the RFA; it was even argued that the rights were more limited in the northern mountains than in Jämtland, which is clearly contradicted by the findings of the Court.

However, the government commission that was appointed in 1982 to examine the legal status of the Saami (the Saami Rights Commission) was fully aware of implications of the judgement. In a report of 1989, the Commission proposed several amendments in the RFA in order to give the Saami more effective protection against such measures of the owners of reindeer herding lands as would encroach upon the reindeer farming. Among other things, the Commission proposed that certain forestry activities that would be detrimental to the reindeer grazing should depend on the permission of the County Forestry Board. Further, the Saami right was defined as a right sanctioned by immemorial use. While it is unnecessary to go into detail here, most proposals intended to strengthen the Saami position were opposed by numerous authorities and organizations commenting upon the report. The result of the discussion was the not too effective protection afforded by the Forestry Act, as earlier mentioned, and the amended RFA. As appears from the above, it implies rather modest changes in the legislation, although it is underlined that the government considers the protection of the reindeer herding rights as important.

In this connection, mention should also be made of the recent legislation concerning Saami hunting and fishing rights, which the Saami regard as a serious menace to their legal position. As mentioned before, the Saami have a right to hunt and fish on the whole-year areas, although they cannot dispose freely of the right. At the same time, the state has the hunting and fishing rights in the capacity of owner of the land. The relationship between the rights of the state and the rights of the Saami is not quite clear; the Saami claim, with some support in the findings of the Court in the *Taxed Mountain* Case, that they originally had the exclusive hunting and fishing right which has gradually been reduced by various legislation and acts of the authorities. However, according to the Game Act of 1987, the Saami shooting right is not on the same level as the rights of owners and tenants on the land, and the new legislation implies a considerable extension of the possibility for local authorities to grant hunting and fishing rights in the whole-year herding regions; the foremost aim is to satisfy the increasing need for these kinds of spare time occupations among tourists as well as local people. Of course, this state of law is incompatible with the idea that the Saami originally possessed an exclusive right to these natural resources.

The general attitude of the Swedish government to the Saami rights is further illustrated by the statements in the Saami bill regarding the ILO convention (no. 169) concerning indigenous and tribal peoples in independent countries. According to article 14 in the convention, "The rights of ownership and possession of the peoples concerned over lands which they traditionally occupy shall be recognized." In the Saami bill, the responsible minister stated that this article was obviously incompatible with the Swedish state of law and that it probably was founded on relations altogether different from those applicable to the Swedish Saami. Hence, it would not come into question to ratify the convention. The parliament, too, rejected proposals to ratify the convention, although in less definite terms.

Recent Historical Research

So far, the position of the Saami does not seem very promising from a juridical point of view. The efforts of the Saami Rights Commission to improve their legal status by creating a more efficient protection have only partly succeeded, and the detrimental effects of forestry and tourism upon the Saami activities are likely to increase. However, these negative traits may be partly compensated by the development in legal history, provided that proper attention is paid to the implications of the findings. Above all, the historical investigations of Kaisa Korpijaakko concerning the legal status of the Saami in the 17th and 18th centuries should reasonably have a considerable impact on the opinion among jurists and politicians. Moreover, in her doctoral thesis (Korpijaako-Labba 1989) she demonstrates in a convincing way that, as far as concerns the Lapp areas of Torne and Kemi in the northernmost parts of Sweden and Finland, the Saami were treated by courts and other authorities as owners of the land (or, more precisely, possessors of taxpayers' rights that later developed into ownership) at least until about 1740; this holds true of Saami villages as well as individual Saami. Further, it is shown that all the requirements for a Saami ownership specified by the Supreme Court in the *Taxed Mountains* Case were fulfilled in these parts: an intensive, long standing, and essentially undisturbed use by Saami villages in areas with comparatively fixed boundaries. The Court had pointed out the possibility that such areas may have existed in the North; now, Kaisa Korpijaakko has apparently proven their existence.

If these results are accepted (as is the case at least among Finnish legal historians) it might have far-reaching consequences for the Saami rights in northern Sweden. As mentioned before, the royal regulations of 1683 on which the claims of the state on the northern mountain districts are based only concerned land without owners; if the Saami possessed the mountains in the capacity of owners, their rights should not have been affected by the regulations. In that case, it is not clear how the state between 1740 and the first reindeer farming legislation in 1886 would have acquired ownership to these regions. In any event, the state cannot refer to any of the ordinary ways of acquiring good title to land originally owned by others. Still more surprising is that the hunting and fishing rights that the Saami must have exercised as owners now have been degraded to second-class rights that they are not even permitted to dispose of. Even if there are only small chances for the Saami villages to be declared rightful owners of the mountains today, the mere possibility of such ownership will strengthen their legal position to a considerable extent.

Kaisa Korpijaakko's thesis (originally in Finnish) was only translated to Swedish in 1994 (Korpijaakko-Labba 1994), which may partly explain why it has not made any particular impression upon the government and the parliament; however, it has been referred to in the legal discussion, and important parts of her findings were presented in shorter papers during the eighties. The Saami Rights Commission mentioned her investigations, and when the Law Council in 1990 examined an earlier draft to a Saami bill (containing essentially the same proposals as the bill of 1992) the Council called attention to the fact that according to recent historical research the ownership of the state had been called into question as far as concerned the northernmost parts of Sweden and Finland. However, the Law Council did not wish to express any doubt concerning the essential basis of the legislation proposed; if new lawsuits concerning the reindeer herding right should lead to other conclusions as to the nature of the right, it was assumed that the legislation—if necessary—would be reconsidered. In the bill of 1992, the statement of the Law Council was shortly mentioned; the reference to Kaisa Korpijaakko's research was passed over in silence. According to the minister, the statement did not give cause for any particular comments on his part. As mentioned before, the government as well as the parliament have not considered it possible to adopt the ILO Convention guidelines concerning indigenous and tribal peoples. Apparently, the possibility of Saami ownership to certain mountain regions also was not seriously considered in this context. One explanation for this might be that the mere idea of such a right was too disturbing from a political standpoint in view of the legislation proposed.

Conclusion

It appears from the above that the legal position of the Saami varies according to the point of view from which it is discussed. The government (independently of political color) prefers to leave Saami ownership out of account. The existence of a rather strong usufructuary right based upon immemorial usage is acknowledged, but concerning the protection of that right, the interests of forestry, other industry, and tourism often outweigh the Saami interests. The same, of course, is true of the opinion among forest companies and other property owners in the North, as well as among the local population on the whole. The general idea seems to be that the judgement of the *Taxed Mountains* Case has finally settled the question of Saami rights in all the mountain regions of Sweden.

On the other hand, one need not be a Saami, nor even particularly partial to the Saami, to feel a strong doubt concerning this somewhat lighthearted attitude to this complex legal problem. Kaisa Korpijaakko's research continues and may add some important facts to what we already know about the historical background. It would be an exaggeration to speak of the Saami having a strong case insofar as concerns the ownership question. However, the arguments that they can adduce seem to have sufficient weight to give the state a lot of worry, should the question of ownership of the northern mountains be brought before a court. As for the limited rights included in the reindeer herding right, the new legislation hardly does full justice to the Saami standpoint. The possibility that the rights of the Saami are far stronger than the legislation has assumed should call for some caution on the part of the government and other owners of the land in question. However, the solutions given by the amended RFA are probably not so manifestly unconstitutional that the rules can be put aside by a court.

REFERENCES

Bengtsson, Bertil. 1987. *Samernas egendomsskydd*, 9–40. Stockholm: Statsmakten och äganderätten.
Korpijaakko-Labba, Kaisa. 1989. *Samelaisten oikeusasemasta Ruotsi-Suomessa.* Helsinki.
Korpijaakko-Labba, Kaisa. 1994. *Om samarnas rättsliga ställning i Sverige-Finland.* Helsinki: Juristförbundets Förlag.
Jahreskog, Birgitta, ed. 1982. *The Saami National Minority in Sweden.* Stockholm: Rättsfonden (The Legal Rights Foundation).
Proposition 1971:51. Rennäringslag.
Proposition 1992/93:32. Samerna och Samisk kultur m.m.
SOU 1986:36. Samernas folkrättsliga status.
SOU 1989:41. Samerätt och sameting.
Taxed Mountains Case. 1981. *North Frostviken Saami Village and Others v. the State*; "skattefjallsmålet." Reported in *Nytt juridiskt arkiv*, 1.

Heikki J. Hyvärinen   Chapter 12

The Legal Status of Rights to Resources in Finnish Lapland

Introduction

The right of indigenous peoples to land, water, and natural resources is a topical and difficult issue throughout the world. This issue is also current in Finnish Lapland. Our indigenous people are the Saami (formerly called Lapps). They number seven thousand among a total Finnish population of five million; their language, culture, and traditional livelihoods distinguish them from the population at large. The term "Saami" comes from the Saami language and was adopted in our legislation with the Decree on Saami Parliament (1973). There are three different groups of Saami in Finland: Northern Saami, Inari Saami, and Skolt Saami. A Saami is defined in the 1991 Act for the utilization of the Saami language before authorities as a person who considers him- or herself a Saami and who has learned Saami as his or her first language or who has a parent or grandparent who learned Saami as his or her first language. According to the new Act on Saami Parliament of 1995, the Saami have their own advisory body—the Saami parliament elected by and among the Saami, whose function is to protect the rights and interests of the Saami people.

Most of the Finnish Saami inhabit and use the northernmost part of Finland, which has been referred to since 1973 as the Saami homeland in the legislation named above. This area encompasses Finland's three northernmost municipalities—Enontekiö, Inari, and Utsjoki—and part of Sodankylä municipality. It is 35,000 square kilometers in size and represents 10 percent of the total surface area of the country. The Saami homeland has a total of twelve thousand inhabitants, of whom four thousand are Saami.

The traditional Saami livelihoods include reindeer herding, fishing, and hunting. Present legislation in Finland does not grant land title to Saami engaged in these livelihoods. At various times in history homesteads were established in the area for farming and cattle raising. Owners of these were eventually granted title to the land they occupied. Both Finns and a part of Saami own homesteads. In the Saami homeland, homesteads account for some 10 percent of the total land area. The remaining 90 percent is land which the state regards as its property.

Earlier, the position of the Saami in the whole area was totally different. Their traditional land use gave them the status of masters on their own lands with all the attendant rights and responsibilities (see chapter 8).

The Saami—unlike many other indigenous peoples—had the status of full citizens hundreds of years ago in Sweden-Finland. Their rights and responsibilities were spelled out in laws and statutes. It was at that time that the principle existed that all land must have an owner. If land had no owner, it was considered as belonging to the state (terra nullius). In the 1900s Finnish legislation rested on the notion that nomads cannot acquire ownership rights. This reasoning deemed the lands which the Saami had used as their own for centuries as "ownerless," meaning that ownership belongs to the state.

The legal basis for state title to the land (terra nullius) in the Saami homeland was shaken substantially back in 1981 in a land title dispute between the state of Sweden and the Saami. In deciding the case, the Swedish Supreme Court accepted the premise that according to the law of Sweden-Finland a nomad could have acquired title to land without engaging in farming or having a permanent dwelling (see chapter 11). Later, in a doctoral dissertation published in 1989, Kaisa Korpijaakko-Labba proved that the Saami as nomads, fishermen, and hunters enjoyed ownership of the land recognized by state officials in the northernmost parts of Finland and Sweden (see chapter 8). The scientific community has not challenged this finding (Bengtsson 1991; Hyvönen 1991; Klami 1990; and Ylikangas 1989).

From the point of view of international law, the land rights of the Saami were brought up before the Finnish parliament in 1990 when it was considering the Governments Bill to Parliament (no. 306/1990) containing a proposal not to ratify the ILO Convention no. 169 on indigenous and tribal peoples. Finland could not become a party to the convention at that time because our legislation did not conform to the provisions of the agreement concerning Saami land rights. In fact, the government of Finland stipulated that the agreement could only be ratified if Finland would better recognize the rights of the Saami to the land they traditionally occupy and own and to the use of the natural resources on these lands.

The Legislation Concerning Land and Water in Lapland

The material foundation of Saami culture consists of the lands and waters of the Saami homeland and the right to use these areas in pursuing the traditional livelihoods of reindeer herding, fishing, and hunting. The Saami do not, however, enjoy any special rights to these areas or livelihoods in accordance with our modern legislation.

The Laws of the State's Right to Land and Water

In the Saami homeland 90 percent of areas are lands which the state regards as its property. According to act on forests of 1886, woodlands not belonging to village communities and other areas outside boundaries to which no one can prove a better right belong to the crown as state woodlands.

According to the act on the right to sell real property and income-producing rights of the state of 1978, state authorities are allowed to sell and let on lease land and water of the state under certain conditions. The main part of Saami homeland belongs to the state. The basis for the administration of state woodlands is the 1993 Act on National Board of Forestry. According to this act the national board manages, uses, and protects state forests. It also handles official duties—assigned to it in special acts—related to fishing, primary livelihoods, hunting, off-road traffic, and the Skolt (Saami) population. Moreover, legislation has been enacted to safeguard the interests of the local populace for provincial and regional advisory councils which assist the National Board of Forestry when deciding on land and water areas. Section 11 of the National Board of Forestry Act—under the heading "Safeguarding Saami Culture"—specifies that the management, use, and protection of the natural resources under the control of the National Board of Forestry in the Saami homeland must be harmonized so that the traditional livelihoods of the Saami and the requirements of the culture are safeguarded. However, according to this legislation the Saami parliament does not have the right to decide who may monitor the "safeguarding of Saami culture."

According to the Nature Protection Act of 1996 and the Wilderness Act of 1990, nature reserves and wildernesses can be established mainly in areas owned by the state. Most parts of Saami homeland are nowadays by the law either nature reserves or wilderness areas. However, in dealing with the government's bill of the Wilderness the parliamentary committee on constitution was not sure whether the ownership of these areas in the Saami homeland belongs to the state or to the Saami (Constitutional Committee of Parliament 1990).

Rights to the Resources in the Underground

According to the 1965 Mining Act, the rights to ore and minerals do not belong to the owner of the land but rather to the person establishing a claim on it. The landowner nevertheless has the right to take part in mining and to receive compensation for mining activity. Mining operations cannot begin until the mining area has been finally demarcated. The committee on the elimination of racial discrimination considered Finland's periodical reports in 1996 and then adopted its concluding observations. As principal subjects of concern, the committee said: "As regards the land rights of the Saami people, concern is expressed over the mining and other economic interests of national and international companies which may be threatening the way of life of Saamis" (CERD/S/SR 1154).

The Right to Reindeer Herding

The right to own and breed reindeers in Finland is regulated in the Reindeer Husbandry Act of 1990. The law permits reindeer herding—with some exceptions—through the province of Lapland and also in parts of the province of Oulu (*poronhoitoalue*). Within this area reindeer herding is allowed independently by landownership. Any Finnish citizen living permanently in this area could own reindeer before 1993. The allowed number of reindeer in Finland is about 230,000.

About 40 percent of the reindeer are living in the Saami homeland and the Saami own 85 percent of these reindeer. Although reindeer husbandry is of great importance to the economy and the culture of Saami people, they do not—according to the law—have any exclusive rights to own or breed the reindeer.

The Right to Fishing

When the Fishing Act of 1982 was passing the parliament in 1981, the three northernmost municipalities—almost the whole Saami homeland—were left out of the new law. The law would not be valid in these municipalities because the fishing rights of the Saami could not without difficulties be regulated in this act. From 1978, the constitutional committee of parliament had stated in its reports that the Saami have had and still have a right to fishing which is protected in accordance with the Finnish constitution (Constitutional Committee of Parliament 1981). For this reason the fishing in three Saami communities was regulated in the elder Fishing Act of 1951, which does not protect the Saami's right to fish, either.

The main rule in Finland is that the right to fish belongs to the owner of a water area, who usually is the landowner of the shore. In accordance to the elder fishing legislation of 1951, the right to fish in state-owned waters in the three northernmost municipalities belongs to the state. Fishing is further allowed to all residents of these municipalities.

The Right to Collect Natural Products

The right to collect berries and mushrooms is usually free. In municipalities of the province of Lapland the Ministry of Agriculture and Forestry can forbid nonresidents of the municipality to collect natural products (usually cloud berries). This right is regulated in the 1955 Act on Certain Restrictions Concerning the Collection of Natural Products.

The Recent Development

In practice, the application of ILO Convention no. 169 and the safeguarding of Saami rights in new internal laws have varied.

A Saami Bill

The land and water rights of the Saami can be established in two ways. First, a court can confirm the rights of Saami on state land where a dispute arises between Saami and the state. Second, these rights can be substantiated through laws enacted by parliament. The question of old Saami title to state land in Finland has yet to be decided in court. On the other hand, there have been numerous attempts to settle the issue by legislative means. In 1952 and 1973 state committees proposed bills which would have guaranteed Saami rights to land (State Committee report 1952, 12; 1973, 46). However, the Finnish government did not bring either proposal before parliament for consideration and both lapsed. In 1990, a permanent state committee—the Advisory Board for Saami Affairs—drafted a legislative proposal entitled "A Saami Bill" (State Committee report 1990, 32).

According to studies done by the advisory board, state officials at one time had recognized in established practice the ownership rights of the Saami to their lands for the purpose of reindeer herding, fishing, and hunting. The Saami are still using these same areas for the same purposes but the land is called state land. The title of the Saami to this land has never been legally terminated, and it should still be in effect today. By contrast, no adequate legal basis for state title to these lands has ever been produced. For this reason, the advisory board considered that the present status of the state with respect to state lands violated the Saami's legal protection of property. Moreover, this situation amounts to a structural barrier causing inequality among different groups of citizens: the Saami are in an inferior position with respect to other citizens because of their special means of livelihood. Finally, the present position of the Saami conflicts with the provisions of international agreements.

To rectify the situation, the advisory board proposed that the rights of the Saami to land, water, and the traditional livelihoods should be safeguarded by enactment of a special Saami law. The law would not give the Saami new rights; it would restore their previous ones. The legislation would also promote the development of the Saami language and culture, improve social conditions, as well as foster sustained growth in the area. According to the committee's proposal, these provisions would neither encroach upon anyone's property nor affect the practice of any established livelihood.

The constitutional committee of the parliament mentioned in its 1990 report (no. 3) the advisory board's proposal to a new Saami bill. The committee considered it wholly plausible that the Saami have title to land. The committee also urged that a government Saami bill be brought before parliament as soon as possible. There would then be legislation on the books to rectify the currently unclear situation with regard to Saami title to state land and the Saami livelihoods like reindeer husbandry, which are an integral part of Saami culture.

The government has not been in any hurry with a Saami bill. In the spring of 1993 the ministry of interior decided to enact further drafting work on a Saami bill to Saami parliament, although no funds were appropriated for the purpose. The drafting work goes very slowly. Despite the unresolved issue between the state and the Saami people of the right to land ownership, the Finnish state is selling land, logging forests, building roads, and carrying on the tourist trade in the Saami homeland without the consent of the Saami people. The Saami receive no compensation for these activities.

Legislation Concerning Reindeer, Fishing, and Hunting

In conjunction with the Treaty on the European Economic Area, section 4 of the 1993 Reindeer Herding Act was changed, permitting citizens of EEA countries who permanently reside in Finland to own reindeer on the same terms as Finnish citizens can. When the bill came up before parliament, the Saami parliament proposed an alternative whereby the right to own reindeer in Saami homeland would be specified in the act as an exclusive Saami right. The constitutional committee of the parliament regarded in its report this alternative proposal as confirming to the EEA Treaty, but the decision was not within this committee's jurisdiction (Constitutional Committee of Parliament 1993).

In dealing with this bill, the committee on agriculture and forestry of the parliament referred to an earlier report of the constitutional committee of parliament (no. 3, 1993) emphasizing the importance of reindeer herding to Saami culture. The committee on agriculture and forestry also urged speedy drafting of a Saami bill. When a government's Saami bill comes up before parliament, the Reindeer

Herding Act can be changed, if necessary (Committee on Agriculture and Forestry of Parliament 1993).

The government's proposal (1993) to parliament on a new fishing act would have abolished the old Fishing Act of 1951 in the municipalities of Enontekiö, Utsjoki, and Inari. In its statement the parliamentary committee of constitution maintained that no legislation should be passed which would change the 1951 Fishing Act to the detriment of Saami rights. The committee cited ILO Convention no. 169 in support of this position (Constitutional Committee of Parliament 1993). The new Fishing Act was passed in 1993 in the form proposed by this committee. In 1997 the Fishing Act of 1982 was extended to apply to the municipalities of Enontekiö, Inari, and Utsjoki. This act does not in any way protect the fishing right of the Saami; it guarantees that right to any permanent resident of any of these municipalities.

On the other hand, the Saami were not heard and their rights to hunting were not mentioned when the government's proposed hunting act was being drafted in 1993. It was the view of the parliamentary committee on agriculture and forestry that the provisions of the bill did not change the hunting procedures in use on state land in a way which would significantly affect legal protection of ownership among the Saami or their right to practice their culture. On the confusion surrounding Saami land ownership and hunting rights, the committee on agriculture and forestry referred to previous statements by the parliamentary committee of constitution (Committee on Agriculture and Forestry of Parliament 1993). When the Hunting Act was passed in 1993 by parliament, the provisions in it pertaining to the Saami were those originally proposed by the government—with no special rights accorded to the Saami. In addition, parliament passed the new Hunting Act prohibiting hunting in nature reserves (Reply of Parliament to Government's Bill 1992). This provision, which applies to hunting by Saami as well, contradicts—in my opinion—article 23 of the ILO Convention no. 169, which guarantees indigenous people a special right to hunt.

The New Basis of the Rights of the Saami

Finland joined the European Union at the beginning of 1995. Part of Finland's treaty of accession is protocol no. 3 which applies to the Saami people. In the protocol the parties recognize the obligations and commitments with regard to the Saami people under national and international law. It also notes the commitment by Sweden and Finland to preserve the means of livelihood, language, culture, and way of life of the Saami people. The protocol also takes account of the dependence of traditional Saami culture and livelihood on primary economic activities, including reindeer husbandry in the traditional areas of Saami settlement (The Treaty of Finland 1994).

The protocol is legally binding for the parties. Regardless of the EC Treaty, article 1 permits that Finnish legislation grant exclusive rights to reindeer husbandry to the Saami people in the traditional Saami areas. In accordance with article 2, the granting of exclusive rights to the Saami people is subject to the consent of the EU organs. Such a right is, for instance, the Saami people's right of ownership to land in the traditional areas.

In the same year of 1995 The Finnish Constitution act was amended by provision (section 14) according to which the Saami; an indigenous people, the Roma; and other groups have the right to maintain and develop their own language and culture. This provision authoritative recognizes the indigenous character of the Saami. Another amendment (section 51a) to the Constitution act recognizes the cultural autonomy—culture and language—of the Saami in their domicile area. (See Hannikainen 1996, 45.) The word "culture" in this connection includes even the traditional Saami livelihoods as reindeer herding, fishing, and hunting (see Government Bill 1993, 65).

In reform of section 14 of the Finnish Constitution act, ordinary laws were left untouched by any of those necessary or essential changes that the new Consitution acts would have required. It is intended that legislation is amended later in order to bring it to harmony with the Constitution act (Jyränki 1996, 749). At the same time, the Finnish parliament enacted a law on the Saami parliament, which specifies the content of the cultural autonomy by giving the Saami the right to consult with state organs in matters affecting their interests. The Saami parliament act of 1995, which entered into force at the beginning of 1996, broadens the concept of "Saami." Except for the lingual basis, the act makes it possible also for the descendants of persons marked in certain old registries as "Lapps" to register themselves as "Saami," even if they have no relationship to the Saami language. Finns whose forefathers were registered as Lapps several centuries ago traditionally are very much opposed to any special rights for the Saami (Hannikainen 1996, 49-50). Some of these Finns have organized into an anti-Saami group, which is not interested in the Saami language and culture. Instead they try to gain economic benefits and nullify the Saami cultural autonomy (Finnish Saami Parliament 1996, 51).

Evaluation and Conclusions

The rights to land, water, and resources in the Saami homeland are unclear. The ministry of agriculture and forestry in Finland wrote about "forests, indigenous people and local communities" in its report to UNEP in 1996 as follows: "The issue of Saami ancient land rights, including title to land, is so far still unsolved in the legislation as is their land rights as an indigenous people...The rights of Saami to lands, waters and traditional livelihoods—reindeer herding, fishing and hunting—have not been recognized nor put into effect in the Saami region. In this juridically unclear situation the material foundation of the Saami culture rests on an uncertain basis. Thus, Saami become estranged from their ancestors' lands and waters and from their use in the ways specific to Saami culture" (Finland's report to UNEP 1996).

Each year the Finnish government provides parliament with a report of its activities. Similarily, Saami parliament draws up an annual report of the activities of the Finnish government, enumerating measures which are of specific relevance to the Saami. In its annual report to the parliament the Finnish government included the report from the Saami parliament unaltered. The government report states on the Saami rights as follows:

> In 1996, the rights of the Saami people to practice the livelihoods that are part of their culture had not yet been harmonized by law with their fundamental rights. . . Reindeer husbandry was integrated into EU agriculture. . . In the preparation of these legislative, administrative and financial changes, Saami parliament was not heard, not consulted and not informed of them. . . It is the opinion of Saami parliament that integration of reindeer husbandry has been carried out without taking into account of the importance of reindeer husbandry as the foundation of Saami culture. Saami parliament also believes that efforts should be made to reach the goals of fundamental rights for the Saami people, of cultural autonomy and of EU protocol no. 3 also as regards those fundamental rights for the Saami people which as yet are not expressed in the law (Government's Report to Parliament 1996, 354 and 356-357). The committee on the elimination of racial discrimination suggested in 1996 "that the Finnish government draft and implement a clear policy on Saami land rights in order to better protect and preserve the way of life of this minority group. The committee also recommends that the government ratify ILO Convention no. 169" (CERD/S/SR. 1154).

To date, the government has not submitted to parliament any bill to secure the rights of the Saami people to land or livelihoods. For that reason, the Saami's rights remain unclear. For the same reason Finland cannot ratify the ILO Convention no. 169.

REFERENCES

Bengtsson, Bertil. 1991. "Skattefjällsmålet och dess efterverkningar." In *Samesymposium*, edited by Marjut Aikio and Kaisa Korpijaakko. Lapin yliopiston Hallintoviraston Julkaisuja 15, Rovaniemi 1991, 34-46.
Committee on Agriculture and Forestry of Parliament. 1993. Maa- ja metsätalousvaliokunnan mietintö, no. 7/1993 vp. Statement no. 7 on revision of Hunting Act.
Committee on Agriculture and Forestry of Parliament. 1993. Maa- ja metsätalousvaliokunnan mietintö, no. 8/1993 vp. Statement no. 8 on amendment to Reindeer Husbandry Act on rights of EU citizens to take up reindeer husbandry.
Committee on the Elimination of Racial Discrimination. 1996. "Concluding observations of the Committee on the Elimination of Racial Discrimination, Finland." Forty-eigth session 26 February-15 March 1996 (CERD/S/SR 1154).
Constitutional Committee of Parliament. 1981. Eduskunnan perustuslakivaliokunnan lausunto, no. 5/1981 vp. Statement no. 5 on revisions of the Fishing Act.
Constitutional Committee of Parliament. 1990. Eduskunnan perustuslakivaliokunnan lausunto, no. 3/1990 vp. Statement no. 3 on Reindeer Husbandry bill.
Constitutional Committee of Parliament. 1990. Eduskunnan perustuslakivaliokunnan lausunto, no. 6/1990 vp. Statement no. 6 on Wilderness Areas bill.
Constitutional Committee of Parliament. 1993. Eduskunnan perustuslakivaliokunnan lausunto, no. 8/1993 vp. Statement no. 8 on amendment to Reindeer Husbandry Act on rights of EU citizens to take up reindeer husbandry.
Constitutional Committee of Parliament. 1993. Eduskunnan perustuslakivaliokunnan lausunto, no. 30/1993 vp. Statement no. 30 on revision to the Fishing Act.
Finland's report to UNEP. 1996. "Finland: Forests, Indigenous people and Local communities" by Ministry of Agriculture and Forestry, Department of Forestry to Secretariat of the Convention on Biological Diversity, United Nations Environment Programme, 6 March 1996.
Finnish Saami Parliament. 1996. "Land Rights, Linguistic Rights and Cultural Autonomy for the Finnish Saami People," by The Finnish Saami Parliament. *IWGIA Indigenous Affairs* no. 3/4 July-December 1997, 48-51.
Government bill 1993. Hallituksen esitys Eduskunnalle perustustuslakien perusoikeussäännösten muuttamiseksi no. 309/1993. Bill no. 309 on revision of Fundamental Rights of the Constitution.
Government's report to Parliament 1996. Kertomus hallituksen toimenpiteista vuonna 1996. Annettu Eduskunnalle 1997 vuoden valtiopäivillä no. K 4/1997 vp. Government's report no. 4 of its activities in 1996 to Parliament.
Hannikainen, Lauri, 1996. "The Status of Minorities, Indigenous Peoples and Immigrant and Refugee Groups in Four Nordic States." *Nordic Journal of International Law* 65, 1-71.

Hyvönen, Veikko O. 1991. "Jaollisesta omistusoikeudesta oikeusjärjestyksessämme." *Oikeustiede/Jurisprudentia* 24, 171-187.
Jyränki, Antero. 1996. "Perusoikeuksien historiasta Suomessa." *Lakimies* 5-6, 739-752.
Klami, Hannu Tapani. 1990. "Käsitteet ja historiantutkimus." *Historiallinen aikakauskirja* 2/1990, 132-135.
Reply of Parliament to Government's bill of 1992. Eduskunnan vastaus hallituksen esitykseen no. 300/1992 metsästyslain muuttamiseksi.
Reply 2 June 1993 on revision of Hunting Act.
State committee report. 1952. 12 "Saamelaisasianin komitean mietintö". Helsinki.
State committee report. 1973. 46 "Saamelaiskomitean mietintö". Helsinki.
State committee report. 1990. 32 "Saamelaisasiain neuvottelukunnan mietintö I. Ehdotus saamelaislaiksi ja erinäisten lakien muuttamiseksi." Helsinki.
The Treaty of Finland on accession to the European Union. 1994. Signed in Korfu 24 June 1994.
Ylikangas, Heikki. 1989. "Kirjallisuutta: Korpijaakko Kaisa: Saamelaisten oikeusasemasta Ruotsi-Suomessa." *Lakimies* 8/1989, 1163-1169.

TORGEIR AUSTENÅ	CHAPTER 13

The Proposal of the Norwegian Government Commission on the Rights of the Saami to Land and Water in Finnmark

The Saami Rights Commission handed over its proposal to the minister of justice in late January, 1997 (NOU 1997). The Commission proposed to enact a law concerning the ownership and administration of land and water in Finnmark. The basis for the proposal is the current legal situation (NOU 1993) and the historical development of Finnmark and international law concerning indigenous people. The main goal for the Commission was to propose rules and regulations to secure the culture of the Saami people inclusive of the material basis for that culture. Another goal was to propose an act which would not give the inhabitants of Finnmark preferential rights to land and water based on ethnic origin. Only 4 percent of the surface of the county has been transferred to private ownership. The proposal from the Commission concerns the rest of the county.

The title to dispose of land and water in Finnmark as an owner was conveyed to a company called Statsskog SF in 1993. This society is owned by the Norwegian state. The right to exercise ownership today is delegated to the local office for administration and selling land in Finnmark (see chapter 10). The Saami Rights Commission proposed to convey the title to dispose of land and water in Finnmark to a new corporate body called *Finnmark Grunnforvaltning* (Finnmark Ground Management). The majority of the Saami Rights Commission further proposed that this body should be organized as a legal entity independent from the rest of the state administration. The minority of the Commission proposed that the new body should be a part of the state administration, but with an independent position.

According to the proposal from the majority of the Commission, the Finnmark Ground Management Organization would have an advisory board consisting of eight persons. Four members would be appointed by the *sameting* (the Saami parliament) and the other four by the chief administrative body of the county. The minority of the Commission proposed that five of the board members be appointed by the County and three by the Saami parliament. Five members of the Saami Rights Commission proposed that there should be a possibility for people in a municipality or *bygd*[1] to decide by a majority vote that Finnmark Ground Management be replaced by Saami Ground Management. The difference between these two bodies is mainly the composition of the board. The board of Saami Ground Management would have seven members, five of which would be appointed by the Saami parliament. The intention behind this proposal is to give the Saami people a stronger impact on disposition over land and water in Finnmark.

The Finnmark Ground Management Organization would have a position as a full owner and the authority to decide over land and water, minerals, sand, gravel, drinking water, and the production of hydroelectric power. The intent is that the Ground Management Organization use its power to the benefit of all inhabitants of Finnmark. The body should not act in a "businesslike" way. Any company's commercial interests and even strong interest of the Norwegian state in exploitation of land and water resources should be set aside. Instead, the long-term interests of the people in Finnmark should have priority. In the evaluation of different interests, the interests of the Saami people should be taken specially into consideration.

The Saami Rights Commission proposes that the right to and disposition of the resources in the outfields should be transferred to the municipalities.[2] In this context, "resources" mean fishing salmon with some types of fishing tackles; fishing salmon in the so-called small rivers (rivers excepting the Alta, Tana, and Neiden); general fishing; hunting; picking cloud berries; grazing land for domestic animals; as well as gathering winter food, firewood for household use, peat, and eggs and down from birds of special kinds. The inhabitants in the municipality have equal right to take part in harvesting from these resources with the exception of firewood and peat. These two resources are reserved for people outside densely populated areas. The outfield resources as mentioned above would be administered by a special outfield body in every municipality. This body would be given a relatively broad competence in its administration of the resources.

The right to and the disposal of the outfield resources from a part of the area within a municipality would be given to a local community (a *bygd*). Such an outfield of a local community is called a *bygdebruksområde* (area to be used by the *bygd*). It would be up to people with permanent addresses in the community to decide whether or not to establish a *bygdebruksområde* and take over the right to and administration of the outfield resources in their area. The decision must be supported by the majority of people of voting age in this community. The majority of the Saami Rights Commission would give the people in a *bygd* an absolute right to decide on establishing a *bygdebruksområde*. The minority of the Commission (four members) would propose a right for the majority in the area to demand establishing a *bygdebruksområde*, but the municipality could turn down such a proposal. The rights of the reindeer herding Saami to use the land and water in Finnmark are unaffected by the proposals for the new organization of the administration of the disposition and use of the resources in the county.

It could be of a certain interest to compare the proposed model for governing land and water in Finnmark with the governing of state commons elsewhere in Norway. In state commons the company Statsskog SF holds the title to ground and the remainder in trust for the state. This company is, as mentioned above, the title holder of Finnmark today. Statsskog SF is a state company. The Finnmark Ground Management Organization would be structured as an autonomous legal entity. The board of Statsskog SF is appointed by the ministry of agriculture. The board of Finnmark Ground Management would be elected partly by the Saami parliament and partly by the chief administrative body of the county. In a state common the rights of commons are held by all legitimate farms in the *bygd*. Rights of commons to timber and firewood are regulated by an act of June 19, 1992 (no. 60). A board elected by commoners and the local chapter of Statsskog SF co-manage the wood resource. Resources other than wood are managed by a board elected by the municipality.

As mentioned earlier, a central goal for the Saami Rights Commission was to secure the Saami culture. This includes the material basis for the culture; land, water, and other natural resources are a part of this basis. To secure the possibility to use all these resources as a part of the Saami culture, the Commission proposed rules and regulations hampering encroachments harmful to the basis of the Saami culture. The Commission has three proposals on this matter. Firstly, the Commission unanimously proposed as a general rule that in all forms for considering applications for allowance of encroachments, securing the Saami culture should be considered.

The majority of the Commission (fourteen members) also proposed a special rule concerning "considerable encroachments." In such cases the Saami parliament would have the opportunity to give its statement on the application. Thirdly, nine members of the Commission proposed a postponing veto for the Saami parliament. A veto would postpone the encumbrance for at least six years. A special minority of the Saami people is living in the east of Finnmark in an area called Neiden. For this special group the Saami Rights Commission proposed special rules and regulations concerning the right to reindeer herding as well as fishing salmon in the river Neiden and the fjords in the area. This is the only proposal from the Commission giving rights to natural resources on an ethnic basis.

The proposals from the Saami Rights Commission will be sent by the Department of Justice to counties, municipalities, and organizations of different kinds asking for their opinions. Thereafter the Department of Justice will redraft the proposed act in their proposals to the Norwegian parliament. In the Norwegian parliament a committee will discuss the proposal, and possibly amendments will be proposed before the committee's proposal goes back to parliament in its usual two-chamber procedure for enacting law. A new act on the resources in Finnmark can not be expected for several years.

NOTES

1. In relation to rights of common, *bygd* is a key word which doesn't translate too well to English. Its original meaning is something like "local community." Also, current usage of the word would suggest some kind of local community independent of more formally defined units such as school districts, parishes, or municipalities. Earlier in our history the *bygd* would be used for the smallest administrative unit, the local law district, and later the parish. In Sweden the word would mean the same. But in conjunction with commons, this translation will not give the right associations. Because the areas burdened with rights of common throughout our history were usually tied to users from some specific local community, the *bygd* became tied to a certain area recognized as "their" commons. During the past eight hundred years the original usage of the word *bygd* has turned around in the legal language, and today the *bygd*, in relation to commons, is defined as those who have rights of common. The *bygd* is defined as comprising those farm enterprises which have rights of common in the area recognized in law as a "commons" (both state and *bygd* commons). Since translation to English in this case is seen as inadequate, the word *bygd* will be used.

2. The resources are listed in the proposed act and can be described as resources to which the existence of rights of common could be argued for.

REFERENCES

NOU 1997:4. "Naturgrunnlaget for samisk kultur" ("The natural resources foundation of the Saami Culture"). Oslo: Statens Forvaltningstjeneste.
NOU 1993:34. "Rett til og forvaltning av land og vann i Finnmark i historisk perspektiv. Bakgrunnsmateriale for samerettsutvalget" ("Rights to usage and management of land and water in Finnmark in historical perspective". Background materials for the Saami Rights Commission.) Oslo: Statens Forvaltningstjeneste.

PART FOUR

The Fisheries of the Barents Sea

Introduction

Outside the near shores of most coastal states the fish used to be a resource with access for all. It was a divisible, non-excludable resource. But technological change made it possible for humans to fish more than the rate of renewal allowed. In several places around the world, the over-fishing has made the total catch so small that only the existence of large subsidies to the fishers has made it possible to continue. The subsidies have further exacerbated the situation. If fish species have not been brought to the brink of extinction, they have at least been fished so low that they have lost most of their value for human societies. They no longer contribute to feeding the growing human population.

However, during this development states have acted to improve on the international management institutions so that they can be used to regulate the catch effort. The most notable result is the two-hundred-mile extended economic zone where each coastal state claims management responsibility. It is the result of this effort, as manifested in the Barents Sea, we are turning to here.

There is an intimate relationship between what happens in the Barents Sea and along the coast of northern Norway. The interdependence of coastal fisheries and the pelagic species is perhaps stronger here than elsewhere. Even if there is evidence of local species of cod in the fjords, the big seasonal fisheries are based on the pelagic species, and the prosperity of the coastal population depends on these seasonal fisheries. This means that one needs to look at both the international and the national management of the fish to understand the pressures forming the policy and the consequences following.

In the first chapter in this section, "The Legal Status of Rights to Resources in the Barents Sea," Geir Ulfstein gives an overview of the situation in the Barents Sea. There are problems in the two-hundred-mile zone around Svalbard. There are problems in the pockets of high seas outside the two-hundred-mile zones of Svalbard, Norway, and Russia. And there are problems related to the management of whale and other sea mammals. There are also considerations originating in problems other than resource management which affect the management regime. Most notable are security considerations and environmental pollution. The problem most in need of action seems to be the problem of verification of agreed upon management goals and the integration of national management objectives with international management decisions.

In chapter 15, "Managing the Barents Sea Fisheries: Impacts at National and International Levels," Alf Håkon Hoel goes into the management of the Barents Sea in more detail. In the Barents Sea, Russia and Norway as coastal states are obliged to manage the fish stocks and their exploitation by their own and foreign fishermen. The main agreements were entered into in the period 1974–1977, ready to be used as the extended economic zone was established. But their interpretation and application at the operational level took time. Even if the total allowable catch was set according to best available advice, over-fishing was extensive and resulted in an all-time low of the fish stocks in 1990. Since then both the regulatory regime and the fish stocks have been recovering. With this recovery, pressure is increasing on the stocks from fishing in international waters.

Part of the problem of management was that fishermen misreported their catches, distorting the data used for management. Additionally, climatic variations and unknown aspects of the dynamic of a multispecies ecology may have contributed to the collapse of the fisheries. Thus it is difficult to evaluate the extent of failure of the fisheries regime. For the same reasons the later improvements cannot unequivocally be attributed to the management. An important part of the current regulatory regime is played by the research on fish stocks carried out by scientists in Norway and Russia, often in joint programs. As for oil fields, equitable sharing of the valuable resources below the surface of the sea needs reliable and trustworthy data. Within Norway the regulatory regime rests on a complicated negotiation procedure centered around the Regulatory Council with the Norwegian Fishermen's Association (NFA) as a major participant with six out of thirteen representatives. NFA's organization combines geography and the labor-capital relationship as organizational principles, thus incorporating conflicts between fishermen in various regions, between fishermen using different types of gear, and between the ocean-going fleet and the coastal fleet. The discussions within the association, as well as within the council, have considerable distributional consequences. Usually the members discuss until they reach an agreement; only in very important and contentious issues do they resort to voting. This is important for the legitimization of the fisheries policy.

The question of legitimization is at the core of chapter 16. In "Management under Scarcity: The Case of the Norwegian Cod Fisheries," Bjørn Sagdahl looks at the political processes generated by the problems encountered by Norwegian fishermen because of declining fish stocks. The government has been faced with confronting demands from different groups of fishermen. The political-administrative response has been a problematic political balance between equity and inequality. The solutions adopted have had severe impacts for the state of the fish resource in the past. The problem of establishing legitimacy to resource management seems to be prevalent. Sagdahl discusses the political limits for conducting a policy for sustainable development by the existing administrative institutions, and concludes that general models offering managerial solutions often neglect such facts.

Geir Ulfstein   Chapter 14

The Legal Status of Rights to Resources in the Barents Sea

Introduction

The Barents Sea has long supported one of the world's major commercial fisheries. In the mid and late 1970s there were peak catches of about 4 to 4.5 million metric tons a year, which was about 6 to 7 percent of the total world marine fish catch. The catch in recent years has, however, been in the order of 1 to 1.5 million metric tons a year, representing only about 1.5 to 2 percent of the world's total. The decline in catches reflects natural fluctuations in fish stocks, but it also reflects over-fishing. The challenge is to design a more effective management regime.

This article will focus on the rights to the fish resources in the Barents Sea at the international level, and the legal aspects of their management. It will be demonstrated that the fish stocks to a large extent are international—common—resources. The current management regime will be presented. Possibilities and problems involved in establishing a more effective management regime will also be discussed. Finally, it will be pointed out that fisheries management in this region is affected by some particular problems: disputed maritime areas, special management regimes for some living resources, and nonfisheries issues such as national security and pollution. These problems must also be taken into account when devising new management arrangements.

General Legal Framework

The general legal framework establishing rights to the living resources and management responsibilities in the Barents Sea is set out in the 1982 Law of the Sea Convention (LOS).[1] Although this Convention is not in force, its general principles on coastal states' rights regarding fisheries management are assumed to reflect customary international law, which is legally binding for all states.

In the 1960s and early 1970s, international fisheries management in the Barents Sea was undertaken on a regional basis under the Northeast Atlantic Fisheries Commission (NEAFC). Article 56 (1) of the LOS Convention establishes, however, the coastal states' sovereign rights for the purpose of "exploring and exploiting, conserving and managing" the living resources in the two-hundred-mile exclusive economic zone. The first point to note is thus that the management regime has changed from a regional regime to management vested in the coastal states (Churchill and Ulfstein 1992, 91–126). Accordingly, the two coastal states of the Barents Sea—Norway and Russia—have exclusive jurisdiction over the living resources in their two-hundred-mile zones. Since the LOS Convention merely establishes vague conservation duties (article 61), Norway and Russia are relatively free to decide which conservation measures to implement.

The most important fish stocks in the Barents Sea are shared between the two coastal states. Shared—or common—fish stocks necessitate cooperation between the owner states. Such cooperation is also an obligation under the LOS Convention article 63. This means that fisheries management in the Barents Sea is still an international problem, but it is now a bilateral rather than a regional problem. Norway and the USSR (now Russia) entered into a treaty on cooperation on fisheries in 1975,[2] and a treaty on fisheries management in the two-hundred-mile zones in 1976 (Churchill et al, 348).

Additionally, management is undertaken by the two coastal states, not by local entities such as the Norwegian county of Finnmark or by national fisheries organizations. States may have other aims than merely serving local entities or fishermen—for example, promoting oil activities on the continental shelf or protecting national security.

Yet another aspect of the coastal states' management is that the principle of balance has been essential in the adoption of management measures. It is obvious that balance is applied when exchanging quotas. But this principle is also relevant for the introduction of other conservation measures, because such measures may have effects upon the fishing opportunities of the two states. Russia will, for example, be more affected than Norway by an increase in the minimum size of fish since the Norwegian Arctic cod migrating in the Russian two-hundred-mile zone is younger than the cod migrating in the Norwegian two-hundred-mile zone. The need for balance may complicate the adoption of agreed conservation measures and thus make management less effective. There may also be a balance between the two states but an imbalance at the national level, in the sense that certain fishermen may be more obstructed by the measures than others. This demonstrates the need for an integration of fisheries policy at the international and national levels.

Lastly, no fishing rights for additional states in the two-hundred-mile zones in the Barents Sea can be founded upon the LOS Convention. Article 62 (2) provides that the coastal states shall allow other states access to "the surplus of the allowable catch" (that is, the balance between the total allowable catch and the coastal state's own catch capacity). The Convention also deals with who gets what in articles 62, 69, and 70. In the Barents Sea, due to the high catch capacity of Norway and Russia, there is, however, no surplus of fish stocks of commercial interest. The two coastal states are thus free to decide to what extent additional states shall be given access. As a consequence, other states' share of the catches of cod in the Barents Sea region dropped from approximately 20 percent in 1976 to around 10 percent in 1987.

An important challenge lies in establishing more effective management of the fish stocks—especially of shared stocks—in the Barents Sea. To what extent could management regimes applied for other international commons be introduced in the Barents Sea? A fundamental aspect of management regimes is their effectiveness and the principles applied in the management. Neither the LOS Convention nor the 1975 or 1976 bilateral treaties between Norway and Russia establish precise management goals. In order to assess the effectiveness of the management, the bilateral treaties should state more precisely which effects are intended and which principles are to be applied. Relevant principles known from environmental treaties are the precautionary principle[3] and the principle of cost-effectiveness.[4]

Another feature of a management regime is the decision-making procedure. Management measures in the Barents Sea are based on recommendations from the International Council for Exploration of the Sea (ICES). Such a scientific body has a great significance for the legitimacy of the measures adopted.[5] The fisheries agreements between Norway and Russia could formalize procedure that management should be based on recommendations from the ICES. Similarly, it could be determined that decisions made by the mixed Norwegian-Russian fisheries commissions should be binding, instead of mere recommendations. But it is not to be expected that such amendments would improve management in the Barents Sea. It could, however, be considered to integrate local interests—northern Norway, for example—in the decision making.

Verification has become an increasingly important aspect of international treaties. The LOS Convention article 73 provides that each of the coastal states enforces its own two-hundred-mile zones. Such coastal states as Norway and Russia have, however, no guarantee that effective enforcement is undertaken in fishing for shared stocks in the two-hundred-mile zone of the neighbor state. Verification procedures are primarily known in treaties on arms control and disarmament, but certain supervisory techniques are also known from environmental treaties (see Birnie and Boyle 1992, 166–8). Such techniques should also be considered in fisheries management in the Barents Sea.

The bilateral treaties between Norway and Russia do not contain any provisions on dispute settlement. So far, there have been no disputes about interpretation of the respective rights and duties under the two fisheries agreements, and it does not seem that dispute settlement procedures are essential in improving fisheries management in the Barents Sea (See on dispute settlement in environmental matters, Birnie and Boyle 1992, 177–86).

In concluding, it would seem that the most urgent subjects to be considered in improving management in the Barents Sea are to establish more precise management goals and principles, to integrate local interests in decision making, and to establish verification procedures.

Disputed Areas

Norway and Russia have not agreed upon the delimitation between their continental shelves and between their two-hundred-mile exclusive economic zones (Churchill and Ulfstein 1992, 54–91). The first contact between the parties took place as early as 1967, but so far they have not succeeded in reaching a solution. Pending a final delimitation, the so-called "Grey Zone Agreement" was entered into by the two states in January of 1978.[6] This agreement provides an interim arrangement for fisheries in the part of the disputed area lying within two hundred miles of the mainland. This agreement has worked well for the area it applies to. There is, however, no such arrangement for the disputed areas further north between the Norwegian Svalbard archipelago and the Russian Novaja Zemlja and Frans Josef Land. This creates problems regarding reporting of fishing activities and enforcement in these areas.

Svalbard's status was disputed for centuries, but eventually this question was one of the territorial disputes addressed by the 1919 Peace Conference in Versailles, which resulted in the 1920 Svalbard Treaty.[7] Norway was granted sovereignty, whereas other states' parties were granted extensive rights, especially nondiscriminatory rights to certain activities. The application of the Svalbard Treaty in the two-hundred-mile zone around Svalbard is, however, disputed (Churchill and Ulfstein 1992, 23–54). Norway claims that, on the basis of sovereignty, it has the right to establish a two-hundred-mile zone around Svalbard and that other states' rights under the Svalbard Treaty do not apply in this zone. On the other hand, Russia contends that the treaty prevents Norway from establishing measures in the Svalbard two-hundred-mile zone on a unilateral basis. Other states (except for Finland, which supports Norway's position) have either claimed that their rights under the treaty apply in the Svalbard two-hundred-mile zone or they have reserved their position. The main effect for fisheries of the treaty applying in the Svalbard two-hundred-mile zone would be a prohibition against Norway adopting discriminatory fisheries regulations (see the Svalbard Treaty article 2 (1)). This dispute has meant that Norway has been careful in introducing conservation measures, and while there have been violations of regulations by foreign vessels, Norway has been reluctant to arrest these vessels. Fisheries management in this zone has been under reasonable control, but further violations may make stricter management and enforcement necessary.

There is still a remaining pocket of high seas outside the two-hundred-mile zones in the Barents Sea (*Smutthullet*), where uncontrolled fishing by third states has recently occurred. General international law provides that coastal states have jurisdiction in the two-hundred-mile zones, whereas the flag states have exclusive jurisdiction on the high seas. This means that conservation measures on the high seas, to be effective, would have to be agreed upon by all states fishing in such an area. There is some basis in the LOS Convention article 116 for claiming that coastal states have regulatory control over fish stocks straddling between a two-hundred-mile zone and the high seas, but this question is controversial (see Burke 1989, 285–310). Norway has tried to gain control by bilateral arrangements with the relevant states, but more firm arrangements may be needed in the future. It remains to be seen whether such arrangements will be adopted on a global, regional, or bilateral level. An international conference on this issue was convened under United Nations auspices in August 1993 without reaching agreement; subsequent negotiations were undertaken in 1994.

Other Management Regimes

Most fish stocks are managed bilaterally between Norway and Russia. There are, however, also examples of marine living resources managed on a regional or a global level. Article 66 of the LOS Convention contains special regulations on anadromous species, the most important such species in the North Atlantic being salmon. The state in whose rivers anadromous species spawn is primarily responsible for the management of these stocks. In general, fishing for such species is prohibited beyond the two-hundred-mile zone. Salmon in the

Barents Sea is managed through the regional 1982 Convention for the Conservation of Salmon in the North Atlantic Ocean.[8] The management of salmon does not, however, seem to create international problems in the Barents Sea.

The LOS Convention article 65 allows coastal states to limit or prohibit the exploitation of marine mammals. States are to cooperate in the conservation of marine mammals. Whales are managed through the global International Whaling Commission (IWC).[9] In 1982 the IWC adopted a prohibition on all commercial whaling. Japan, Korea, Peru, Norway, and Russia filed objections and the decision was thus not binding for them. However, these states later affirmed that they would cease whaling and 1989 began a period free of commercial whaling.

Marine mammals eat a considerable amount of fish and compete with fish for food. A ban on whaling makes multispecies management in the Barents Sea difficult. Since the minke whale stock was at a sufficiently high level, Norway decided to resume commercial whaling in 1993. The catch of marine mammals, however, may be met by import restrictions from other states—especially the United States—and by actions by environmental organizations.

Nonfisheries Issues

Several nonfisheries issues have an impact on fisheries management. The Kola Peninsula bordering on the Barents Sea contains the largest naval base in the world and harbors the most important of Russia's four fleets. This has meant that security issues have played a major role in cooperation between Norway and the former Soviet Union. This has especially affected cooperation in enforcement. With the disappearance of the cold war, cooperation between Norway and Russia should be easier, but security aspects will still be taken into account by the two states.

The Barents Sea may contain considerable quantities of oil and gas. These prospects may also influence fisheries management. For example, there is reason to assume that Norway can accept nondiscrimination in the Svalbard zone for fisheries purposes, but not for oil and gas exploitation. But since there is a connection between the legal regime in the two-hundred-mile zone and the continental shelf, Norway may not readily accept nondiscrimination in the two-hundred-mile zone. The result may thus be that oil and gas interests prevent effective fisheries management in the Svalbard zone.

There have been reports of Russian dumping of nuclear waste in the Barents Sea. If fish in this area become contaminated—or if the consumers get such an impression—this may have disastrous effects on fish exports. Consequently, there is a link between effective management of nuclear contamination and fisheries management.

Another example of the connection between fisheries management and export is the recent agreement on the European Economic Area (EEA). The European Community got increased quotas in the Barents Sea in exchange for better market access.[10] Similarly, Norwegian fishermen fear that the European Community may get even higher quotas under the negotiations on Norwegian membership in the Community.

Conclusions

The two-hundred-mile system has made it easier to adopt adequate management measures in the Barents Sea. Before the introduction of the new ocean regime, all states fishing in this area had to agree in the regional fisheries organization, the NEAFC, on which conservation measures to implement. Now management is left to the two coastal states, Norway and Russia (except for whales and salmon). But because of the shared stocks, fish in the Barents Sea are still an international common resource.

Fisheries management in the Barents Sea is primarily a question of the effectiveness of the bilateral cooperation between Norway and Russia. This cooperation may be developed, inter alia, by drawing upon experiences in the management of other international commons. As was previously stated, it has been concluded that in this respect the following matters should be considered: to establish more precise management goals and principles, to integrate local interests in decision making, and to establish verification procedures.

The effectiveness of the management arrangements, will however, also be influenced by questions related to the disputed areas in the Barents Sea: the unsolved delimitation of the two-hundred-mile zones, the two-hundred-mile zone around Svalbard and the remaining area of high seas in the Barents Sea. Secondly, problems with the management of marine mammals may make multi-species management less effective. Thirdly, nonfisheries issues such as security, oil exploitation on the continental shelves, pollution, and export conditions will also limit the number of options available. The most urgent problems to be solved are related to the management problems in the two-hundred-mile zone around Svalbard and in the remaining area of high seas. If unregulated fishing increases in these areas, it could undermine the whole management regime in the Barents Sea.

NOTES

1. XXI International Legal Materials 1245 (1982).

2. United Nations Treaty Series, CMLXXXIII, p. 3.

3. See the 1992 Rio Declaration on Environment and Development, Principle 15 (31 International Legal Materials 874 [1992]).

4. See, *inter alia*, the 1992 United Nations Framework Convention on Climate Change, art. 3 (3) (31 International Legal Materials 849 [1992]).

5. Reference may be had to the Intergovernmental Panel on Climate Change (IPCC), which provided the scientific basis for the 1992 United Nations Framework Convention on Climate Change, op. cit.

6. The Norwegian text can be found in *Overenskomster med fremmede stater*, 1978, p. 436.

7. League of Nations Treaty Series II, p. 8.

8. Text in British Government Paper *Cmnd*, 8830, 1983.

9. The IWC was established by the International Convention for the Regulation of Whaling, Washington, 2 December 1946. Text in United Nations Treaty Series, CLXI, p. 72.

10. See Norwegian Government proposition, St. prp. no. 102 (1991–92).

REFERENCES

Birnie, P.W. and A.E. Boyle. 1992. *International Law & the Environment*. Oxford: Clarendon Press.

Burke, W.T. 1989. "Fishing in the Bering Sea Donut: Straddling Stocks and the New International Law of Fisheries." *Ecology Law Quarterley* 16, no. 1.

Churchill, R., M. Nordquist, and S. Houston Lay. 1977. *New Directions in the Law of the Sea*, V. Dobbs, Ferry, New York: Oceana Publications.

Churchill, R. and G. Ulfstein. 1992. *Marine Management in Disputed Areas. The Case of the Barents Sea*. London and New York: Routledge.

ALF HÅKON HOEL   CHAPTER 15

Managing the Barents Sea Fisheries: Impacts at National and International Levels

Life in northern Norway has always been—and to a considerable extent still is—based on the rich fish resources off its coasts (Brox 1989; Jentoft 1991). A characteristic of these resources is their internationality—the most important fish stocks are shared between Norway and Russia. Hence, the bilateral fisheries regime set up by Norway and Russia in the mid-1970s is of crucial importance to northern Norway, as the region's welfare depends upon how well the Barents Sea fishery resources are managed.[1]

Purpose and Perspective

Taking this as the point of departure, the purpose of this article is to describe the Barents Sea fisheries regime and assess its performance. Following some introductory remarks on resource management in general, the legal basis for the regime and the regime itself are analyzed; its performance with regard to management and distribution of resources among various groups of users is discussed.

According to conventional wisdom, two characteristics of fishery resources necessitate their use being subject to management. First, they are conditionally renewable resources, which require that exploitation should not exceed the resources' carrying capacity if a stable long-term yield is to be expected. Second, it is assumed that ownership rights to fish resources are non-existent, leading to a competitive race for scarce resources with disastrous ecological and economic consequences. In order to avoid such "tragedies of the commons," ownership rights to resources must be established, or the right to manage resources must be vested with a public authority (Hardin 1968; Gordon 1954).[2]

The enormous expansion of international fisheries after World War II, based on dramatic changes in technology and scale of fishing operations, led to overfishing in many areas within a few decades. In the North Atlantic, for example, the Northeast Atlantic Fisheries Commission (NEAFC) was not able to manage fisheries in an appropriate manner. The failure of international regulatory bodies to manage fishery resources according to their sustainable yield was the impetus for the establishment of extended coastal state jurisdiction in the late 1970s. The two-hundred-mile exclusive economic zone concept emerged from the third United Nations Law of the Sea Conference, and essentially entailed that the right and duty to manage marine resources within the two-hundred-mile zone—which contained most of the world's fishery resources—was shifted to the coastal states. At the international level, the approach to resolving the "tragedy" thus was one of vesting the control over natural resources with the coastal states (Eckert 1979), thereby abandoning the "public authority" approach represented by the international fisheries commissions established in the 1950s and 1960s.

The Barents Sea Fishery Regime

The Resources and the Economy

The Barents Sea—some 1.4 million square kilometers of shallow waters between the European continent and the Arctic Basin—is among the world's most productive ocean areas. The Barents Sea ecosystem is based on stocks of pelagic fish species on which other species (most importantly, cod) feed. The ecosystem stretches southward along the Norwegian coast and westward into the Norwegian Sea.[3]

The total catch from the biological production of the Barents Sea used to be considerable. In 1980 it amounted to some 2.4 million metric tons of fish, or 3.75 percent of the world catch. After 1985 the percentage declined sharply due to the reductions in major fisheries in this area, most of which were at an all-time low in 1990 when the total yield was a fraction of that in 1980. Since then, most fisheries have improved dramatically. Cod quotas for 1997 amounted to 890,000 metric tons, a sevenfold increase from the 1990 low of 120,000 metric tons.

To the North Norwegian economy, the Barents Sea fishery resources are crucial. Of a total population of 460,000 in the three northernmost counties of Nordland, Troms, and Finnmark, 13,500 are fishermen and 6,700 work in the fishing industry onshore. The processing plants are located in some 250 local communities, most of which have less than one thousand inhabitants (Brox et al. 1989). Since the work force constitutes about half of the population, about 10 percent of the northern Norwegian population depends directly on the fisheries for their income. Indirectly, a far larger share of the northern Norwegian population depends upon fisheries for their income.

A Variety of Legal Bases

The Barents Sea fishery resources are found in a number of areas where different legal conditions apply.[4] In international waters the fishery resources are subject to the open access rule—anyone may fish what he wants. Since 1991, fishing in international waters in the Barents Sea has increased, peaking in 1994 with 42,000 metric tons of cod taken in the "loophole" situated between the Spitzbergen Archipelago and Novaya Zemlja. A Greenlandic fishery here in 1991 was halted by a bilateral agreement between Norway and Greenland of September 1991 and EU fishing in 1992 was brought to an end with internal EU regulations. Since 1993 vessels from Iceland have dominated the fishery.

The two-hundred-mile exclusive economic zones of Norway and Russia are the most important in terms of weight and value of fish caught. The fishery regime here is based on part 5 of the 1982 Law of the Sea Convention (LOSC), the essence of which is that the coastal state decides on the management and use of the resources within the zone. As for resources that are shared (i.e., between the zones of two countries), article 63 of the LOSC states that "these States shall seek... to agree upon the measures necessary to coordinate and ensure the conservation and development of such stocks." It is left to the states to decide how they will share the resources.

The existence of such a disputed area created problems for the regulation of fisheries, especially third-country fishing. In 1977 a practical arrangement for handling these problems was negotiated: the "gray zone." In this zone, which is situated in the southern part of the disputed area and to some extent to the west of it, Norway and Russia regulate and control their own fishermen and the third-country fishermen each of them has licensed. As a fisheries arrangement, the zone appears to have worked well.

According to the Svalbard Treaty of 1920, Norway holds sovereignty over the four-mile territorial waters around the Spitzbergen archipelago, but all treaty parties are subject to equal treatment. Also, in the two-hundred-mile Spitzbergen Fishery Conservation Zone which Norwegian authorities erected in 1977, fisheries are regulated on a nondiscriminatory basis. When establishing the zone, the aim of Norwegian authorities was to get the fishery under control while at the same time avoid conflicts with other states over the jurisdictional status of the zone (Frydelund 1986).[5] Doubts have, however, been cast as to the long-term effects of this arrangement.

Fishermen from Iceland in particular have challenged the Norwegian rule in these waters.

The upshot of this is that the legal basis for regulating the Barents Sea fisheries is very complex. The patchwork of different legal conditions renders management difficult as the rights and duties of the states concerned change according to where in the Barents Sea their vessels fish. This legal complexity used to be compounded by international security concerns, but this is less important now than it was a decade ago (Schram Stokke and Hoel 1991). Other policy concerns are important, particularly the prospects for petroleum resources in the area, which render the resolution of the jurisdictional issues difficult. In relation to this, environmental concerns are becoming a major policy issue (Brubaker 1991), not least because of Russian handling of radioactive material and nuclear test detonations at the Novaja Zemlja archipelago.

The International Fishery Agreements

The international fishery agreements covering the Barents Sea are of two types: multilateral and bilateral. Although the latter are the more important, multilateral regimes are also of some significance. The Northeast Atlantic Fisheries Commission (NEAFC) was the major management body until the establishment of economic zones in the late 1970s.

The International Whaling Commission (IWC) has a decisive say in the management of large whales. Since 1976 it has set quotas for the northeastern Atlantic minke whale stock. Furthermore, due to the adoption of a temporary moratorium on commercial whaling in 1982 and pressure from the United States, Norwegian authorities decided to halt commercial whaling from 1988 onwards, awaiting the completion of a comprehensive assessment of whale stocks to be carried out by the IWC (Hoel 1989a). In 1992 the Scientific Committee of the IWC agreed on a stock estimate of 86,700 animals. This implies that the northeastern Atlantic minke whale stock could sustain commercial exploitation; as a response to that Norway supposition, Norway resumed minke whaling in 1993.[6] In 1996 the IWC Scientific Committee adjusted the stock estimate upwards to 112,000 animals.

There are a number of bilateral fishery agreements relating to the Barents Sea. The most important of these are the Norwegian-Russian agreements.[7] The task of the Joint Commission is to manage resources and to ensure the long-term yield from the ecosystem. In 1990 the total yield from the Barents Sea fisheries was in the order of some 3.7 billion NOK.[8] The potential yield however, is considerably higher, as most stocks in 1990 were at an all-time low. Management on a multispecies basis may result in a total yield three to four times higher than this (Flaaten 1989:40).

The basis for negotiations is the scientific advice and management strategy options given by the Advisory Committee on Fisheries Management (ACFM) of the International Council for the Exploration of the Sea (ICES). The ACFM advice for the Barents Sea fisheries management is the result of research carried out mostly by Norwegian and Soviet scientists, to some extent in joint programs. The international scientific screening and elaboration of national scientific work provides the legitimacy ICES management advice carries with administrators and fishermen in its member states.

According to the 1976 reciprocal fishery agreement, quotas are to be set according to "the need for rational management of the living resources, catching methods, and the traditional catch levels of the contracting parties and other relevant factors." The three joint stocks of cod, haddock, and capelin are shared on a 50-50 basis in the case of cod and haddock, and 60-40 in favor of Norway for capelin.[9] Thus each party's quotas follow automatically when the total quotas (TAC) are set.

The delegations to the meetings consist of government officials, research personnel, and representatives from the fishermen's organizations. The negotiations proceed as follows: scientific advice regarding catch levels is reviewed, TACs for the shared stocks are established, quotas of the joint and exclusive stocks are exchanged, and various types of technical regulations relating to fishing seasons, gears, and areas are established. The exchange basically consists of the Russians giving Norway some of their cod quota, while Norway gives Russia a share in her quotas of the exclusive stocks of red fish, herring, and most importantly, blue whiting. The outcome of the negotiations is adopted in a protocol.

The most contentious issues over the last ten years have been the questions of mesh size regulations and Norwegian coastal fishing. The former relates to the claim of Norwegian scientists—supported by the ICES—that the mesh size in trawls should be increased in order to utilize the growth potential of fish. The Russian counter-argument is that mesh size is not as important to the exploitation pattern as commonly believed.[10] The trawl-free zones set up by Norway in the mid-1970s have been a bone of contention, as Russian fishermen claim that the zones prevent them from taking their quotas. The issue of Norwegian coastal fishing is rooted in the 1974 Tripartite Fisheries Agreement between Norway, Russia, and Great Britain. At that time the overriding concern was the advent of economic zones, and Norway advocated the coastal state preference principle for distribution of fishery resources among nations. To this end the ocean-going fleet of all nations was regulated, while the coastal fleet remained unregulated. A sentence in the Tripartite Agreement which allows for unlimited coastal fishing—even when the total quota was taken—remained in the subsequent bilateral agreement between Russia and Norway, and its essence was not changed before 1984. Russian resistance to this section of the agreement stems partly from over-fishing of the TAC by the Norwegian coastal fleet in the early 1980s; they also maintain that Norwegian coastal fishing is a wasteful way of exploiting the resources, as it disturbs fish in the spawning grounds.

The two parties give resources to other states in separate bilateral negotiations. The Faroe Islands is the more important third-country for Russia, while the European Union is the major recipient of the Norwegian share of third-country quota.

Internal Aggregation in Norway

At the national level in Norway there is considerable legal complexity as to the status of fishery resources and fishermen's rights in relation to these. While commonly regarded as a resource which belongs to the nation ("fish is a national resource"), the actual content of the "common property" rights concept is very difficult to handle in legal practice (Fleischer 1990; Ørebech 1989). Beyond doubt, however, is the right of the fisheries authorities to regulate entry into the fisheries as well as fishing itself.

The formal point of departure for the internal decision-making process on distribution of northern fishery resources is the result of the bilateral negotiations with Russia. These negotiations are normally concluded by late November. On the basis of TACs agreed-for shared stocks in such negotiations, the Fisheries Directorate works out a proposal on how various fisheries are to be regulated. Within a few weeks from the presenting of the regulatory proposal, the director of fisheries meets with several of the fishing industry's organizations, the Marine Research Institute, and environmental authorities in the Regulatory Council to discuss the proposal. The Council has eleven members, six of which belong to the fishermen of the Norwegian Fishermen's Association (NFA) and Norwegian Seamen's Union. The shore side of the fishing industry has three members, while the Fisheries Directorate has one member. Saami fishing interest recently got one representative. The fishery interests thus hold a majority in the Council. These are, however, more often than not fairly divided among themselves, and for that reason the fishermen do not always win acceptance for their proposals for distribution of fish quotas. In addition to the regular members of the Council several observers are admitted.

The Regulatory Council was established in 1973 as a forum for administrators and fishermen to prepare for the negotiations in the international fisheries commissions. Given the importance of international resources to Norwegian fisheries—then, as now—it is evident that Norway's approach to these negotiations is crucial to the welfare of its coastal population. The task of working out the strategy for international negotiations, however, shifted to a working group under the Sea Boundary Committee when the two-hundred-mile zones were established. This working group has held a very low public profile, considering the importance of its role to coastal Norway. The Regulatory Council in the late 1970s took on tasks corresponding to those formally vested with it in 1983, when its role was defined in the Marine Fisheries Act. According to this, the Council shall on the basis of the information given by the Marine Research Institute, "consider which regulations of the fishery which are required and how they may be appropriately implemented."[11] In practice this involves discussions in Council on which seasonal, temporal, and technical restrictions should apply to the quotas set, as well as distribution of quotas on various user groups. In the case of national stocks, the setting of quotas is also discussed.

The Regulatory Council meets three or four times a year; their meetings are preceded by a bargaining process within and among the organizations. The major actor is the NFA, which organizes both labor and capital in fisheries: most fishermen in the coastal fleet and the ocean-going groundfish/shrimp fleet hold a rank-and-file membership through the regional (county) departments of the organization.[12] The regional departments in turn have numerous local divisions.[13] There also exist four independent organizations associated with the NFA, three of which organize ship owners. These may also hold membership through the regional departments, and thereby have several channels of influence with the NFA. Thus the NFA combines geography and the labor-capital relationship as organizational principles, thereby incorporating conflicts between fishermen in various regions, between fishermen using different types of gear, and between the ocean-going fleet and the coastal fleet.[14]

Fisheries regulations always have distributional implications, and very often these center on the coastal-ocean, north-south, and gear controversies. In the NFA the Directorate's regulatory proposal is subject to a thorough examination and debate on its board, yielding compromises which leave the Associations' Council members with their hands tied to particular solutions regarding how resources are to be managed and distributed.[15] The major lines of conflict in the Council correspond to those within the NFA. The coastal fishermen's interests are represented by the NFA, while the ocean-going fleet's interests are represented by the NFA and by the Seamen's Union. Moreover, the latter interests are often allied with the fish processing industry which prefers the ocean-going fleet due to the volume and regularity of its landings. It follows that power relationships within the NFA are crucial for its position on various issues as well as the outcome of the Council's deliberations (Hoel, Jentoft, and Mikalsen 1991).

Distributional decisions in the Council are mostly made with reference to gear types, and the north-south and coastal-ocean dimensions are indirectly affected as various types of gears are not evenly distributed along the coast. As a result of these contradictions, the deliberations in the Council sometimes result in conflicts being solved by raising the quota for exclusive stocks, as has happened with saithe. The Council does not make such decisions for cod, however, as its TAC is decided in the proceeding negotiations with Russia.

The Council decides mainly by way of debate, while resorting to voting only when matters are contentious. This is significant for the outcome of the deliberations, as the observers are allowed to take part in the debate. Their views are included when the director of fisheries sums up the deliberations and formulates his advice to the Fisheries Ministry. The fishermen's influence is now more than before balanced with environmental considerations, due to the observer status of the environmental groups, and also because the fishermen themselves and the fisheries administration have become increasingly concerned with this aspect. The advice provided by the Council is to a large extent adhered to by the fisheries authorities, especially when the Council is unanimous in its recommendations. The Ministry has, however, introduced additional measures from time to time, in particular with geographical redistribution in mind.

This organization of the regime, with cooperative structures both in the preparations for and in the delegations to the international negotiations and the Regulatory Council, leaves the NFA with a considerable influence over the Norwegian fisheries policy, and thereby also over coastal community development, in that it has a decisive say over the distribution of resources. In return, Norwegian authorities benefit from qualified technical advice concerning the complex details of fishing. And many conflicts are resolved by the NFA, thereby relieving the authorities of the task of setting up compromises. In addition, when the fishermen are participating in the formulation of fisheries policy from the outset, policy holds greater legitimacy among the fishermen and its implementation may be more successful (Hoel, Jentoft and Mikalsen 1991; Jentoft 1991).

What emerges from the above, then, is a picture of a two-tiered decision-making system where the important decisions regarding management and distribution of the Barents Sea resources are made at the international level, while the distribution of those resources among various groups in Norway is decided on by a corporatist body, comprising regulators as well as those to be regulated. The next question, then, is how this organizational setup has functioned, in terms of how well resources are managed and how they are distributed.

Fisheries Policy: Resource Management and Distribution

International Management and Distribution

Both groundfish and pelagic fish are important in the Barents Sea fisheries. As mentioned, cod is the most important species economically, and its management therefore also the more important: the total catches were mostly between some five hundred thousand and one million metric tons between 1960 and 1977, when extended coastal state jurisdiction was established. After that time the total catches sharply declined before increasing again during the 1990s.

The sharp reduction in the catches of cod stemmed from the decline in the cod stock and consequent quota regulations from 1977 onward. In 1977 a TAC of 810,000 metric tons was set, while the 1990 TAC of 120,000 metric tons was at an all-time low. By 1997 the total quota was up to 890,000 metric tons. As for the pelagic species, during the highly intensive herring fisheries at the Norwegian coast in the 1960s, herring was brought almost to extinction. During the 1970s the capelin fisheries expanded enormously, reaching their peak with almost three million metric tons taken in 1977. In 1985 it was discovered that the stock was about to collapse, and the fishery was halted early in 1986. In 1991 the fishery was restarted, with a TAC of 850,000 metric tons. This quota marked the real start of multispecies fisheries management in the Barents Sea, as the single-species recommendation from the ICES was at 1 million metric tons. 150,000 metric tons was thus set aside as food for other fish species. Herring is now replacing capelin as the key species in the Barents Sea ecosystem, and is the basis for a major international fishery in the Norwegian Sea. The 1997 TAC here amounted to 1.5 million metric tons.

Underlying the improvement in various fish stocks is a change in regulatory philosophy in the Joint Commission. Management is based on increasingly strict principles. Another important measure in rebuilding the fish stocks has been a system of regular surveillance of fishing ground and automatic closures of areas with a high percentage of immature fish in the catches. However, the failure to control especially the operations of foreign fishing vessels is (in the long term) threatening the successful management efforts.

In addition to the fishery, there are also whaling and sealing industries which have been cut back, albeit for other reasons than those related to biology. In recent years, the latter has been in the order of 40,000 animals, and is carried out both by Norwegian and Russian sealers. Whaling is conducted by Norwegians only in this area, and has averaged about 1,800 animals per year since World War II. As mentioned, the commercial catch was halted in 1988, removing an important fishery to some fifty vessels in northern Norway. Only a few animals were taken for scientific purposes in 1988–1990, while ninety-five were taken in 1992. In 1993 some 225 animals were taken in the traditional coastal whaling. In 1997 the catch had increased to more than 500 animals.

As to the distribution of catch between countries, there is a dividing line before and after 1977. Before 1977 these waters were international outside a twelve-mile fishery zone. Most of the fishery resources were therefore subject to international management under the auspices of the Northeast Atlantic Fisheries Commission (NEAFC) and a more or less untamed international exploitation, as reflected in the share of the total catch by third countries. After 1977 Norway and Russia kept most of the resources to themselves. The third-country quota of cod has been steadily reduced from 19 percent in 1977 to 10–11 percent since 1984. By the late 1980s Norway and Russia had established as a firm principle that they retain about 90 percent of the total quota of cod for their own fishermen. A slight increase in EU quotas has followed from Norway's entry in the European Economic Area Agreement, and may continue to increase as a consequence of an eventual Norwegian EU membership.

As to the bilateral results of these negotiations, Norwegian government officials claim the outcome of the negotiations are fairly balanced and that the negotiations with Russia have been far more businesslike and practically oriented than the case is with the EU (Paulsen 1989). However, some have argued that the final distribution appears to be skewed in Norway's disfavor (Hoel 1989b; Schram Stokke and Hoel 1991), at least when measured by conventional western price indices. Moreover, Russia annually gets the afore-mentioned 40,000 metric tons of "Murmansk cod" in addition to their 50 percent share, which probably is a part of the Northeast Arctic cod stock.

On the other hand, taking Norway's catch of marine mammals in Soviet waters and the over-fishing of the cod quotas in the early 1980s into account,[16] as well as the fact that the quotas Russia receives from Norway eventually are not taken,[17] the distribution appears somewhat more balanced. It should also be taken into account that Norway's major goal in these negotiations has always been to secure as large a transfer of cod as possible from its counterpart, in exchange for fish species of less interest to her. This strategy has definitely been in the interest of the coastal population in northern Norway, and since substantial amounts of cod have been obtained this way the strategy may, from that perspective, be characterized as successful. The Norwegian share of the cod quota increased from 41 percent in 1977 to 86 percent in 1984, then leveling off to between 50 and 60 percent annually. On average, in the 1977–1991 period Norway has had 52 percent of the quotas of cod.

Internal Distribution Results in Norway

The total Norwegian catch of all fish species in northern waters has been reduced from 2.5 million metric tons in 1977 to 470,000 metric tons in 1990, a reduction of 80 percent (Hersoug and Hoel 1991). Northern waters thereby became less important to Norwegian fisheries in general. The reductions in available resources have of course led to considerable overcapacity in the fishing fleet, adding to the economic difficulties fishermen and their communities face. With the resumption of the capelin fisheries in 1991 and the increase in cod quotas, the importance of Northern waters has increased again.

Scarcity has served to intensify the distributional conflicts between the North and South and between various gear types. The actual distribution of catches of cod between conventional gears and trawlers appears, however, to be fairly constant, although there have been considerable deviations for certain years (Hersoug and Hoel 1991). The variations stem mainly from shifts in the migration pattern of cod, which is unavailable for coastal fishermen when it stays too far off the coast as happened in the years 1986–1988. The political decisions of the Regulatory Council are not of much help to the coastal fishermen when natural phenomena intervene in its distributional scheme. Over the last decade the distribution of the most important species, cod, has been about 65–35 in favor of the coastal fleet. This distributive key has been a recurrent source of tension. In 1989 the NFA suggested a scheme for making the distribution of cod on gear types dependent on TAC size. This implies basically that when cod quotas are low, trawlers will have about 25 percent of the total quota, and when quotas are high their share rises to 35 percent. By establishing such a fixed distributive scheme the annual conflicts may be softened, as the various groups' shares will not be subject to bargaining each year.

The great redistribution in Norwegian fisheries during the 1980s appears to be that between the North and the South. While taking about a third of the total catch in 1977, in 1990 North Norwegian fishermen were down to a fifth (Hersoug and Hoel 1991). The basic reasons for this development are scarcity of resources in the North, and, in the pelagic sector, a considerable transfer of fishing licenses from the North to the South.

Just as important are the impacts of these cutbacks for the fishing industry. While half of all Norwegian fish catches were landed in North Norway in 1977, only 20 percent of the catches were landed in the North in 1990. This decline stems to a large extent from the closure of the capelin fishery. As to cod, landings were in absolute terms in 1990 down to a third of the 1977 level (Hersoug and Hoel 1991). The supply to the North Norwegian fishing industry thereby became dramatically reduced. In the groundfish sector this has increasingly been compensated for by deliveries from foreign, particularly Russian, fishing vessels, which in 1993 landed more than 100,000 metric tons of cod in Norway.

Explaining the 1990 Crisis

The Barents Sea fisheries regime as described above is the institutionalization of an attempt to avert a "tragedy of the commons" in the area. The three joint stocks the regime is to manage were all sharply reduced during the 1980s, as were several exclusive stocks owned by either Norway or Russia. By 1990 it was therefore no bold conclusion that the regime's success was at best qualified, while a few years later management policies appeared largely successful. It should be noted, however, that it is open to discussion exactly how much of the development in the resource situation may be attributed to the regime. Explanations can therefore be grouped into two categories: "natural" and "political."

As to explanations relating to natural phenomena and science, scientific advice has in some instances been inferior, as in the case of the collapse of the capelin stock (Tjelmeland 1989). Fishermen misreporting their catches have compounded these problems in that the data on which management is based is faulty. In addition, climatic variations have not been taken fully into account and may be an important explanatory factor (Loeng 1986). Thus not only inferior advice, but also neglect of factors which contribute to stock development are features here.

As for the political aspect of management, the more popular explanation is the corporativist hypothesis stating that fishermen's greed in combination with fisheries authorities' lack of understanding is the cause of the crisis (Brox 1989; Nilsen 1991). However, the quotas have by and large been set in accordance with the scientific advice given (ICES 1989/1991). Summing the total quotas for cod set during the 1980s yields a lower total than the sum of advised TACs in the same period. This suggests that the vagaries of nature are more important in explaining the resource situation than management policies. As the decline in resources only to some extent can be explained with reference to political factors, so the improvement in stocks may stem from other factors as well. In addition to such issue-specific, fisheries-related explanations, come those related to the complex legal basis for the regime and other policy concerns which have a direct bearing on the regulation of fishing in the Spitzbergen area, for example.

The upshot of this is that, while the introduction of two-hundred-mile economic zones which conferred resource rights to coastal states did not result in an immediate improvement of resource management, this change of ocean law has been instrumental in the distribution of resources. The partial phasing-out policy regarding foreign fishing after the introduction of two-hundred-mile zones has left the two coastal states with about 90 percent of the cod quotas. This pattern appears to be challenged now, as witnessed by the admission of new fishing nations into the area and the increase in the EU's share here. Thus, the turn-around trend witnessed in the development of resources is accompanied with a certain tendency towards international redistribution towards third countries. In this context it is the establishment of rights by formal agreements and the abjuration of well-established principles by coastal states that is important, not the actual quantities of fish involved.

As regards the distribution of quotas between the two coastal states in the area, overall power relationships appear to have little explanatory value (Schram Stokke and Hoel 1991).[18] More issue-specific explanations, such as the parties' interests for different fish species and bargaining dynamics as the salience of focal points, are more important explanatory factors. An example of the former is the Norwegian interest in highly valued species as opposed to the Soviet interest in quantity. An example of focal points is the 50–50 division of resources and the stability of the annual transfer of blue whiting from Norway to its counterpart.

Turning to the national part of the regime, the mobilization of a broader public interest in fisheries management has undermined the legitimacy of both the corporativist regime and its policy. As long as the fisheries were a matter for the fisheries authorities and the fishermen's organizations, the NFA was very useful to the authorities. Firstly, in that it functioned as an information central, providing the technical knowledge required in the international negotiations, and secondly in its role as a clearinghouse, in which the directorate's proposal is melded into a politically feasible regulatory scheme. This is no longer true: the increasing scarcity of resources has intensified conflicts not only among fishermen, but also among regions as the economic repercussions of scarcity have been felt onshore. There has been a growing concern among other groups in society—politicians and environmentalists—for how fisheries are managed and distributed. The extent to which the distributional pattern can be explained by the corporativist organization of the fisheries regime is difficult to assess, however. In general, the management aspect is not that important in the decision-making process at the national level, as the economically most significant stocks are stocks shared with other nations. The basic reason for resource shortage in North Norway is the decline of the fishery resources in the North, which, as we have seen, is due to a mix of factors where the more important probably are beyond the realm of national politics.

Prospects for the Coastal Populations

Social Disruption Following the 1990 Crisis

The quota reductions for cod until 1990 meant that the inputs to the North Norwegian economy were reduced with some NOK 3 billion, relative to the early 1980s catch levels.[19] As the North Norwegian economy is, to a large extent, based on fish and the fishing industry, one should expect that the economic upheavals resulting from catch reductions can be measured along traditional social indicators. It is, however, difficult to assess precisely how much of the scores on these indicators can be attributed to the variations in the fisheries.

The decline in the population in most municipalities in Finnmark and the northern part of Troms (the two northernmost counties) in the latter half of the 1980s, as well as the generally stagnant population in most coastal communities, is basically a consequence of long-term changes in the age structure of the population.[20] Imbalance in sex composition in most communities is also of relevance here (Jentoft 1991). However, the declining in-migration (Eikeland 1991) which traditionally has contributed significantly to the population may, to some extent, be ascribed to the fisheries crisis.

The rise in unemployment to levels far above the national average—in 1990 at 13 percent in Finnmark and at 17 percent in northern Troms—was evidence of the decline in the fisheries sector and in dependent industries. The same applied to the soaring number of private and company bankruptcies. The biggest vessel owners in the North were the banks, whose troubles in turn stemmed not least from the problems in the fisheries. Following the improvements in the resource base, unemployment levels have now dropped considerably, and are now generally low compared to other parts of Norway. In addition to such measurable social indicators, more intangible changes are also occurring: people's general outlook on the prospects of staying in their home district, and young people's attitudes toward the fisheries industry are being negatively affected.[21] Such attitudes do not co-vary with the fluctuations in fish stocks, and may take considerable time to change.

The General Outlook

A basic feature of the public debate on the Norwegian fisheries policy is that it is almost devoid of reference to the international context necessary for fisheries management. This is reflected also in the debate on regime change. It follows from the above that the prospects for the North Norwegian population depend in large part upon the development of the resources in the Barents Sea—that is, on the performance of the bilateral fisheries regime. It is at the international level that significant advances can be made in management that renders the coastal population better off. It is obvious that considerable improvements in management are already made, as witnessed by the increase in the cod and capelin stocks. With a more directed effort at multispecies management there is a great scope for deriving more benefits from the resources.

This raises two questions: what are the obstacles to further improvement of resource management, and who will benefit? The fishing pattern has to be improved—a difficult thing to achieve due to the complex legal situation in the area that prevents the coastal state from exercising control over fisheries throughout the migratory range of fish stocks.[22]

This brings us to the second question: who will benefit from the results of increased stocks—those who have carried the costs by tight management, or others? With the European Economic Area agreement negotiated in 1991, the EU is set to increase its share of the cod quotas in northern waters.[23] In addition, Greenland has obtained a share in Norwegian waters, and irregular fishing in the international waters in the Barents Sea has been on the increase the last three years.

As the ownership entitlement approach, which implies that the coastal states are given control over resources, is being challenged at the international level,[24] with other nations enjoying new privileges in coastal states' waters, an opposite tendency emerges in fisheries management at the national level. A prominent solution suggested for improving management is to vest ownership rights with single actors, such as companies or persons (Hannesson 1984; Strukturutvalget 1989). A consequence of this is the privatization of fisheries resources, leaving fishing rights in the hands of a privileged group of persons. Efforts to this end, however, were not successful. The Norwegian government, following an extensive public debate, rejected the idea of individually transferable quotas. It is evident that the coastal population as well as the fishermen have been left with an uncertain legal foundation for claiming ownership of the resources off their coasts, a legal structure that offers no protection from outside interests seeking to reap those resources.

NOTES

1. This is the perspective of the Norwegian government in two consecutive Reports to Parliament: 32 (1989–90), *Framtid i nord* and 32 (1990–91), *På rett kjøl*.

2. This menu of choice has been increasingly contested as the commons paradigm assumes an open-access situation which seldom is found in reality. It is suggested that there may also be a third way, commonly termed "co-management," in which the existence of common property is viewed as a solution to, rather than the cause of, resource management problems (McKay and Acheson 1987; Berkes 1989).

3. The catch figures referred to here therefore include not only the catches in the Barents Sea proper, but also the areas to the west and south where the fish stocks are exploited. The statistical reference areas are ICES areas I and IIa,b.

4. The most thorough analysis of the legal conditions in the Barents Sea is found in Churchill and Ulfstein's *Marine Management in Disputed Areas: The Case of the Barents Sea.*

5. Russia has lodged an official reservation to the zone, while several nations have reserved their positions with regard to the Spitzbergen Treaty. Only Finland recognizes the Norwegian approach here.

6. In response to the abdication of the IWC from its treaty-based management responsibilities and the development of multispecies fisheries management requiring the role of marine mammals in ecosystems to be taken into account, the Faeroe Islands, Greenland, Iceland, and Norway have set up an alternate marine mammals regime: the North Atlantic Committee for Cooperation on Research on Marine Mammals. As the name indicates, this organization is concerned with research only (see Hoel 1992).

7. The Norwegian-Russian agreements are as follows: the 1974 agreement on trawl-free zones, the 1974 agreement on fisheries cooperation, the 1976 agreement on reciprocal fisheries relations, and the 1977 gray zone agreement.

8. The basis of this calculation is as follows: the 1990 cod catch of 171,000 metric tons, priced at NOK 10 per kilo, has a firsthand value of 1.7 billion NOK. The 23,000 metric tons of Haddock is given the same price per kilo, being worth 0.2 billion, while saithe, Greenland halibut, and red fish amount to 183,000 tonnes and are given an average value of NOK 5, totaling NOK 0.9 billion. In addition, the Norwegian shrimp fishery in this area amounted to 85,000 tonnes in 1990. The total value, based on an average price of NOK 10 per kilo, amounts to NOK 0.9 million. Marine mammals are not included.

9. The basis for this distribution is the zonal attachment of stocks.

10. The Soviet experiments show that most small fish that escape through the meshes in the trawl die anyway. It is also argued that bigger mesh sizes force the fishermen to increase trawling time, causing a higher mortality rate among young fish, precisely what one wants to avoid.

11. This mandate applies to the articles 4, 5, and 8 in the 1983 Marine Fisheries Act, which authorizes a variety of regulations.

12. In addition, some fishermen are organized in the Norwegian Seamen's Association, and some in the Coastal Fishermen's Organization.

13. In Troms County, for example, there are about eighty local departments.

14. Due to the remuneration system in Norwegian fishing, the labor-capital conflict manifests itself not so much between ship owners and crew as between coastal vessels with labor as the major economic input and ocean-going vessels having capital as the major input.

15. The NFA members meeting in the Council reflect the membership profile: two coastal fishermen from the North, one representative from the trawling interests in the North, one representing the purse seine interests in western Norway, and one representative for the North Sea fishermen in southwestern Norway.

16. In 1982, for example, the Norwegian quota was over-fished with some 120,000 metric tons, the same quantity as the 1990 TAC.

17. The blue whiting quota, which has been varying between 290,000 and 385,000 metric tons, has never been taken in its entirety.

18. This seems to be the case also for other instances of international fisheries cooperation. (See Underdal 1980 for the case of NEAFC.)

19. A reduction in the cod catch of 100,000 metric tons corresponds in 1990 prices to roughly 1 billion NOK. The Norwegian cod quota, three-fourths of which is taken by North Norwegian fishermen, has been reduced by more than 300,000 metric ton since 1977.

20. See the Report to Parliament, no. 32, 1990-91, page 16.

21. This is by no means a feature of North Norway alone. On Canada's Atlantic coast the experiences are similar (Andersen 1989).

22. Spanish fishing vessels, for example, landed catches in August 1991 consisting of fish averaging about 300 grams each, less than half the legal minimum size in Norway (700 grams/47 cm) (Fiskeribladet 1991).
23. There are two components in this: first an increase in the EU's general TAC share in cod in the Norwegian Economic Zone from 2.14 to 2.9 percent and secondly an additional amount (also in the Economic Zone) increasing from 7,000 metric tons in 1993 to 11,000 metric tons in 1997. Given a TAC of 700,000 metric tons in 1997, the EC quota in northern waters will therefore constitute 3.46 percent (Spitzbergen Zone) + 2.9 percent (Economic Zone) + 11,000 metric tons. This amounts to 55,500 metric tons, or 7.9 percent of the TAC and 15.8 percent of the Norwegian share of the TAC.
24. For example, the EU's fisheries policy shifts the competence to manage fish resources from the member states to the EU. Thereby management strategies may change. It is to be noted that the EU approach to fisheries management represents a reversion of the process of transferring ownership and management rights to coastal states which resulted in the establishment of the two-hundred-mile economic zone principle during the United Nations Law of the Sea negotiations in the 1970s.

REFERENCES

Andersen, R. 1989. *2JKL Cod Stock Allocation and the Inshore/Nearshore Fishing Sector*. St. John's: Government of Newfoundland and Labrador.
Berkes, F., ed. 1989. *Common Property Resources*. London: Belhaven Press.
Brox, O. 1989. *Kan bygdenæringene gjøres lønnsomme?* Oslo: Gyldendal.
Brox, O., et al. 1989. *Mot et nytt ressursregime for nordlige bestander*. Tromsø: Troms fiskarfylkning.
Brubaker, D. 1991. *International Management of Pollution. Problems in the Barents Sea*. Tromsø: Institute of Law, University of Tromsø.
Churchill, R. and G. Ulfstein. 1992. *Marine Management in Disputed Areas: The Case of the Barents Sea*. London: Routledge.
Eckert, R. 1979. *The Enclosure of Ocean Resources*. Stanford: Hoover Institution Press.
Eikeland, S. 1991. "Kva fortel livsløpanalysar om utviklingsproblema på Finnmarkskysten?" *Tidsskrift for samfunnsforskning* 32, no. 3.
Flaaten, O. 1988. *The Economics of Multispecies Harvesting*. Berlin: Springer Verlag.
Flaaten, O. 1989. "Økonomi og flerbestandsforvaltning." *Et Barentshav av muligheter*. Tromsø: Fiskerikandidatenes forening.
Fleischer, C. A. 1990. "Eiendomsretten til fisken i havet." *Perspektivanalyser til langtidsplan for fiskeriforskningen* (1990–1994). Trondheim: Norges Fiskeriforskningsråd.
Frydelund, K. 1986. *Lille Land - hva nå?* Oslo: Gyldendal.
Gordon, S. 1954. "The Economic Theory of a Common Property Resource." *Journal of Political Economy* 62, no. 2.
Hallenstvedt, A. 1982. *Med lov og organisasjon*. Tromsø: Universitetsforlaget.
Hamre, J. 1991. "Barentshavets fiskeressurser—utvikling og potensiale." In *Forvaltningen av våre fellesressurser*, edited by N. Chr. Stenseth and G. Kristiansen Oslo: Ad Notam.

Hannesson, R. 1984. "Markedsmekanismens positive egenskaper." *Fra forhandling til marked*, edited by H. Jervan. Oslo: Gruppen for ressurstudier.
Hardin, G. 1968. "The Tragedy of the Commons." *Science* 162.
Hersoug, B. and A. H. Hoel. 1991. "Hvem tok fisken?" Mimeograph. Tromsø Norwegian College of Fisheries Science.
Hoel, A. H. 1989a. "Norwegian Marine Policy and the International Whaling Commission." *Journal of North Atlantic Studies* 2.
Hoel, A. H. 1989b. "Ressursforvaltningen i nordområdene." *Norsk Utenrikspolitisk Årbok*. Oslo: NUPI.
Hoel, A. H. 1992. "Regionalization of International Management of Whales: The North Atlantic Committee for Cooperation on Research on Marine Mammals." *Arctic*, summer 1993.
Hoel, A. H., S. Jentoft, and K. Mikalsen. 1991. "User Group Participation in Fisheries Management." Proceeding from World Fisheries Congress, edited by S. Jentoft and B. McCay. Athens, 1992.
ICES 1989/1991. Extract of the Report of the Advisory Committee of Fisheries Management. Copenhagen: Council for the Exploration of the Sea.
Jentoft, S. 1991. *Hengende snøre*. Oslo: Ad notam.
Loeng, H. 1986. "Havklimaets betydning for fiskeressursene." *Barentshavets ressurser*, seminarrapport. Trondheim: Norges Fiskarlag.
McKay, B. and J. Acheson. 1987. *The Question of the Commons*. Tucson: University of Arizona Press.
Nilsen, R. 1991. "Yttersideværi og fjordbygder i Finmark." In *Forvaltningen av våre fellesressurser*, edited by N. Chr. Stenseth and G. Kristiansen. Oslo: Ad Notam.
Ørebech, P. 1989. *Om allemannsrettigheter*. Dissertation. University of
Paulsen, T. 1989. "Norges samarbeid med EF innen fiskerisektoren." In *Norge, EF og fiskeriene*. Tromsø: Norges Fiskerihøgskole.
Ressursoversikten. 1991. Bergen: Fiskeridirektoratet
Strukturutvalget: 1989. Oslo: Høringsnotat. Ministry of Fisheries.
Sætersdal, G. and G. Moore. 1987. "Managing Extended Fisheries Jurisdiction: A Decade's Modest Progress." *Ceres* no. 119.
Shram Stokke, O. and A. H. Hoel. 1991. "Splitting the Gains: the Political Economy of Barents Sea Fisheries." *Cooperation and Conflict XXXVI*.
Tjelmeland, S. 1989. "Hva skjer i Barentshavet?" *Et Barentshav av muligheter*. Tromsø: Fiskerikandidatenes forening.
Ulfstein, G. 1987. "Reguleringsproblemer i fiskevernsonen ved Svalbard." *Lov og rett*, no. 3.
Underdal, A. 1980. The Politics of International Fisheries Management. Oslo: Universitetsforlaget.

BJØRN K. SAGDAHL   CHAPTER 16

Management under Scarcity: The Case of the Norwegian Cod Fisheries

The introduction of the two-hundred-nautical-mile economic zone on January 1, 1977, gave promising prospects for resource maintenance and economic growth in the Norwegian fishing industry. Some few years later the scene was dominated by crisis in the arctic cod fisheries, an almost breakdown of the stock, and continuous quarreling about the allocation of the diminishing quotas. The 1980s ended in the lowest quotas since regulations of the stock started some fifteen years earlier.

This experience with management of a common property resource is hardly specifically Norwegian. It is shared by most communities and nations depending on common pool resources (CPRs). The reasons why such situations have developed seem not to be due to the lack of scholarly advice. The "tragedy of the commons" is thoroughly described and analyzed by numerous scholars in the field. Yet it seems hard to find a politically accepted recipe among the recommendations, ranging from market solutions and self-government to traditional top-down governmental administration. At least this could be said to be a prevalent problem in the Norwegian resource policy context.

Two main concerns have to be met in the policy applied: Resource maintenance is the overall concern, not only because of internal economic considerations but also because of our international judicial obligations. This is expressed by setting the MSY (multiple sustainable yield) standard and the yearly TAC (total allowable catch) on the basis of scientific advice. And as Norwegian-arctic cod is a stock we are sharing with Russia, close cooperation is needed to reach this goal.

The second concern is one of allocation of the negotiated TAC. This implies negotiations both on the bilateral and the national level. While an accepted allocative key is used for the allocation on the bilateral level, it has been far more difficult to agree upon the allocation on the national level.

Different sciences approach the above mentioned problems in different ways. But in the search for general bio-economic and management models it is easy to overlook the social-political and institutional context management is depending on. In other words, there are certain social-political limits for the application of approaches and models in a given society. A functional model of management is the one that stands the test of political scrutiny.

This study could be read as a warning against the search for general models in managing CPRs. If managerial models shall work, they have to be developed on the basis of the social-political realities they are meant to affect. We are not managing only economic actors, but also social, political actors. Our case study on the management of the Norwegian cod fisheries underlines this simple lesson—the necessity of having a profound understanding of the affected social-political setting in lining out workable managerial strategies.

The character of the problem is, of course, closely related to the actual status of the resource system and the amount of the national quotas. In situations with insufficient and shrinking quotas, the question of rights and favoritism enters the agenda. Negotiated obligations to let so-called third countries exploit the resources despite the hardships national fishing communities are facing make the allocative decisional process even more politically delicate.

We are here confronted with the problematic political balance between legitimacy, allocation, and effective resource management. Legitimacy is a key word for understanding the relationship between government and the governed. As such, it is a fundamental concept in political analysis concerning characteristics and relations between input and output in any political system as the polity itself (Easton 1957, Beetham 1991). In a Weberian tradition we may speak of different sources to authority. Although resource management is based on formal laws, other sources to legitimacy may turn out to be important implementing managerial schemes. The affected actors' own perception of what is legitimate in a certain context constitutes a political reality that often has to be dealt with. Legitimacy for social scientists is always legitimacy-in-context, rather than absolutely, ideally, or abstractly (Beetham 1991). As an analytical concept it is rather vague and general and has to be specified contextually; this will be undertaken in later paragraphs.

The balance between legitimacy and political outcomes is essential for any government. So also in our case. What we shall focus on is the problem of shaping a national policy for sustainable resource development in this context. What are the options and the political limits? To what degree do the national allocative problems influence the political outcomes? The underlying problem we are facing is first and foremost one of access to the resources and priorities under shifting resource situations. We are here dealing with a fundamental problem in the literature about CPRs, the open access character of such resources, and its implications for ecological and economic maintenance (Hardin 1968; Pearse 1981; Keen 1988).

Access and Enclosure

The open access to the fish resources in the Norwegian waters has been considerably modified in the last fifty years. And even if we expand the span of time there have been different kinds of restrictions in taking up fishery as a way of living, both formally and informally. Since the Raw Fish Act was passed in 1938 the entrance to the raw fish market was more or less controlled by the organized fishermen. And in order to be registered as a fisherman in the public files to obtain professional rights and benefits, a minimum of documented fishing activity was needed. Only registered fishermen were allowed to have boats over fifty tons registered as fishing vessels, and this was a necessity if the boat was to be used for that purpose. By organizing and making use of the open political channels to the government in the 1930s, so-called outside, private capital, and opportunistic speculations in the natural ups and downs of the fisheries were stopped by laws and organizations (Hallenstvedt 1982). This policy was even strengthened in the post-war period, although some exceptions had to be made to develop a limited fleet of trawlers and deep sea fishing vessels. The renewed Trawler Act of 1951 nevertheless gave protection to fishermen drifting with traditional, passive gear. Trawling continued to be licensed, limited in number with restrictions on fishing areas (Sagdahl 1982a).

Despite the substantial reduction of the number of fishermen during the 1950s and 1960s the pressure on the resources turned out to be too hard. The first fishery to experience the increasing scarcity was the rather industrialized herring fishery. The technological development had increased the efficiency far beyond the limit for sustainable development of the Atlantic-Scandic stock of herring. The traditional open access to the fishery for those belonging to the enclosure of fishermen and the lack of proper legal backing made the introduction of licensing in this fishery belated and inadequate as a managing instrument to maintain the stock in time. The result was a total breakdown with a following ban on industrialized exploitation of this resource. Some fifteen years later the stock is still too small to be normally exploited, although there has been some recovery during recent years.

While the access structure to the herring fisheries could be formed by a national public policy, this was not the case with access to the cod fisheries in the North. The extension of the fishing border to twelve nautical miles in 1961 gave rather minor protection for the pressure on the resource from the growing fleet of foreign trawlers. With a transition period of fishing up to six nautical miles for a time period of ten years for foreign trawlers, little seemed to be gained by the extension of the border. The coastal fishermen who had pressed for an extension were consequently dissatisfied with the solution and feared a coming breakdown of the stock (Sagdahl 1982b). Additionally, there was no limit for fishing cod in the nursery areas of the stock in the Barents Sea. Any restraints on Norwegian fishing on the stock would therefore just bring negative socio-economic impacts for the industry and the affected communities without any certain positive effect for the enhancement of the stock. Or if it did, it was a high price to be paid by the Norwegian fishing industry and almost impossible politically to bring about. The dilemma of this situation is a rather classic one in the study of collective action. Its logic leads to such tragic situations as the one described by Hardin. Although it was feared the cod stocks were diminishing, the fishing effort of the international trawler fleet in the Barents Sea continued to increase.

But increased fishing effort and the lack of proper jurisdiction and managing tools to limit the effort was only one dimension of the problem. Norwegian purse seiners gradually increased their fishing on capelin in the Barents Sea after the breakdown of Atlantic-Scandic herring. Later other nations followed. The food chain then became disrupted with severe consequences for cod, seals, and other species belonging to the top of the chain. At least, this could be maintained to be one of the reasons why severe imbalances in the ecological system became a fact at the turn of the 1970s. The growing exploitation of the shrimp stock probably also had some effect on the balance of the ecological system, along with circumstances in nature itself beyond human control. The need for improved management of the ecosystem in the northern waters became evident. This also implied restrictions on the Norwegian fisheries in the North. A new fishing policy had to be formed and adopted and the question of formal, expanded limitations to the access structure in the cod fisheries became urgent in this respect. Allocative policy on the international and the national levels could no longer be avoided.

Allocative Policy

Even before the economic zone was implemented, a policy of bilateral cooperation with the former Soviet Union to restrict fishing practice in the northern waters was adopted. Since 1975 total quotas of cod and other species were yearly negotiated under the scientific advice from the International Cooperation for the Exploration of the Sea (ICES). The maintenance of the cod stock nevertheless turned out to be unsuccessful. Despite growing restrictions on access to the resource, the improved management failed to give the necessary results. By the end of the 1980s the biological condition of the stock was worse than ever and in 1988 a situation of crisis was officially stated. Comprehensive restrictions both on inshore and offshore Norwegian fisheries in the northern waters were introduced under a bitter political struggle. The old issues of who to blame for the situation and who should pay the costs dominated the public agenda. Policies for the fisheries became a national matter; massive regional political mobilization with both organizational and political impacts occurred. This situation highlighted the political limits for solving CPR problems with public policy and the shortcomings of the policy in the past. Our intention is to sort out these limits and reveal their consistence, thereby shedding light on some of the causes of resource management failure in the North. We will especially focus on the national allocative policy in this respect. Our thesis is that due to the national allocative problem, sustainable management of the stock became neglected. Why so?

Allocative political processes often take place in situations with formidable political pressure from affected interests and under considerable political noise. But whether such situations occur or not is, above all, dependent on the allocative object. If scarcity is the problem as often is the case in allocative public policy, there is a difficult process of legitimizing the allocative pattern (Salisbury and Heinz 1970), especially if the situation at hand has the character of a zero-sum or a minus-sum game. If so, some will become winners and some will become losers in the allocative processes. Not only the character of the allocative object is important in this respect. Groups dependent on the allocation will have different needs and claims addressing the allocating body. In the fisheries we often find that local or regional dependence on the resources leads to political pressure for unequal access. Equal treatment could be conceived as political-administrative favoritism if some of the affected interests reject the blame for the scarcity of the allocative object and are not willing to bear the burden of what other actors have caused. This implies that not only the character of the allocative object is important for the degree of political noise, but also the historical setting and the involved interest structure.

In the process of allocating the Norwegian quota of Norwegian-Arctic cod, the government is confronted with different demanding groups and socio-political considerations. First and foremost is the case of the coastal fishermen in the North and the belonging communities. The cod fishery is the backbone of the economy in the North and especially at the coast. Problems in the cod fishery will easily affect most of the economic structure of the dependent regions and above all the labor market. Both local and regional public authorities will therefore take a strong interest in fishing policy and how it is executed. Some communities are dependent on trawlers and freezing plants, but these are few in number compared to conventional processing. A split of interests between active and passive gear and the belonging industry can be noticed in this connection, but by and large the coastal fisheries constitute the main interest in the North with a rather great potential for rallying political support. This implies regional departments of political parties, members to parliament, and organized interest groups, as well as general public support.

Allocating cod quotas also affects the fishing industry at the west coast. This industry is generally more capitalized than in northern Norway, both at sea and shore. With a general decrease in availability of fish resources at the West and in the South, ground fish fisheries performed by private trawlers and factory trawlers, and the unlicensed, growing fleet of auto-liners has become more and more dependent on the cod resource in the North. As the west coast has a more complex economic structure than northern Norway, the possibility of gathering a similar degree of political support is less. But the image of having a modern, competitive fishing fleet and its importance for coastal communities as well as the national level has been turned into a political asset both within the fishing industry and on the national level.

Allocating scarce resources and securing a sustainable development by governmental policy and administration seems, with this background, to be a risky political project. Government is here confronted with considerable political tension. The tension lies between those fishing with active (trawl) and passive (nets, lines, etc.) gear—capital-intensive versus labor-intensive forms of fishing in a period with growing unemployment, especially in the coastal areas in the North—and a developing regional conflict imbedded in regional political networks with open channels to the national political level. Members of parliament are above all regional representatives. Besides, one of the traditional political bases for the Labor Party in northern Norway is the coastal areas. And the party has as such had a rather close cooperation with the Fishermen's Union, being part of the same social-political movement in the North (Hallenstvedt 1982).

With the exception of a department within the Labor Union offering membership to the crews of the industrialized fishing fleet, the rest of the Norwegian fishermen are more or less organized through the Fishermen's Union. Whereas the union on the regional level channels the interests of the coastal fishermen of the North, it turns out to be a more complex organization on the national level. As an umbrella organization it adjusted to the differentiated structure of the fishing fleet that developed and comprises today the above mentioned tensions. The negotiated decisions that follow from such an organizational structure may differ considerably from the more homogeneous interests advocated by its northern members. The organizational voice and dissatisfaction with its way of functioning has become a dominant issue in the organizational debate in the North. In 1988 it led to a split as a discontented group of coastal fishermen formed an alternative or rather supplementing organization, the Coastal Fishermen's Union. So far it has not succeeded in getting a formal status within the Fishermen's Union. Nor has it been accepted by the government and the Ministry for the Fisheries as a functional actor in the governmental organizational network (Sagdahl 1992a).

To form and implement a resource policy and allocate scarce quotas in such a context easily challenges the political authority of the responsible government minister and the legitimacy of the decisions. In addition, fishing policy is formally linked to the general district policy with a responsibility for employment and the general economic well-being for communities and regions linked to the industry. These are officially stated political goals along with narrower industrial goals such as resource maintenance and economic industrial efficiency. The potential for goal conflicts are therefore manifold as the one or another is activated. The character of the blend decides the political reactions. But in situations where the allocative goods are scarce and diminishing, the allocative decisional process will easily be lifted out of the quiet scene of routine policy to one afflicted with political noise and contending parties. Since the final outcome has severe consequences for economic maintenance and stock enhancement—as in our case—the room for political action is limited. International and bilateral obligations complete the political scene in this respect.

Our question is how to legitimize allocative decisions within this framework. Legislative backing is, of course, a prerequisite, but is not necessarily sufficient to give political room for decisions without severe political costs. Giving co-influence and co-responsibility to the affected parties by corporate political-administrative bodies is a well-known governmental technique in such situations (Olsen 1983; Cawson 1985). An advisory body for resource regulations was put up long ago in the early 1970s, but with decreasing quotas and growing concerns its representativity has been questioned by the coastal fishermen in the North.

Legitimizing the decisions by sticking to scientific advice is another source, but the scientific validity of this advice has been challenged by the fishermen's own experiences and impressions of the present state of the resource. Later admittance of inaccurate prognoses has weakened the political functionality of this legitimacy source. Almost paradoxically, it has become more politically important both nationally and internationally as situations of resource crisis have come about. But it is also in such situations that the problem of legitimacy is stressed and challenged by the affected parties. Unacceptable political costs for the responsible government will therefore easily follow, a situation any government seeks to avoid in a parliamentarian system such as Norway's.

The above mentioned sources of legitimacy could be systematized as procedural legitimacy and scientific legitimacy. A third source should also be mentioned—what we here according to input/output analysis choose to label "outcome legitimacy." If the affected actors are more or less content with the political-administrative outcomes, the two other sources will become less activated. But in allocative situations where discontent, protests, and considerable political noise dominate the scene, all the sources of legitimacy will easily be challenged by the affected actors. A situation of reduced quotas of cod with no escape route for the affected actors (zero-sum, minus-sum game) illustrates such a scene. Allocative decisions in such situations will easily imply considerable political costs unless the character of the situation can be redefined in some way or another.

The Allocative Pattern

Regulatory policy in the cod fisheries on this background seems to be a political challenge where any government may easily become unpopular by those affected. The coastal fishermen in the North have not defied regulations as such, but have maintained that those who caused the situation should also pay the price of restrictions to the resource. Since the 1950s they have pressed for access limitation to the resource by extending the fishing border and thereby limiting national and foreign deep sea trawling. The extension of the Icelandic fishing border in 1972 reactivated their demand. They feared a breakdown of the resource if the capital intensive fishing effort was not restricted and above all the consequences for themselves and their communities.

Until 1980 the coastal fishermen avoided being a target group for the expanding regulations of the fishing effort on Norwegian-Arctic cod. To solve the national allocative problem, thus preventing political noise and a possible compliance problem, Norway had negotiated an exclusive right for those fishing with passive gears in the newly established bilateral commission with the former Soviet Union. Passive gear fishing could continue, although the national quota had been reached. The result was a massive over-fishing of the total quota of cod for most of the regulative period up to 1988. And when the resource situation made more comprehensive regulations neccessary from 1980 on, including restrictions on the coastal fisheries, the government got its first lesson on what was to come. Believing not to have caused the depletion of the resources, the coastal fishermen in the North regarded it illegitimate to have to bear the burden of the suggested time-limited fishing ban. Considerable outcry, as well as threats of civil disobedience, soon ensued (Sagdahl 1991).

What the coastal fishermen consider to be fair management and acceptable distributional solutions in times of scarcity seems to be closely linked to their perceptions about causal explanations to depleted resources. Since it began in the 1930s, deep sea trawling has been considered to be a threat to resource maintenance—not only because of its efficiency and the amount of total catches, but also because of the gear's unselective character when used in typical nursery areas. H. P. Saxi argues that the resentments against deep sea trawlers have become a part of the cultural pattern of the coastal fisheries, especially in the county of Finnmark, the county with the longest and strongest experiences of national and foreign trawling (Saxi 1988). It should also be added that the Norwgian trawlers were initiated and owned more or less by the fish producers, and that most of them still are closely integrated with the fish processing industry. In addition, over time they have been considered to constitute a supplementary fishery to the coastal fleet. This position became the rational behind the Trawler Act of 1951 that regulated the development of Norwegian deep sea trawlers (Sagdahl 1982a). The coastal fleet's major position was later stated in several policy documents. So when distributional questions of scarce resources came to the fore in the 1980s, the coastal fishermen's perceptions of fair management and allocation of quotas were nourished by a traditional cleavage at the coast and backed by a causal model of thinking about the occurrence of the breakdown of the resource system. And the resource crisis itself was the final evidence to the validity of their explanatory model of thinking.

Forming and implementing a policy that favors sustainability of the resource system is easily impeded by the distributional problems in such a situation. Table 1 reflects some of the regulatory political problems the government has been facing. The discrepancy between TAC and the total catches, quotas given to third countries in situations of national scarcity, and the over-fishing for years of the Norwegian quota convey a message of an underlying political landscape not easy to handle for any government. But in 1988 there was a change in the problem structure, when a near state of emergency was declared due to the reported status of the cod stock. Improved models and new data had changed the former optimistic message from the marine biologists to one of crisis. Drastic reductions of the quotas and a sudden stop in the over-fishing by the Norwegian coastal fishermen became necessary. While extended regulations with prospects for enhancement in the latter part of the 1980s had seemed a boon to resource conservation, a contrary situation had arisen where all groups of fishermen had to share the extended burdens of regulation. The

TABLE 1 Quotas and Catches of Norwegian-Arctic Cod 1977-1991 (in 1000 ton)

Year	QUOTAS					CATCHES						
	TAC	Third count.[a]	Soviet Union	Norway[b]	Types of gear		Total	Third count.	Soviet Union	Norway	Types of gear	
					Active	Passive					Active	Passive
1977	890	150	37*	370	180	190	945	146	370	429	161	268
1978	890	130	38	380	195	185	733	69	267	397	151	246
1979	740	90 (15)	32	325	135	190	485	40	119	326	132	194
1980	430	48 (8)	19	191	80	111	420	33	115	272	89	183
1981	340	35 (5)	14	158 (5)	60	98	448	38	83	327	76	251
1982	340	35 (5)	9	208 (55)	60	148	406	36	40	330	69	261
1983	340	35 (5)	6	240 (88)	60	180	328	33	23	272	68	204
1984	260	20 (4)	4	195 (75)	55	140	311	25	22	264	55	209
1985	260	20 (4)	5	185 (65)	55	130	336	34	63	239	63	176
1986	440	40 (15)	15	250 (50)	93	157	457	47	151	259	102	157
1987	600	56 (24)	20	342 (70)	177	165	553	51	204	298	174	124
1988	491	41 (18)	20	250 (25)	120	130	456	40	169	247	122	125
1989	340	28 (12)	13	178 (22)	62	116	353	39	134	180	65	116
1990	200	14 (6)	7	113 (20)	28	85	287	14	75	118	29	89
1991	225	18	10	129	29	98	251	19	79	153	32	121

a. The numbers in parentheses denote quotas supposed to be caught in the protection zone at Spitzbergen.
b. The numbers in parentheses denote quotas transfered from the Soviet Union to Norway.

political costs of enforcing detailed regulations in the coastal fisheries could no longer be avoided. In the years prior to 1988 we find that the regulating authorities had met the compliance problem of the coastal fishermen by following at least four supplementary strategies, as outlined below.

First, the size of the total quota was increased somewhat over the biologically recommended one, or the maximum quota was chosen in situations where options were recommended. By doing so the formally zero-game situation gave better opportunities to avoid national allocation conflicts.

Second, the negotiated right to have the opportunity of exceeding the quota by fishing with passive gear turned the zero-game allocation situation into a plus-sum situation for the most numerous group of fishermen. In reality no fixed quota existed for this group until 1989, where the stated resource situation made this negotiated right impossible to go on with.

Third, the Soviet Union was willing to transfer a considerable share of its cod quota in exchange for Norwegian quotas of other species. Thereby the allocation situation improved and conflicts could more easily be avoided.

Fourth, the negotiated quotas for third countries were considerably reduced, although not to the size demanded by the coastal fishermen in the North. Giving away shares of the quota was regarded as unacceptable when Norwegian fishermen had to bear the burden of reduced fishing.

This strategy was regarded as politically functional until 1988. Except for two periods with bans on fishing—lasting some few weeks—there had been no set quotas for the coastal fishermen in the North. This does not imply there was a general agreement on the policy. The fishery closure during the seasonal fishery of spawning cod at Eastertime had been heavily criticized since it was introduced in 1980, being very economically important for those fishing with passive gears. Additionally, the extended weekend restrictions on fishing that were introduced in the midst of the 1980s were fought. One important underlying reason was the general decline in the availability of cod at the coast. The important seasonal fisheries at the Lofoten Islands and at the coast of the nothernmost county (Finnmark) some months later had shown a decline since 1984. Spawning cod were reported to be meager, the growth of the immature part of the stock slowed down, and seals invaded the coastal waters in the North with severe consequences for the availability of any species of fish in affected areas and hence the economic sustenance of the fishermen involved. The ecological system in the Barents Sea seemed to be out of order; the coastal fishermen feared for their future and pressed unsuccessfully for a sudden reduction of the deep sea trawling of immature cod. The backing from marine biologists was lacking in this respect. Their prognoses reported stock improvement and "better times" at the turn of the 1980s (Sagdahl 1992a). In the spring of 1988 these experts, on the contrary, had found the situation to be alarming. Severe administrative measures had to be taken to avoid a complete breakdown of Norwegian-Arctic cod. The most important measure in this respect was the suggestion of the use of boat quotas for almost all kinds of vessels. All groups of fishermen would be affected. The reactions followed immediately.

Regional Political Mobilization

The new recognition of the state of the cod stock led to a renegotiated reduction of the quotas for 1988 and a major reduction for the following years. The coastal fisheries also became an important target group for extended regulations at this time. Their right to over-fish the national quota of cod was dropped. Access to their main resource was utterly reduced by administrative measures. The introduction of individual boat quotas for all parts of the fishing fleet was heavily disputed. Some former participants were even closed off as the quotas of cod were allotted on the basis of average catches over the preceding three years. Economic sustenance became difficult and led to a reduction of crew and for some even to bankruptcy and selling off the boats. The economic and social fabric of many coastal communities became endangered and led to a comprehensive social and political mobilization.

While the previous resistance against restrictions for coastal fishing for cod had been largely maintained by coastal fishermen and their local and regional organizations in the North, heavier political actors now entered the political scene. Local politicians and mayors of coastal municipalities with national and regional political networks came into the foreground. Wide support was rallied among different groups and professions. Formal movements were established and environmental interest groups got unexpected allies demanding a new policy for resource maintenance and the fisheries in the North. Mass meetings of fishermen and other coastal citizens demanded that the responsible minister leave his post. The conflict was covered by the national media and coastal problems in the North were highlighted.

The regional mobilization that was triggered off in the wake of the resource crisis led also to a political focus on the regional allocation of fish resources in general and how the capital-intensive fishing fleet from the west coast had increased their share of the available resources during the 1980s. The northernmost regions' dependence upon the resources in the Barents Sea and the northern waters became a hot topic. A policy of regionalization of access to the resources was advocated by influential actors, leading county politicians, and the public county assemblies in the North. Preparatory steps were taken to form an alternative fishing policy based on a regionalization of fishing rights by regional quotas and a licensing system. This represented a severe challenge and a political attack on the present fishing policy and also caused political mobilization in the western part of the country.

Another momentum should also be noted in this respect. Coastal fishermen in the North had been dissatisfied for a rather long time with their national interest organization, the Fishermen's Union, and its way of functioning. One criticism held that its heterogeneous character had prevented it from being an efficient advocate for the coastal fisheries in the North. This judgement became utterly nourished by the disputes over regulations and the pattern of quota distribution. A new organization was formed, challenging the established interest structure of the industry and its political-administrative network.

Forming and implementing a policy to meet the reported resource crisis strained the traditional base of legitimacy under the conditions mentioned above. As the Ministry of Fisheries was forced to abandon the former strategy due to the biological status of the cod stock and Norway's international responsibility as a co-manager of the stock, the policy was from then on strictly derived from advice from ICES. The zero-sum situation that rapidly developed into a minus-sum situation at the turn of the 1980s made the national allocation an extremely difficult administrative task. The regional challenge from the north would easily lead to comprehensive political costs for the governing political party and especially for the responsible minister. The Labor party, which had just taken over the governmental responsibility, was in particular politically vulnerable to political pressure from the North. Additionally, the new minister for fisheries was an elected parliament member from Finnmark, the nothernmost county.

Formally the allocative decision was an administrative and not a political matter. The political implications were nevertheless unquestionable—a negotiated order was needed. The advisory body, the Council for Resource Regulations, formally had a mandate to suggest a solution, but without the consent of the Fishermen's Union it would not work politically. The organization possessed the key to the problem of allocation. What we find here is a typical corporate solution to a political problem (Lembruch and Schmitter 1979; Cawson 1985). Framing the factual policy was left to a private organization outside the government, thereby obtaining a sufficient legitimacy base to solve the allocative question. The top executives of the organization had their meeting close by, giving advice to the council. The press from the local organizational level in the North and the organizational split gave a recommendation that favored the coastal fisheries in the present situation. No other option seemed politically possible. The coastal fisheries came out with 75 percent of the quota, but this relative share was to be reduced if the Norwegian quota increased in the following years. If so, the trawlers' share of the quota was to be increased.

Although this could be regarded as a temporary victory under the present situation, the scientific justification for the extremely low quota was questioned by the affected fishermen. At that time they were experiencing a growing availability of cod despite the scientifically stated status of the stock. The fish also seemed to be in a good condition. The food base had been improved. Both the herring and the capelin stocks were in a state of recovery. Indeed, the marine biologists and ICES proved to be mistaken in their calculations. The scepticism towards their science and advice became stronger than ever. Consequently there was a demand for an immediate increase in the quota of cod.

While the 1980s began with access limitation to the resource by regulatory measures and prospects for a gradual deregularization when the stock had recovered, the decade ended in a situation of crisis—biologically, economically, and politically. The policy that until then had met the situation as it developed was now ad hoc. It was meant to be temporary. The need for a long-range policy to avoid the resource fluctuations and the political costs of administration became apparent as time went on. However, to form a functional policy under the current circumstances was more than a challenge. What could be regarded as functional for sustainable management of the resources could easily turn out to be politically unfunctional.

Functional Policy Solutions

The introduction of the economic zone in 1977 gave an impetus for long-range planning and development of the fishing industry. The policy document that passed parliament in 1978, forwarded by a social democratic government, concerned further access limitations to the ground fish fisheries in the North as neccessary. Deep sea trawling was to be reduced and the coastal fleet to be favored. The rapid and rather unexpected decline of the resources, especially the cod stock, made the policy document obsolete even before its implementation. The revised plan that passed parliament in 1983 under a nonsocialist government differed from the previous one by favoring market solutions to hierarchial management. This new policy direction was later followed up in 1989 when the Ministry of the Fisheries presented a preliminary working document where the access problem and the classic tragedy of the commons was to be solved by the introduction of privatization of fishing rights and individual transferable quotas (ITQs).

The influence of fishery economists and other coastal states such as New Zealand and, above all, Iceland was noticeable. But the political setting was different. To launch a policy based on privatization of the fishing rights would represent a fundamental shift favoring those with access to financial backing. The capitalized part of the fleet, regardless of regional affiliation, would profit from such a policy. If such a policy was carried through without any modification, in the long run the coastal fleet would be the loser and the districts in the North marginalized. The regional conflict, as well as the other conflict dimensions in the industry, became activated. Hence the political institutions in the North took an interest in the shaping of a new policy, a policy that should favor the region.

The regional conflict dimension in Norwegian politics is the oldest and probably the most fundamental of those structuring Norwegian politics. This dimension does not follow the lines of the political parties, but exists within the parties (Rokkan 1987). The crisis in the cod fisheries had activated this latent conflict dimension. The reorganization of the county administration in the 1970s and the following development of both the political and administrative levels throughout the 1980s had also led to new political arenas eager to be activated as regional political instruments, constituting meeting places for problems, participants, and solutions. In this, the regional consequences of the resource crisis were a perfect example. Landsdelsutvalget, an advisory regional body for northern Norway which developed at this time, used its budget and organizational network to supply the documentation needed. It was established that natural resources that belonged to the region should benefit actors and communities in the North and not the distant fishing fleets belonging to other regions or even foreign countries. There was broad political support for these positions across the lines of the political parties.

The highly disputed Norwegian relation to the European Community also contributed to the activization of the regional conflict dimension. Additionally, the EC question activated all the conflict dimensions in Norwegian political life. Although shaping a new fishery policy and the government's aspirations for a future membership in the EC were different political processes with different backgrounds, they coincided in time and were regarded as closely linked by influential groups in the coastal areas. Political resistance could easily be rallied in this climate, especially in the North. The fear for increased market solutions and growing pressure on the resources of the North was widely shared among the inhabitants, especially at the coast. The general political frame for launching a shift in the fishery policy to management by market mechanisms was, in fact, the worst thinkable. Public opinion polls gave discouraging results for the Labor Party government, especially in its northern stronghold. The public hearing of the preliminary policy document returned the message of an increasing loss of voters in the North if this policy would be carried through. Both local and regional departments of the Labor Party in northern Norway rejected the proposition, which politically ended the ITQ suggestion. Another policy had to be lined out.

The rewritten policy document returned to the principles laid down by the Labor Party government at the turn of the 1970s. The coastal fleet was the one to be favored due to its positive impacts for economic maintenance of the coastal districts in the North. The overcapacity of the fishing fleet, it was argued, was found in the bigger, deep sea fishing vessels and expanded licensing was recommended for this part of the fleet.

The former discussion of the access problem and the use of market mechanisms was not in the fore any longer, obviously for political reasons. The political problems of legitimizing such a policy under the prevailing circumstances had been too great a challenge. This could be discerned from the policy document itself. There was, however, a considerable discrepancy between the general analysis and its policy recommendations, although the urgency of the matter gave no time for a complete rewriting of the document. Besides, the analytical model of thinking could also probably be said to mirror the prevailing analytical approach found in the ministry. Although the ITQ question was left in the dark, some of the propositions could be linked to the market model of thinking found in the ministry. The introduction of a resource fee could be said to fall within the established analytical frame. Whether this remedy will have any effect for entry limitation or not is dependent on the size of the fee. What is more important as a political signal is the principle involving the open-access structure of fish as a common property resource. The proposed fee reflects a new way of thinking in this respect.

The proposition of making a new public record for registration of fishing rights should also be noted. Formal qualifications, not only experience in fishing, should be demanded as entrance tickets. Both of these propositions may have impacts for the coastal fleet for access to fishing rights. Over time the enclosure of the commons will probably narrow if these policy recommendations are implemented. None of these proposals were justified by referring to any access structure as a problem for biological and economic maintenance. They were more or less presented as practical propositions to reduce management costs and improve unreliable public data on registered fishermen.

Making a split between the coastal and deep sea fishing fleet by direct limitations to fishing rights is above all justified by its political functionality. The experiences from the 1980s show that limiting resource access by a detailed regulatory system for the coastal fleet will be perceived illegitimate (Sagdahl 1992a). The new policy document underlines the importance of perceived legitimacy of the political-administrative measures for its efficiency. An important political lesson apparently seems to have been learned. The document also stresses the importance of control and the improvement of this variable for successful resource management. Here we are facing another limitation to sustainable resource management by policy solutions on the national level.

Legitimacy and Control

While the legitimacy of restrictions on fishing rights is questioned by those believing themselves free of blame for the situation that has arisen, the gravity of the situation may demand comprehensive action is taken. Not only is the allocation of benefits important in this respect, but also the allocative pattern of burdens, for perceived legitimacy and compliance to the administrative measures. This is important within industry at the national level as well as the international level—as in the case of the resources in the Barents Sea. The motivation for subjection to national or group limitations is closely linked to the perceived compliance by other nations. The coastal fishermen in the North have consistently complained of suspected illegal fishing by vessels from the EC and especially from Russia. These suspicions are rather widespread in the North, although insufficiently documented. Several reported cases of illegal fishing indicate that the problem of over-fishing seems to have far greater proportions than earlier expected.

But it is the belief, whether justified or not, which constitutes the political reality. Reports of uncontrolled fishing by EC vessels outside the economic zones in the Barents Sea, a situation similar to one on the east coast of Canada (Sullivan 1989), have also nourished the criticism of the insufficiency of the control regime. Former irregularities of fishing and shortcomings of the surveillance system in the fishing protection zone around the isles of Spitzbergen have also constituted a management problem. These events and the shortcomings of policing the implementation of rules laid down in the resource policy and negotiated treaties have undoubtedly influenced the compliance problem in the Norwegian fisheries in the North. What is more, the uncertainty of impacts for a sustainable resource management is an even bigger problem. Both stock estimates and prognosed impacts by the set quotas will be affected by unreported catches. An improvement of management control is therefore decisive for improved legitimacy and the efficiency of the regulatory measures.

The motivation for abiding regulatory statutes and sticking to low quotas in the domestic fisheries will naturally be influenced by the above mentioned momentums. Especially for those believing that there is unfair treatment and that the TAC has been set too low. The road to overfishing and the use of black markets for selling the illegal catches is not long under such circumstances. Individual benefits to solidaric misery could easily be preferred in such situations. We are here confronted with the well-known problem of free-riding, where organizational bounds turn out to be insufficient to control and discipline self-interested actors (Olson 1971). Solidaric behavior will, in this case, favor only those who break away, who go on fishing regardless of given quotas and regulatory prescriptions. The fact that part of the cod stock and the vessels operate in international waters makes solidaric behavior even more difficult to come by. The need for improved management and especially for monitoring and control seems today to represent a major challenge to the sustainability of the resource system in the area.

The transition to multistock management models as signalled by the government could also be said to necessitate a better policing system. Such a management scheme implies easily disputed decisions in a fishing industry made up of specialized and differentiated fisheries and its supporting economic activities ashore. Industrialized capelin and herring fishing has to be balanced against the bio-economic considerations of the cod fisheries and whaling. What interests to be favored are not merely just economic and biological questions in the light of professional models, but also questions of political networks and political realities. The political pressure from economic actors in the industry is diverted from their own investments and economic needs, not the well-being of the industry and the resource system as a whole. While the trawler interests pressed for increased quotas of pollock in the fall of 1992, the coastal fishermen protested, referring to the observed depletion of the stock. Multistock management when implemented may therefore turn out to be otherwise than intended. And if carried through by stressing political and administrative authority under heavy protest from discontented fishermen, the probability for illegal fishing and monitoring efforts will increase. This dilemma can hardly be avoided if multistock management is ever to be designed and implemented.

Towards a New Management Regime?

An improvement of just the Norwegian control system would not suffice to solve the problems of legitimacy and compliance. The resources in the Barents Sea are bilateral resources with Russia, and to improve the efficiency of the control system Russia must be included. Besides, there are special problems of control in the protection zone at Spitzbergen and the jurisdictional problem of fishing activities outside the economic zones. These are special challenges that need special solutions on the international level.

The big question is how to organize an efficient surveillance and controlling system. The prevalent model of thinking is diverted from the national judiciary system. Besides, the character of the former regime that Russia was a part of supported this way of thinking and did not invite cooperative solutions on the bilateral level. The political presuppositions for closer cooperation in policing the northern waters were lacking.

To solve the compliance problems brought on by public deregulations and self-governing systems of the affected parties, the formalization of local informal systems of cooperation have been advocated by a number of scholars (McCay and Acheson 1987; Jentoft 1989; Pinkerton 1989; Ostrom 1990). But the Norwegian fishing industry consists of contending actors not easy to reconcile, and there is a long political tradition of regarding the public authorities as the natural problem solvers. Environmental pressure groups have also taken an interest in the management of the resources; adding to this are the bilateral and international aspects. On this background top-down management seems to be the most plausible organizational approach (Sagdahl 1992b). Especially when dealing with matters concerning legal authority and bilateral questions, no other approach seems legitimate or functional. Still, it remains a question of how best to organize to improve the compliance problem.

To regard the ecological system of the Barents Sea as an undivided unity regardless of national economic zones could be said to be a legitimate starting point. The present administrative institutions involved in the administration of this ecological system are divided both on the international and the national levels. They are parts of different political-administrative networks partly stemming from their functional differentiation and historical backgrounds. Both conflict and cooperation is found within and between these networks. Their functional potential for securing a sustainable development of the ecosystem is limited as to the complexity of the problems they are facing. Their ability to handle policing functions has been questioned for a number of years and the need for improved efficiency has become a politically recognized fact in the Norwegian context.

If the ecosystem of the Barents Sea was the only consideration to be taken regardless of national economic zones and borders for forming functional institutions for sustainable resource management, then bilateral co-administration by one organization located in the area could be said to be preferable. Such an institutional framework for the policing functions will undoubtedly give improved possibilities. The political reality of such a solution can of course be questioned. Not at least will the mere existence of established institutions form a bar for such a development. The imbedded interests of their present localization and networks will easily make any transformation unrealistic. Institutional transformations and relocations are heavy political processes not easy to carry through.

The political orientation towards an eventual membership of the Common Market could also be said to constitute an obstacle in this respect. Some of the EC countries have a strong interest in getting extended access to the fish resources in the North. Any institutional change has to include the EC fishing interests in the institutional framework if membership becomes a reality.

The present improvement of the resource situation in the Barents Sea could also be said to be working for institutional conservation instead of institutional development in a regional and ecological context. But increased resources have also led to increased catch efforts in the area, as well as in the international loophole beyond the control of the nation state. Both international law and the present organization of management and control seem therefore to be out of correspondence with the problem structure as it has developed. And the long-range effects for the ecosystem become consequently harder to foresee. That is why the importance of solving the problems of control will probably become the greatest challenge in the years to come if a sustainable development of the ecosystem of the northern waters shall ever be reached.

REFERENCES

Beetham, D. 1991. *The Legitimation of Power*. Atlantic Highlands, New Jersey: Humanities Press International, Inc.
Cawson, A., ed. 1985. *Organized Interests and the State. Studies in Meso-Corporatism*. London: Sage Publications Ltd.
Easton, D. 1957. "An Approach to the Analysis of Political Systems." *World Politics*, no. 3.
Hallenstvedt, A. 1982. *Med lov og organisasjon*. Oslo: Universitetsforlaget.
Hardin, G. 1968. "The Tragedy of the Commons." *Science* 162.
Jentoft, S. 1989. "Fisheries Comanagement: Delegating Government Responsibilities to Fishermen's Organizations." *Marine Policy*, April.
Keen, E. A. 1988. *Ownership and Productivity of Marine Fishery Resources: An Essay on the Resolution of Conflict in the Ocean Pastures*. Blacksburg, Virginia: The McDonald and Woodward Publishing Company.
Lembruch, G. and G. L. Schmitter. 1979. *Trends Toward Corporatist Intermediation*. London: Sage Publications Ltd.
McCay, B. J. and J. M. Acheson, eds. 1987. *The Question of the Commons: The Culture and Ecology of Communal Resources*. Tucson: The University of Arizona Press.
Olsen, J. P. 1983. *Organized Democracy*. Oslo: Universitetsforlaget.
Olson, M. 1971. *The Logic of Collective Action: Public Goods and the Theory of Groups*. Massachusetts: Harvard University Press.
Ostrom, E. 1990. *Governing the Commons: The Evolution of Institutions for Collective Action*. Massachusetts: Cambridge University Press.
Pearse, P. H. 1981. "Fishing Rights, Regulations, and Revenues." *Marin Policy*, April.
Pinkerton, E., ed. 1989. *Co-operative Management of Local Fisheries*. Vancouver: University of British Columbia Press.
Rokkan, S. 1987. *Stat, nasjon, klasse*. Oslo: Universitetsforlaget.
Sagdahl, B. 1982a. "Teknologisk endring og interessekonflikt. Trålfiskets innpasning i torskefiskeriene." In *Fiskeripolitikk og forvaltning-*

sorganisasjon, edited by K. H. Mikalsen and B. Sagdahl. Oslo: Universitetsforlaget.

Sagdahl, B. 1982b. "Kystfisket og fiskerigrenseutvidelser. Noen trekk ved beslutningsprosessene." In *Fiskeripolitikk og forvaltningsorganisasjon,* edited by K. H. Mikalsen and B. Sagdahl. Oslo: Universitetsforlaget.

Sagdahl, B. 1991. "Balancing the Brink." In *Regional Development around the North Atlantic Rim,* vol. 1, edited by M. Leroy. Swansea, G. B.: International Society for the Study of Marginal Regions.

Sagdahl, B. 1992a. "Ressursforvaltning og legitimitetsproblemer. En studie av styringsproblemer ved forvaltningen av norsk-arktisk torsk." *NF-rapport* nr. 15/92–20. Bodø: Nordlandsforskning.

Sagdahl, B. 1992b. "Allmenningsressurser og forvaltningsorganisasjon. Statlig overformynderi eller selvhjelp?" *NF-rapport* nr. 16/92–20. Bodø: Nordlandsforskning.

Salisbury, H. and J. Heinz. 1970. "A Theory of Policy Analysis and Some Preliminary Applications." In *Policy Analysis in Political Science,* edited by I. Sharkansky. Chicago: Markham.

Saxi, H. P. 1988. "Innovasjon og politisk motstand. Konfliktlinjer i spørsmålet om tråldrift." *NF-rapport* nr. 10. Bodø: Nordlandsforskning.

Sullivan, K. M. 1989. "Conflict in the Management of a Northwest Atlantic Transboundary Cod Stock." *Marine Policy,* April.

Part Five

Comparisons and Conclusions

Introduction

The utilization of the resources of the complex ecosystems in the Barents Sea and on the plains of northern Fenno-Scandia has created very diverse systems of use rights and powers of management. The legal system defining rights and duties and empowering appropriators is currently in a state of reformation. The situation in the Barents Sea changes through the yearly negotiations among the states with established interests in the fish resource. On land the situation changes through the political process; in particular, the situation of the reindeer herding communities has been changing rapidly. The situation for the Saami in Russia is unknown, but presumably it has been deteriorating with the worsening of general social conditions since the collapse of the Soviet Union.

Both the status and the rate of change for reindeer herders is different in the three Nordic countries. In Norway and Sweden the fate of reindeer herding is closely tied to the fate of the Saami people. One of the more interesting developments is the Saami Rights Commission established by the Norwegian government in 1980 after a major incident of civil resistance to the development of a dam for generation of hydroelectric power. (The Alta development, with some modifications, was completed in 1987.) In 1984 the commission presented their first recommendations. One result was to enact elections of representatives to a Saami parliament (*Sametinget*). This process was established in 1989. In 1993 Sweden created a similar *Sameting*. Finland had already created a Saami parliament (*Sameparlamente*) in 1973 by a provisional ruling. New legislation made it permanent in 1995, and it was renamed as *Sametinget* (NOU 1997, 5).

The major task of the Norwegian Saami Rights Commission was to consider the rights to the land and water the Saami had been living in. In 1997 they presented their final recommendations (NOU 1997, 4), briefly summarized by Torgeir Austenå (see chapter 13). Their proposal is to create a new system of commons in Finnmark County modeled as a hybrid between the state commons and *bygd* commons (see part 2). The outcome of this proposal depends on the political processes in Norway and Finnmark in the next few years.

Douglass North (1990) puts the political processes (broadly conceptualized) as the primary driving force determining the long-term track record of various economies. The role of scientific knowledge in political processes will easily be exaggerated by scientists. But there is also reason to believe that on the margin those who heed valid knowledge will do better than those who do not (North 1990, 73–82). The process of change both in the Barents Sea and on the plains of northern Fenno-Scandia will do no worse, and perhaps a lot better by being informed by theoretical developments in resource governance and studies of similar situations in other parts of the world. As a step in this direction two articles are included in this section: one discussing the fisheries of Namibia and the other the range lands of the inner Niger Delta in Mali. But the study of institutions in northern Europe may also contribute to the growing science of resource governance. To this end we have included an article discussing how to conceptualize and study property rights based on experiences in Norway.

"The Namibian Fisheries Resource and the Role of Statutory Law, Regulations, and Enforcement of Law in Its Utilization," by Carl-Hermann Schlettwein and Pierre Roux, establishes Namibia as an interesting control case in the sense that this may be one of the few cases where statutory law designed exclusively to manage the fish resource sustainably is the only institution affecting the resource. The Benguela current flowing north along Namibia's coast is highly productive. Yet, upon independence the sea had nearly been emptied by the extensive fishing of a mostly European fishing fleet. In rebuilding the fisheries, Namibia had to start from the bottom. Based on the common law as existing and applied in South Africa since 1920 (as well as the law of the sea), they had to design a regulatory system which could both rebuild the fish stocks and harvest them in a sustainable manner when they were rebuilt. Schlettwein and Roux provide a survey of the legal history and current status of these regulatory policies. The description provides a baseline for comparison of the development of Namibia's own fisheries. If Schlettwein and Roux are correct in their assumptions, Namibia should benefit from its fish resource without disastrous fluctuations. The analysis may also provide a baseline for comparisons with regulatory systems where local informal institutions and established common law rights as well as international treaties complicate the regulatory regime.

Trond Vedeld's chapter, "State Law Versus Village Law: Law as Exclusion Principle under Customary Tenure Regimes among the Fulani of Mali," is based on fieldwork among the Fulani in the "leydi of Dialloubé" in the north-central inland delta of the Niger River. His concern is chiefly to discover how customary property rights regimes interact with the regimes the state administration (or those who control it) tries to impose. The traditional tenure regime with roots in the Dina State (1818–1862) is today under pressure from a variety of processes originating both in the political and the ecological system. The result is increased pressure on the resources. This is most visible in the enclosure of the flood plain areas suitable for rice growing (these areas are also a critical pasture land for the Fulani during the dry season). Vedeld outlines both the customary system of rights to cattle and access to pasture as well as the existing theory of property rights as enacted by the state. The system is very diverse, with all levels of control ranging from open-access to full private inheritable property. This diversity exists in a societal situation of ambiguity, leading to a variety of conflict areas depending on the choice between state and customary law as well as a choice of interpretation. The consequences of the various actions and inactions chosen by officials and administrators have long-term implications for the legitimacy and authority of the state. (One problem for the state is that the diversity of customary rights is difficult to match in substantive statutory law. Vedeld suggests that the state should put more effort into developing appropriate legal procedures for resolving conflicts equitably rather than trying to impose its own system of property rights.)

In his reflections in "Animal Property, Access Rights to Pasture, and Transhumance Management," Vedeld raises the question of whether there are examples in modern law of access rights to or management controls of range lands being tied to cattle ownership. The answer is yes. In Norway, access to pasture for reindeer herding Saami within the districts set aside for reindeer herders depends only on ownership of reindeers, not on the ownership of the ground—as long as the pastures have been used as such from "time immemorial."

In "The Analytical Importance of Property Rights to Northern Resources," Audun Sandberg examines the role of property rights as a link between the physical world and the social world. His point of departure is the settlement history of the North and the legitimacy of access to resources granted by the old Roman law principles of first occupancy, possession, beneficial use, and effective control *(primus occupat, possessiones, usum fructum, and dominium)*. He then considers six different resources of the North for their potential contribution to our knowledge of design principles for resource management systems. The systems for collecting sea birds' eggs and downs are seen as those most uninterrupted by government intervention. If the management regime of sea fish of Namibia at one extreme is totally government designed, the collection of birds' eggs and downs is at the other extreme with a thousand-year history of uninterrupted development. The management of sea fisheries are mostly government designed with only a couple of fjord fisheries existing with any local tradition of management. But, as is also documented in part four of this volume, the many-stranded interrelations of international as well as national players at different levels make this a rich field to study. The third resource, the coastal waters as environment for fish farming, is an interesting case illustrating the conflict between direct government intervention and regulation of the fish farmers and local coastal zone management. A more consciously designed resource management strategy does not exist. In addition, a focus on wild salmon as a resource shows a great potential of insight into a very diverse set of management rules governing the fish's life from its hatching upstream in a river to migration to the North Atlantic to its return some years later to spawn. A rather recent addition to the range of northern resources is the water power used in generation of hydroelectric power. The abundant and cheap energy has played a major part in transforming the North to modern industrialized communities. Recent efforts to liberalize the market for hydroelectric power and privatize the ownership structure of the power plants according to the stock-owning model sets the stage for a new battle over the control of an important resource between the local communities and central government. In this, property rights to kinetic energy assume new importance. Finally, forests, pastures, and berries are a diverse arena of management regimes. Much of the wasteland in northern Norway is so-called "unmatriculated" state land and state commons (see chapter 7). These unclear property rights are now contested by many interests. Legal battles as well as government commissions are slowly redefining them. The dynamics of these changes warrant closer study.

Sandberg ends his discussion by calling for a more precise use of the concept of property rights in the social sciences and suggests that to make property rights more useful in the comparative study of institutional systems, they should be deconstructed into rights of access, rights of extraction, rights of exclusion, rights of management, and rights of alienation.

A main conclusion of this volume is that even the fairly simple ecosystems of northern Fenno-Scandia need complex institutions for their governance. The rapid growth in complexity of institutions for the governance of the resources of the Barents Sea could be seen as necessary to match the complexity of the ecosystem. But, of course, the complexity of institutions shall not only match the complexity of ecosystems, but also that of social systems. And which complexity is more important to the legal system?

Implicit in this book is that legal systems are a subject in need of better study. A better understanding of the chain of causation from the aggregation of everyday experiences of individual people's resource usage through legal cases as well as the political system until a change in the law is effected—however small—must be integrated with a better understanding of the chain of causation running from the decisions of the lawmaker; the detailed implementations of the regulating agencies; and the interpretations of the local administrations, law officers, and the public regarding their impact on everyday activities and the statuses of resources. This understanding must then be integrated with knowledge of the specific resources as well as the ecosystems generating the resources and the communities making their livings from them.

REFERENCES

North, Douglass C. 1990. *Institutions, Institutional Change and Economic Performance*. Cambridge: Cambridge University Press.

NOU (Official Norwegian Report) 1997:4. "Naturgrunnlaget for samisk kultur" (Natural Resources for Saami Culture). Oslo: Statens Forvaltningstjeneste.

NOU 1997:5. "Urfolks landrettigheter etter folkerett og utenlandsk rett" (Aboriginal peoples rights to land according to international law and the law of other states). Oslo: Statens Forvaltningstjeneste.

CARL–HERMANN SCHLETTWEIN AND PIERRE ROUX   CHAPTER 17

The Namibian Fisheries Resource and the Role of Statutory Law, Regulations, and Enforcement of Law in Its Utilization

Introduction to the Benguela System and the Namibian Marine Environment

The sea off Namibia is known to be highly productive due to the up welling of nutrient-rich waters from the cold Benguela current which flows northward along the coast. As with other coastal up-welling systems, the fauna of the Benguela current is dominated by fish species that can utilize the rich plankton production in the upper water layers. However, relatively few species make up the bulk of the total fish biomass: clupeid, pilchard, and anchovy represent the pelagic inshore fauna; horse mackerel, with smaller and varying amounts of chub mackerel, characterize the offshore small pelagic fish; hake, often termed "demersal," inhabit the whole water column with its main distribution offshore—the juvenile part of the population extends into shallower waters. In addition, there are a number of less abundant fish and shellfish species, in particular, snoek, kingklip, sole, monkfish, squid, tuna, deep sea crab, and rock lobster, which are, however, of significant economic importance.

Traditional Systems of Resource Management and Fishing Rights

Traditional systems of management and historic or traditional rights to fish that involve conservation practices, forms of tenure, and other cultural rules of access were never developed by local fishermen along the Namibian coast. Traditional fishing practices were never established due to the inhospitable nature of the largely unpopulated Namibian coastline which forms the Namib Desert stretching from the Oranje River in the South to the Kunene River in the North. The fishing communities of Walvis Bay and Luderitz, the only two fishing ports on the Namibian coast, have never exercised the freedom to fish unregulated without statutory control of resource management. Although fishing has been carried out in Namibia from the earliest times, the trawling industry was only established at the beginning of this century and it was only during the 1940s that a pelagic and other shoal fish industry developed, whereas the first regulatory control for the better protection of fish had already been introduced in 1922 (Proclamation no. 18 of 1922).[1]

Namibia has a modern management regime where the state has the exclusive responsibility for the management of the resource and the protection of the marine environment.

The Common Law

The common law of Namibia is Roman Dutch law. Roman Dutch law "as existing and applied in the province of the Cape of Good Hope (South Africa) on 1st January 1920" was introduced by Proclamation 21 of 1919 to be the common law of the territory of Southwest Africa. Legal continuity was preserved and it continues to apply in Namibia by virtue of the provisions of articles 66(1) and 140(1) of the Namibian Constitution, but subject to the qualification, however, that it is for the courts of Namibia to interpret and pronounce on the content and development of such common law in Namibia.[2]

In accordance with such common law the right of fishing in the open sea was common to all, subject to any established local customs amongst fishermen such as the custom that the first arrival would have the first trek.[3] With the introduction of increasing regulatory control over marine resources, inroads were made from time to time into such common law by the statutory regimes that applied.[4] Under the provisions of article 66 of the Namibian Constitution only so much of the common law and customary law that does not conflict with the constitution or any other statutory law still remains in force.

The common law right of unrestricted access to fishing was accordingly modified in 1922 when the first regulatory control was introduced and progressively repealed by subsequent legislative developments that replaced it with a regime where the state controls the utilization and access to the resource. Under the present legal regime that is governed by the Sea Fisheries Act of 1992, read together with the Sea Fisheries Regulations (Government Notice no. 1 of 1993 published in Government Gazette no. 1 of 4 January 1993), no person may utilize a marine resource without having been granted a right of exploitation in terms of the act, and the exploiter is restricted to a quota allocation. No fishing vessel may operate in the territorial sea and exclusive economic zone of Namibia without a license with the imposition of fishing conditions.

Traditional Foreign Fishing Interests

Regarding the issue of traditional foreign fishing interests, Namibia does not recognize any right of access for any foreign fishing fleet on the grounds of traditional fishing. There is no special provision contained in the national fishing legislation which affords a preferential right of access to foreign fishing interests on the basis that they have long fished the waters.

The foreign fishing vessels that engaged in fishing in Namibia's waters before independence were all ordered by the Namibian government to leave when it generated the exclusive economic zone. According to evidence given in *Redondo v. The State*, as many as two hundred and fifty foreign vessels fished in Namibian waters immediately before independence. Since these foreign vessels had severely damaged the stock and in particular the hake resource through dramatical depletion by over-exploitation, Namibia was under no obligation in international law to negotiate a phasing-out agreement or arrangement. It cannot be argued, as Portugal did in reply to the Canadian twelve-mile fishery proposal at the 1958 Geneva Conference, that the coastal state should be obliged to respect the rights of foreign fishermen who had been engaged in fishing for a long period of time without damaging the stock. Scientific assessments show that in 1969, when foreign fishing commenced, the total hake stock in Namibian waters was approximately 2,385 million metric tons, and had decreased to approximately 486 million metric tons in 1990.

In any event, true historical rights to access have not been established by these foreign fishing fleets. The foreign fishing interest was only relatively recently established in 1969 when the pelagic fishing stocks in western European waters became depleted and Namibia's fishery resources became a major focus of attention of foreign fishing vessels. It was also during this period (1969) that the International Commission for Southern Eastern Atlantic Fisheries (ICSEAF) was formed under its founding treaty. No treaty rights to traditional fishing were derived from the ICSEAF treaty or any other treaty binding on Namibia. Namibia was never made a party to the ICSEAF but was merely referred to by the member states as the coastal authority. (O'Connell 1976, 536–538). (Nor is Namibia under any legal obligation at international law to recognize any traditional foreign fishing rights.)

We wish to point out at this stage that an outstanding feature of the Namibian Constitution is that it is "international law-friendly." Article 144 incorporates the general rules of public international law and international agreements binding upon it into the law of Namibia. Under article 96 an "international, fully law-abiding" framework is established which "fosters respect for international law and treaty obligations" and "encourages the settlement of international disputes by peaceful means" (Erasmus 1989–1990, 81; Szasz 65).

The Law of the Sea Convention of 1982 clearly provides that the rights of the coastal state in the exclusive economic zone (EEZ) are exclusive with respect to the limits of its harvestable capacity. Article 62 establishes that having declared an allowable catch, the coastal state is free to determine its harvestable capacity, and only where it does not have the capacity to harvest the entire allowable catch is the coastal state obliged to give access to its EEZ to other states in respect of the surplus through agreements or other arrangements.

Article 62(3) of the Convention, which establishes the "general surplus rule" that governs access to the EEZ in respect of the surplus to other states, is carefully worded. It does not refer to "traditional rights." Without indicating any priority it sets forth a list of relevant factors and foreign fishing interests that the coastal state has to take into account in giving access to other states to its exclusive economic zone in respect of any available surplus.

One of the relevant factors which the coastal state is obliged to take into account is the need to avoid economic dislocation in states whose nationals have habitually fished in the zone. Under this rule traditional fishing interests have not been given priority over other interests, namely landlocked states (article 69) and developing states in the subregion or region (article 70). The provision also acknowledges other national interests (O'Connell 1976, 565–568).[5]

The above consideration will generally not be applicable to Namibia and is particularly not applicable to access to the hake resources—the primary interest of foreign fishing among Namibia's fish resources. There is at this stage a limited (if any) surplus available and the development and growth of the local industry is being promoted. The economic dislocation that may have been caused in some of the states concerned occurred as an inevitable result of overexploitation.

There exists, therefore, no basis in international law upon which it can be argued that the economic dislocation which may have taken place was caused by the establishment of Namibia's exclusive economic zone and the disruption of foreign fishing operations, and that it should now be taken into account when considering giving access to any surplus fishing opportunities when the resource recovers.

Legal History

The first statutory control was introduced in 1922 under the Sealing and Fishing Proclamation[6] to provide for the better protection of fish and seals in the territorial waters of Southwest Africa. The proclamation introduced the requirement that fishing boats had to be licensed, and prohibited their use without a licence or in non-compliance with the conditions of the licence (section 12). The administrator of the territory of Southwest Africa was given the power to appoint closed seasons and prohibited waters; limit, restrict, or prohibit the catching of any species of fish; and to give special protection thereto (section 6). The administrator was further empowered to make regulations and impose penalties for the contravention thereof pertaining *inter alia* to the following: daily returns; the regulation of fisheries; sizes of marketable fish, mesh sizes, methods of catching fish, and the licensing of nets; and the protection and preservation of fisheries (section 5).

Proclamation 18 of 1922 was amended by Proclamation 36 of 1930 and extended the licensing requirement to factory vessels. Proclamation 18 of 1922 was replaced by the provisions of the Sealing and Fisheries Ordinance, 1949 (Ordinance no. 12 of 1949) which consolidated the laws relating to sea fisheries and sealing.[7] Whilst retaining the protective controls and prohibitions established under the provisions of Proclamation 18 of 1922 (set out above), it provided for further conservation practices such as the declaration of marine sanctuaries[8] and established wider regulatory powers aimed at the protection of the marine environment (section 25) with a more extensive list of offences that carried increased penalties and forfeiture clauses (section 18). The Ordinance also introduced a processing licensing and quota system fixing the maximum quantities of fish that may be treated by a factory (section 2). The Sea Fisheries Act of 1973 (Act 58 of 1973)[9] and its regulations[10] in turn replaced Ordinance no. 12 of 1949 and by virtue of section 24 applied to the territory of Southwest Africa.

Article 140(1) of the Namibian Constitution incorporated Act 58 of 1973 into the law of Namibia and it continued to apply to Namibia until it was repealed by the Sea Fisheries Act of 1992. It reenacted most of the provisions that were contained in Ordinance no. 12 of 1949. Among the important provisions that had a bearing upon conservation and management were the following: the establishment, control, and management of fishing harbors; the registration and licensing of fishing boats; the licensing of fishing factories; the stipulation of closed seasons and of quotas; the control of fishing nets and of other methods of catching fish; specific measures to protect lobsters and other kinds of fish; and the control of whaling (Faris 1973).[11]

The provisions of Act 58 of 1973 were inadequate in that they did not cover the following matters:

- There was no provision made for the granting of defined fishing rights and the orderly exploitation of the resource.
- No proper quota and catch control existed. Processing quotas/licenses were granted to established processing houses which protected their interests, and it lead to monopolistic conditions prevailing in the industry.
- Quotas were transferable and that system was abused by the marketing of "paper quotas" to foreign interest groups.

The (unamended) provisions of the Territorial Waters Act of 1963 (the South African Act) governed the maritime zonal regime and prior to independence it established a limited exclusive fishing zone of twelve nautical miles for the territory of Southwest Africa.[12]

Act 58 of 1973 contained no provision in any way limiting or restricting foreign vessels from fishing in the exclusive fishing zone of Southwest Africa. The provisions of section 8 of the act which dealt with the licensing of fishing vessels and factories were not applicable to foreign vessels and confined to the terra firma and to the territorial waters.[13]

It is against this background of legislative shortcomings that the Namibian legislature shortly after independence enacted the Territorial Sea and Exclusive Economic Zone of Namibia Act (Act no. 3 of 1990) to establish an exclusive economic zone and through an act of promulgation by reference incorporated the provision of Section 22A of Act 58 of 1973 into Namibian law.[14] Section 22A then prohibited the unauthorized use of a foreign vessel as a fishing boat or factory within the exclusive economic zone of Namibia with an increased maximum penalty of R 1 million.

The Sea Fisheries Act of 1992 followed and established an advanced system of rights of exploitation—quotas and licensing requirements with provisions designed to exercise proper quota and catch by regulatory control of fishing and factory vessels.

The Contribution of Fisheries to the National Economy

Namibia's policies on fisheries strongly reflect the contribution of fisheries to the national economy and Namibia's dependence on expanding that contribution. Key features of the contribution are discussed below.

The Fisheries Sector as a Major Contributor to the Economy

Whereas in the pre-Independence year 1989 fisheries (excluding fish processing in the enclave of Walvis Bay) contributed U.S. $16.8 million to Namibia's GDP, this figure rose to U.S. $120.7 million for 1991 and U.S. $148.8 million for 1992. The estimate for 1993 is in the order of U.S. $188.1 million. Fish processing in Walvis Bay alone increased from U.S. $14.2 million in 1989 to U.S. $34.2 million in 1991, with estimates for 1992 and 1993 being U.S. $40.2 and U.S. $47.3 million respectively. Altogether this makes an impressive increase. Expressed in percentages the contribution rose from 2.11 percent in 1989 to 8.6 percent in 1991, with projected figures of 9.5 percent and 10.5 percent for 1992 and 1993 respectively. This increase is due to a number of factors, the most important one being the establishment of an EEZ and the resulting control over the offshore resources. Further, the increased portion of white fish landed and processed ashore and the consequent value addition significantly contributed towards the said increase. Lastly, the introduction of so-called quota fees, a form of royalties, for the major species such as hake, horse mackerel, and pilchard generated further revenue.

In the two-year period from 1989 to 1991, fisheries are estimated to have contributed nearly 40 percent of the growth in the Namibian economy. In 1992 the value of fisheries output is projected to overtake agricultural output, and by 1993 the value of fisheries output is projected to reach 60 percent of the value of output of the mineral sector, which would make it the second largest contributor. In terms of growth in employment opportunities the fisheries sector unfortunately did not perform as impressively. It is estimated that the sector currently utilizes a workforce of approximately five thousand persons. Nevertheless, the fisheries industry can already be counted as one of the major industrial employers in Namibia.

Fisheries as Contributors to Growth in Employment, Output, and Incomes in Namibia

The gains recorded so far have been achieved in a period where the focus of the fisheries policy has been on stock recovery. If this policy is successful, and there are already indications that it is working, then there will be scope for further substantial growth. In the medium term (5–10 years) and based on current international market prices the value of the Namibian fisheries is projected to be in the order of U.S. $420 million. Prospects for the other major sectors are less encouraging. Agricultural output is limited by the scarcity of suitable land and water. The mining sector, especially diamonds and uranium, is presently the most valuable sector of output. However, these resources are nonrenewable and their prices are vulnerable and fluctuate, as is the case for most of the base metals. There is still room for future activity from further exploration and exploitation, but additional output gains are likely to bring higher economic and environmental costs. The potential conflict of interests between offshore mining activities with fisheries could be seen as a point in case.

The Need for Sustained Growth in the Fisheries Sector

The first government of an independent Namibia has inherited an economic structure characterized by harsh economic disparities and social inequity. It is estimated that 70 percent of the national income is received by 5 percent of the population, while much of the rural population lives in relative poverty, with little opportunities at hand for employment or income, and with only limited access to basic services. Reducing the disparities requires major expansion of resources to be committed to basic services, especially health and education, and to infrastructural development and maintenance, requirements for economic advancement and an expansion of job opportunities in less developed areas. Government programs are aimed at poverty alleviation with sustained growth in order to address disparities without disruptive redistributive measures. For this strategy to succeed Namibia must have sustained growth in output, jobs, and income from the fisheries sector.

Specific Management Policies for the Main Stocks

Before we turn to some of the sector-specific pieces of legislation and their enforcement, some of the management policies for the main stocks and the reasons for them should be mentioned.

Hake

The average annual total allowable catch (TAC) for hake set by ICSEAF before independence was about 400,000 metric tons, a figure in no way consistent with the available biomass. In order to redress the situation rapidly, the then president-elect of Namibia requested ICSEAF member countries (even before independence) to withdraw their fleets. The new government subsequently radically cut the TAC to 50,000 metric tons for the remainder of 1990. For the 1991 fishing season the TAC was set at 60,000 metric tons, for 1992 90,000 metric tons, and for 1993 120,000 metric tons. The aim was and still is to restrict catches in such a way as to allow the stock to increase to four or five times its present level. In addition, a ban on trawling in shallow water less than 200 meters deep was put into force, pursuant of the policy to establish exploitation patterns capable of improving the protection of juveniles and small-sized fish.

In spite of registering a doubling of the fishable biomass since 1990, the present volume is still only about a quarter of the biomass at which the stock would peak and support a maximum sustainable yield. Consequently, the TAC for 1993 was kept at 120,000 metric tons, which is also in line with keeping fishing mortality below 20 percent. This program is showing definite signs of success as preliminary results for last year indicate a further substantial increase in the hake biomass.

Pilchard

Of the Namibian fisheries resources, the collapse of the pilchard (Sardinops ocellatus) was perhaps the most dramatic. From an estimated biomass of about 6 million metric tons in the late 1960s, the biomass dropped to a meager 50,000 metric tons in 1980. In 1980 all directed catches for pilchard were stopped. A slow process of keeping the canning industry alive and rebuilding the stocks ensued. This resulted in a present stock biomass of about 800,000 tons. This is not yet a total recovery but there are at least encouraging signs.

Until a consistent stock recovery is demonstrable, pilchard fishing is allowed only to supply fish to the labor intensive canning industry and limited amounts to the less labor intensive fish meal plants. So far, this policy appears to yield the wanted results. For the first time in many years the age structure of the stock has recovered to include four age classes. The presence of three- and four-year-old fish bodes well for the future, because recruitment from older fish is thought to be much better than from smaller, young fish.

Anchovy

The initial policy on this species, which is less valuable than pilchard and is used mainly for fish meal production, was to limit catches until the stock had recovered. Experience on the Namibian coast, however, indicated that anchovy biomass build-up fluctuates irrespectively of fishing mortality. In addition, it also appears that increases in anchovy biomass can be seen as a response to the development of the pilchard resource. Taking this into account, it therefore appears to be the most practical venue to extensively fish anchovy when abundant and hold back TACs when they are not. Whether both pilchard and anchovy can be managed in a balanced way to yield substantial catches from both stocks remains to be seen.

Cape Horse Mackerel

This stock is in a healthy state. Since the possibility exists that it constitutes a competition for the increasing pilchard and hake stocks, the policy is to maintain fishing at a high level, allowing fishing mortality at a level of about 30 percent of the total biomass. Recent biomass surveys indicate a stock size of between 1.5 and 3 million metric tons. A TAC of 450,000 metric tons was therefore maintained for the past three seasons.

Purse seining of horse mackerel in the offshore area is encouraged because the risk of taking pilchard as by-catch is diminished. Midwater trawling is restricted to a water depth greater than 200 meters to minimize accidental (or even deliberate) pilchard and hake by-catches. Recent reports from surveys indicate that these measures are proving successful and hake by-catches are now down from 15 percent to about 3 percent of the total landings.

Rock Lobster

This resource has been dramatically depleted and therefore drastic catch restrictions are applied in a programs to rebuild the stock. Prompted by the extremely low catch of 376.4 metric tons for the 1990–1991 season, the TAC for lobster was cut back from the pre-independence 2,000 metric tons to a mere 100 metric tons for the past season. For the current year a TAC of 200 metric tons is made available. The cut in TAC has been quite traumatic for the lobster fishing companies and has significantly reduced employment opportunities in the industry. To compensate for the lost fishing opportunities, temporary hake quotas were allocated to the affected companies. Maintaining low quotas and upholding the minimum size (which allows for females to breed at least twice before recruiting into the fishable population), should ensure a recovery and future sustainability.

Crab

Special measures were introduced to protect the red crab (Chachyon marineta) resource. Thus the number of crab licences have been reduced from five to four, catches are restricted to depth greater than 400 meters to protect females and juveniles, and larger mesh sizes or escape gaps on traps are being introduced. Extensive tagging programs to facilitate the assessment of biomass from tag returns is underway. Spider crab is being exploited at low levels only and research to assess stocks is implemented.

Minor Species

A number of minor species—for example, different tuna species, snoek (Thyrsites atun), and species that are landed as by-catch to the white fish industry such as different species of squid, kingklip (Genypterus capensis), and monk or angler fish (Lophius upsicephalus)—are targeted. With the exception of snoek, these have not been the aim of specific management measures and are now also being looked into for their development potential.

Tuna is targeted and may prove easier to protect and manage than fish that come as by-catch of other species, and considerations are on the way for determining the best catch technology. Also, the possibility of demarcating zones for either long lining or trawling has been considered in order to afford some protection to kingklip and to avoid gear competition. The management of snoek, however, may prove difficult. Certainly the total ban on the use of any form of gill net or drift net in Namibian waters will go a long way to protect the larger pelagic species such as tuna. Joint efforts by countries bordering the south Atlantic will enhance efforts by individual countries to rebuild and protect, especially the migrant species like tuna.

Utilization of the Fisheries Resource and the Right to Fish in Contemporary Namibia

The utilization of Namibia's fisheries resource is foremost governed by the Namibian Constitution, the supreme law of Namibia, and further regulated and controlled by the Sea Fisheries Act (Act no. 29 of 1992) and the Sea Fisheries Regulations promulgated thereunder by Government Notice no. 1 of 1993 (Government Gazette no. 1 of 1993).

The Namibian Maritime Zonal Regime

It is necessary to make reference to the Namibian maritime zonal regime as it determines the extent of Namibia's marine resources and has a direct bearing on the scope and application of the fisheries legislation. No exclusive economic zone existed before independence. Article 100 of the Namibian Constitution refers to an exclusive economic zone, and read with section 4(1) of the Territorial Sea and Exclusive Economic Zone of Namibia Act, no. 3 (the "Namibian Act 3 of 1990") created an exclusive economic zone for the entire Namibian coastline from the middle of the Orange River in the South to the Kunene River in the North including Walvis Bay and the offshore islands extending outside the territorial sea of Namibia within a distance of 200 nautical miles from the low water line (*S. v. Martinez*, 1991, 750 B-E; *S. v. Curraz*, 1991, 6; and *S. v. Redondo*, 1992, 18 and 26).

The Sea Fisheries Act (Act no. 29 of 1992) and the Sea Fisheries Regulations of 1993 apply by virtue of section 4(3) of the Territorial Sea and Exclusive Economic Zone of Namibia Act (Act no. 3 of 1990), read with Article 1(4) of the Namibian Constitution, to the entire exclusive economic zone of Namibia as defined in Act no. 3 of 1990, including the EEZ around Walvis Bay and the offshore islands.[15]

Namibia exercises sovereign rights and jurisdiction under the specific legal regime of the EEZ over the exploration and exploitation of marine resources within the exclusive economic zone in terms of the provisions of the Sea Fisheries Act of 1992 (Section 3(b)).

Conservation and Resource Management

In respect of the utilization of fisheries resources, the provisions of Article 95(l) of the Namibian Constitution stipulates that as a principle of state policy "[the] State shall actively promote and maintain the welfare of the people by adopting, *inter alia*, policies aimed at the . . . maintenance of ecosystems, essential ecological processes, and biological diversity of Namibia and the utilization of living natural resources on a sustainable basis for the benefit of all Namibians both present and future . . .". The legal status of the provisions relating to principles of state policy as defined in Article 101 of the constitution is that they do not constitute legally enforceable norms but are intended as guidelines to the government in making and applying laws. The principles also constitute second- and third-generation human rights—that is, socioeconomic rights and particularly rights to a sound ecosystem. They furthermore have the force of presumptions of statutory interpretation and will, in time, gain the force of law through judicial precedent (Carpenter 1989–1990, 56–57).

Pursuant to this constitutional guiding principle and by giving effect thereto, the Namibian government in its white paper titled "Towards Responsible Development of the Fisheries Sector" (presented to the National Assembly by the Minister of Fisheries and Marine Resources, December 1991) adopted as the government's main objective for the fisheries sector the following policy—namely "to utilize the country's fisheries resources on a sustainable basis and to develop industries based on them in a way that ensures their lasting contribution to the country's economy and overall development objectives." In regard to conservation of stock, the policy reads, "The government is committed to rebuilding depleted fishery stocks to their level of full potential. This will be accomplished through a program of catch restrictions and other regulations over an expected time period of 5–10 years. All stocks will otherwise be exploited on a sustainable basis and at moderate levels, in general below that estimated to give maximum sustainable yields. Regulation measures for the purpose of adjusting the exploitation levels will include TAC specifications by stocks, effort restrictions by fleet limitations, and closures of the fishery by the time periods or areas."

The National Fishing Corporation of Namibia Act of 1991 was enacted to implement the above government policy objective and it makes provision for the incorporation of a National Fishing Corporation as a public limited liability company within the framework of the Companies Act of 1973 for the purpose of exploiting the fisheries and other marine resources of Namibia, whilst at the same time contributing towards the development and efficiency of the industry as a whole. The objective of the Corporation is to carry out normal business of a fishing company for profit and shareholders' gain, such as the catching, processing, marketing, and selling of fish and other marine resources. Provision was also made for the Corporation to facilitate, promote, guide, and assist new businesses and undertakings with a view to promote Namibian interest in the industry, particularly in human resources development in fisheries. The government will retain control of the Corporation by virtue of its capital structure in which the government holds 51 percent of the shares with voting rights.

The Act was designed to attract private investment in the Corporation in favor of Namibian interest while allowing minority participation by foreign companies. An incentive to private investment is the fact that the government's participation in the profits of the Corporation is limited to 15 percent.

The principle of state policy in Article 95(1) of the Constitution was incorporated into and further amplified by the provisions of Section 2 of the Sea Fisheries Act of 1992 regarding the determination of general policy, which provides as follows: "The Minister may from time to time, determine the general policy with regard to the conservation and utilization of the Namibia marine resource to be applied with a view to (a)the protection of the marine ecology, and (b)the promotion, protection and sustained utilization of the sea, its resources, and derivatives thereof *to the greatest benefit of all Namibians, both present and future*" (authors' emphasis). This provision, which determines the primary principles according to which resource management is to be conducted, requires that proper conservation and management measures be taken designed to protect the marine ecology and restore and maintain fisheries resources with the sustainable utilization of the resource at a yield which will produce the greatest overall benefit for Namibia. For the purpose of determining allowable catch, it introduces the concept of the "optimum sustainable yield" as an alternative standard for Namibia, rather than the scientifically related maximum sustainable yield equation.[16]

Such principles of fishery management and conservation practice conform to international law and more particularly with the 1982 Law of the Sea Convention. Article 61(1)–(3) of the Convention, having established the right of the coastal state to exclusively determine the extent of the allowable catch in its exclusive economic zone, proceeds on the basis that the coastal state, taking into account the best scientific verified estimates, is obliged to take measures designed to maintain or restore the population of harvestable species at levels which can produce the maximum sustainable yield. The yield is qualified by relevant environmental and economic factors including the economic needs of coastal fishing communities and the special requirements of developing states. As pointed out by O'Connell (1976, 565), ". . . by qualifying maximum sustainable yield according to relevant environmental and economic factors, including the economic needs of coastal fishing communities, [it] has weakened the scientific character of the determination which the coastal state is obliged to make of the level of exploitability as a step towards the determination of the surplus available for allocation. The formula is a composite one, in which subjective judgements of an economic character modify objective judgements about verifiable ecological facts."

There are, therefore, no constraints on Namibia's right to set a total allowable catch (TAC) at a lower level than that which is scientifically required to maintain populations at a level that can produce the maximum sustainable yield (MSY). It may determine a TAC at lower levels designed to restore populations and rebuild stocks, or it may do so for sound economic reasons. For example, setting a TAC at a lower level than that related to the MSY may, by reducing supply to key markets, contribute to maintaining prices. Limiting fishing activities will also generally help to maintain catch rates and therefore profitability. It would therefore, for example, for economic reasons, be perfectly legitimate under Article 61 to set a TAC less than the MSY in order to maintain catch rates and thereby sustain profitability and the economic viability of the local fishing industry.

Fishing Rights

As pointed out above, traditional fishing rights of unrestricted access to the sea were never established in Namibia and whatever common law rights to the freedom of fishing existed have been replaced by a common property regime where the state exclusively controls the utilization of marine resources and access thereto.

The Namibian Constitution in Article 100 established sovereign ownership of natural resources within the territorial waters and exclusive economic zone. Its provisions proclaim *inter alia* as follows: ". . . natural resources . . . within the territorial waters and the exclusive economic zone of Namibia shall belong to the state if they are not otherwise lawfully owned." The provisions of Article 100 have to be read and applied subject to Namibia's commitment under its constitution to respect international law and treaty obligations.

In relation to the application of international law, Article 144 of the Namibian Constitution provides that: "Unless otherwise provided by this Constitution or Act of Parliament, the general rules of public international law and international agreements binding upon Namibia under this Constitution shall form part of the law of Namibia." In order to give effect to international law, the Sea Fisheries Act of 1992 incorporated in Section 3(b) the specific legal regime that applies in international law in the exclusive economic zone and claimed "the sovereign rights of Namibia with respect to exploration and exploitation" of marine resources within the exclusive economic zone. Within the territorial sea it exercises sovereignty over the resource (Section 3(a)).

A sophisticated quota–licensing system with comprehensive regulatory control measures was adopted under the Sea Fisheries Act of 1992. The system consists of the following primary implementation instruments:

- Granting of rights of exploitation
- Determination of total allowable catch
- Allocation of quotas
- Licensing of fishing and factory vessels
- Licensing of foreign vessels under a fishing agreement in respect of the surplus quota of the allowable catch
- Monitoring of quota holdings
- Regulatory control over fishing vessels and factories
- Law enforcement

Rights of Exploitation

A right to utilize Namibia's marine resources can be granted by the state exclusively. The act created the legal framework for the orderly exploitation of fish and other marine resources by which fishing rights defined as a right of exploitation can be acquired to utilize the resource. In terms of Section 14 of the 1992 Sea Fisheries Act, "Any person who desires to acquire a right to utilize living marine resources, aquatic plants, shells, or guano for commercial purposes may . . . apply to the Minister . . . for *a right of exploitation.*" When considering the granting of a right of exploitation the Minister may have regard to criteria set forth in the act and regulations. The criteria can be grouped under the following headings:

- Commercial viability
- Namibian interest
- Regional development within Namibia
- Multilateral and bilateral co–operation
- Conservation and economic development of marine resources[17]

In order to ensure that the objectives of the act are realized, a right of exploitation is not transferable, except with the approval of the Minister and then only if the quota (or a portion thereof) is also transferred to the same person (Section 14(10)). A right of exploitation confers on the holder thereof, referred to in the act as an "exploiter"

(Section 1), a long-term right to utilize the resource. The exploiter is not entitled per se to access as of right, but must first acquire a quota allocation (Section 16).

Quotas

The Act created a quota allocation system that facilitates the implementation of proper conservation and management measures and policy to ensure that fishing by exploiters is controlled and related to the catching of an allowable catch with an optimum sustainable yield for Namibia qualified by relevant environmental and economic factors. Quotas also provide the instrument by which catch can be effectively controlled. Fishing efforts by quota holders are controlled by monitoring transshipments and landings through inspections and reconciling them with quota holding.

The Minister, after consultation with an advisory council, determines (from time to time) the total allowable catch (TAC) for a particular species which shall be available for the allocation of quotas during a specified period (Section 15). The total allowable catch of the various harvestable species is presently determined annually.

Quotas are denominated in absolute quantities and defined in the act as "the maximum mass or quantity of fish of a particular species allocated to a person which such person may catch during a specified period in a defined area" (Section 1). The act sets about a mechanism to ensure that quotas are distributed fairly amongst exploiters. Quotas are granted to exploiters or refused by the Minister in accordance with guidelines prescribed by him and are granted on such conditions as the Minister may determine (Section 16(1)). Guidelines prescribed in the regulations by the Minister for the allocation of quotas incorporate the criteria applicable to the granting of a right of exploitation (Section 3(2)). The government raises revenue from levies on quotas. A quota fee determined by the Minister in a notice in the Government Gazette is payable in respect of a quota allocation (Section 20). Quotas are not divisible or transferable except with the prior approval of the Minister (Section 18). The Minister may, when the total allowable catch of a resource is to be reduced as a conservation measure, suspend, cancel, or reduce a quota allocated (Section 17(4)).

Licensing

The act, like most fisheries laws, provides for a licensing regime that regulates and controls the catching of fish and the use of fishing and factory vessels in Namibian waters. The licensing of vessels implies a policy of "license limitation" by which the number and sizes of vessels that have access to the resource as well as their overall catching capacity can be controlled and limited in relation to available catch to prevent overextended fleets and over-fishing. No vessel may be used as a fishing vessel or factory in Namibian waters unless it has been licensed in terms of the act (Section 26(1)). The licence is issued by the Minister subject to such conditions as the Minister may determine (Section 26(5)). Through the imposition of comprehensive conditions and the regulations, extensive controls are exercised over fishing and factory vessels (the most significant of such conditions being catch control by quota specification).

Quotas are allocated to fishing vessels fixed in relation to the catching capacity of the vessel as a condition of the licence. With the quota linked to the fishing power of the vessel and monitored, overfishing is more effectively contained. Some of the other important conditions that may be imposed are those pertaining to the area within and the period during which the vessel may fish, methods and fishing gear that may not be used or which may not be carried on board the vessel, species which may not be caught, sizes, discarding, by-catch, transhipment, inspection, and the placement on board of fishery control officers. These conditions may equally be imposed in respect of foreign fishing vessels operating in Namibian waters under a fishing agreement with another state or community of states (Section 27(2)(c)). Failure to comply with the conditions entitles the Minister to cancel the license (Section 26(7)(a)). The license is not transferable (Section 27(8)).

Foreign Fishing

Special provision is made in the act for foreign fishing that puts in place the key element of the 1982 Law of the Sea Convention pertaining to giving access to other states to the surplus of the allowable catch through agreements. The act establishes a legal framework in which Namibia can negotiate on conditions it deems fit—a fishing agreement with another state or international organization representing a community of states—to authorize the operation of fishing and factory vessels of such foreign state, or a member state of such community of states, within the Namibian waters (Section 27(1)). Whenever such agreement has been entered into, the Minister may, upon application by the owner of a foreign vessel to which the agreement relates, issue a permit authorizing the owner to operate it within the Namibian waters as a fishing or factory vessel. The permit is issued for such period, subject to such conditions and restrictions and against the payment of such fees as the Minister may determine (Section 27(2)-(3)). This permit is also not transferable (Section 27(4)).

The white paper also encourages the formation of mutually beneficial joint ventures between Namibian companies and foreign enterprises whereby the latter are expected to contribute in terms of capital investment as well as technology transfer.

Other Control Measures

The regulations contain detailed and sophisticated provisions that make provision for an extensive control system, particularly over foreign vessels operated within Namibian waters. They are the regulations of a modern management regime similar to those of Canada and Norway.

The government of Namibia does not presently have the manpower and surveillance capacity nor the financial resources to fully monitor and patrol the full extent of the area of approximately 187,500 square nautical miles covered by the exclusive economic zone. The state is only able to utilize two ill-equipped patrol vessels, one of which it owns and one which it charters.

The regulations were consequently designed to assist the government in its control efforts and place the burden on vessel operators in respect of reporting at entry and exit from the exclusive economic zone (Regulation 46), notification of off-loading and transhipment times (Regulation 47), logging and reports on catches (Regulations 32–35), compulsory regular port calls and inspections (Regulation 45), etc. For the same reason, unauthorized transhipment of fish or fish products at sea is prohibited and unless authorized by the license may only be carried out in a fishing harbor under the supervision of a fishing control officer (Section 27 read with Regulation 47). Fishing gear has to be stowed away as prescribed in the regulations by a fishing vessel that is not authorized to operate in Namibian waters and when any fishing vessel transmits a marine reserve or closed area (Section 30).

The important regulations that have a bearing on compliance control are the following: the establishment of catch control (a vessel that carries fish or fish products caught within the Namibian waters may not take it out of the Namibian waters unless the catch has been inspected and no further fishing operations have been carried out by the vessel since the inspection); inspections and the carrying of fishery control officers on board (who have to be provided with food and officers' class accommodation and be remunerated by the vessel owner); marking of fishing vessels and gear; and reporting requirements (any fishing or factory vessel that enters or leaves the Namibian waters has to report by radio on *inter alia* the quantity of fish carried on board) (part VII of the Regulations).

In addition to the standard provisions that are normally contained in fisheries regulations such as marine reserves, closed areas, and closed periods, the regulations have sophisticated provisions on methods of trawling, drift nets, trawl nets, and purse-seining that may not be used or carried on board by different vessels, their mesh sizes, maintenance of mesh openings, and prohibited attachments (part III).

The provisions enumerated above are not an exhaustive catalogue; the regulations are, however, comprehensive and cover the whole field of fishing activities including angling from the shoreline.

Law Enforcement

The Namibian fisheries regime, being regulatory, depends largely on enforcement by criminal law through the judicial system. The independent judiciary system of Namibia exercises its judicial power subject only to the Namibian Constitution and the law, and may not be interfered with by the executive branch. Judges are appointed by the president upon recommendation of an independent Judicial Service Commission and have tenure of office. Criminal trials are conducted subject to the due process of the law in accordance with the bill of rights that includes fair trial procedures (chapters 3 and 9 of the Namibian Constitution).

The 1992 Sea Fisheries Act carefully describes offenses that have a bearing on conservation; without detailing these infractions it can be said that post-independence court proceedings have shown the significance of the special set of provisions pertaining to illegal fishing by foreign vessels in Namibian waters. Any owner, lessee, charter, or master that operates a foreign registered fishing or factory vessel in Namibian waters without a license is guilty of a serious offense, which carries a maximum penalty of R 1 million with mandatory seizure and forfeiture of the vessel, fishing gear, and catch upon conviction (Section 33(P) and 35).

The act incorporated the bond and other security procedures established by Article 73(2) of the Law of the Sea Convention, and provides for the release of foreign vessels upon the posting of a bond in an amount equal to the reasonable value of the vessel (Section 38).

An outstanding feature of the act is the presumptions that aid the state in proving offenses. When trying an offense a court may apply any of the following presumptions if it is proved that:

- A fishing vessel was used in connection with an offense. It is then presumed that the offense was committed by the fishing gear carried on board the vessel and in respect of all fish and fish products found on the vessel.
- A net, line, or cable was cut or released from a vessel, or abandoned. It is then presumed that the vessel was fishing at the time.
- A vessel carrying a cargo of fish has, over a period of two or more days, maintained a presence or generally remained in Namibian waters, or covered a particular area or periodically

reversed its course to and from Namibian waters. It is then presumed that the vessel operated within Namibian waters.
- Processed fish or fish products in excess of one metric ton were found in an onboard factory. It is then presumed that the vessel operated as a factory within Namibian waters.
- Samples taken of fish on board a vessel have certain characteristics. It is then presumed that the whole cargo has the same characteristics.

These presumptions may obviously be rebutted by evidence to the contrary (Section 36).

Recent events in Namibian waters have underlined the practical importance of these substantive provisions. Bearing in mind Namibia's limited surveillance and patrol capacity to control the vast area over which the EEZ extends, the fisheries resource, a valuable national economic asset, is particularly vulnerable against unauthorized exploitation. The eminence of these enforcement measures was confirmed by the Namibian Supreme Court in Redondo's case. It held the offence to be "a serious economic crime against . . . Namibia" and stressed "the need to deter potential offenders . . . inasmuch as the unlawful depletion of Namibia's fishing resource affects all the inhabitants of Namibia not only because fishing is a source of food, but an economic resource as well" (Ackerman 1992, 38, 40; see also *S. v. Martinez*, 1991, 762 D-E; *S. v. Pineiro and Others*, 1991).

Policy on Granting Exploitation Rights and Quotas

The Namibian government issued a policy document on the granting of exploitation rights and fishing quotas in June 1993. It supplements government policy on management of the marine resources published as the white paper, "Towards Responsible Development of the Fisheries Sector," in December 1991. If read together with the 1992 Sea Fisheries Act, the following criteria for granting rights and quotas emerge (Stuttaford 1994).

Criteria for Granting Rights and Quotas

According to section 14(6) of the act and Section 2(2) of the regulations, the following criteria are used when the government considers applications for exploitation rights and in the allocation of quotas:

1. Whether or not the applicant is a Namibian citizen
2. Whether the applicant is a company, whether or not the beneficial control of the company is vested in Namibia citizens
3. The beneficial ownership of any vessel which will be used by the applicant

4. The ability of an applicant to exercise the right of exploitation in a satisfactory manner
5. The advancement of persons in Namibia who were socially or educationally disadvantaged by discriminatory laws or practices prior to independence
6. Regional development within Namibia
7. Cooperation with other countries, especially those in the Southern African Development Community
8. The conservation and economic development of marine resources

Ownership of Vessels

Except in specified fisheries, every applicant for a right of exploitation is required to show how there will be investment in vessels within three years of the date from which the right is valid. The exceptions are sectors where the economic viability of fishing operations is such that some level of charter arrangements may be necessary in the medium term, such as the crab and tuna fisheries and midwater trawling for horse mackerel. Even in these cases, however, priority is given to applicants prepared to make investments in vessels and/or shore processing facilities (Stuttaford 1994).

Namibianization and Foreign Ownership

Having for many years seen its fish stocks raped by a host of foreign trawlers whose owners contributed nothing to the country's welfare, Namibia has embarked on a policy aimed at securing increasing benefits for Namibia, especially through onshore development. This approach provides increased opportunities for Namibians to participate in fishing and related businesses, but it also provides scope for foreign investment through joint ventures or wholly owned foreign ventures.

While priority is given to Namibians in accordance with the criteria listed above, there are opportunities for rights to be granted to joint ventures or wholly foreign owned ventures where the foreign investment is shown to contribute to economic and overall development in Namibia. In general, this requires that such ventures make contributions which are beyond the capacity for the existing Namibian fishery for which rights are being sought.

Foreign investors seeking to participate in fishing ventures are expected to cooperate with Namibian businesses through joint ventures. Notwithstanding this policy, rights may be granted to wholly foreign-owned ventures in exceptional cases where this would be beneficial to Namibia, and where there is an appropriate plan for Namibianization of the business. Moreover, in new joint venture applications for exploitation rights, the right is granted to the Namibian partner in the venture.

For existing joint ventures, priority is given to applications from the Namibian partners, taking into account the extent of ownership and control of the Namibian interests. Rights may be granted to existing joint venture rights holders, rather than to the Namibian partners, with the agreement of the Namibian partners where it can be shown that the joint venture business and Namibia will benefit (Stuttaford 1994).

Term of Rights: Ten Years

Ten years' tenure of rights are granted to:

- Ventures at least 90 percent owned by Namibians with significant investment in vessels or onshore processing facilities. For this purpose, 50 percent ownership by the venture of a vessel or an operational onshore processing facility in the fishery, for which rights are granted, is regarded as sufficient for a significant investment. Ten-year rights may also be granted where Namibian rights holders own a smaller share in a larger venture.
- Ventures with more substantial foreign ownership which make or have the capacity to make a major contribution to economic and overall development of Namibia. For this purpose, employment of five hundred Namibians onshore in activities related to the fishery for which rights are sought is regarded as sufficient for a major contribution.

Ten-year rights may be granted to smaller joint or wholly foreign-owned ventures which make an innovative contribution to the development of the fishing industry in Namibia, such as developing new products or new export markets, and where a longer-term right is necessary to secure the investment involved.

Term of Rights: Seven Years

Seven years' tenure of rights are granted to:

- All other majority Namibian-owned ventures having at least 50 percent ownership in vessels or in operational onshore processing facilities in the fishery for which rights are sought, including ventures which only operate in the fishery by chartering vessels or other similar arrangements.
- Ventures with less than 51 percent Namibian ownership which do not have significant onshore investments in the fishery for which rights are sought.

Term of Rights: General

Rights may be granted for shorter terms in particular circumstances, such as in the early stage of development of a new fishery. Rights granted for four or seven years may be extended if the venture fulfills the criteria for longer term rights. For example, a four-year right granted to a wholly Namibian venture without any initial investment may be extended to ten years upon a significant investment in the fishery. Rights may be terminated or downgraded if a venture no longer fulfills the criteria under which it was granted.

Available Rights and Performance

Rights are granted in the following fisheries: crab, demersal hake (both demersal trawl and longline), demersal monk and sole, linefish, rock lobster, midwater trawl, small pelagic purse seine, tuna (longline, pole, line, purse seine), collection of aquatic plants, and others which applicants may specify. Any rights granted may be subject to performance conditions on matters such as investment and employment.

Vessel Quota System

At the time of publication of the policy document it was intended to allocate quotas to individual boats in the crab, hake, horse mackerel, rock lobster, and pilchard fisheries. However, by early 1994 the principle had been applied only in the demersal trawling industry and partially in the pelagic sector (Stuttaford 1994).

The objective of the system is to keep the capacity of the fishing fleets in Namibian waters in line with the medium-term availability of the resources for each fishery. This is necessary for the following reasons:

1. To avoid the situation where rights holders introduce too many vessels, resulting in vessels being tied up unproductively for long periods of the year
2. To provide stable, year-round employment for crews and those employed in fish processing
3. To reduce the cost of fish supplied to onshore processing plants enabling Namibian value-added products to be more competitive in world markets, increasing the prospects of onshore employment growth
4. To increase the profitability of the industry and thus the capacity to invest, to train, to pay taxes and levies, and to pay higher wages
5. To reduce the tendency for vessel owners to fish illegally, to dump fish, or to target by-catches in order to keep their vessels producing when their quota is insufficient

Conclusion

Namibia's fishing resources are potentially large and valuable. They can support a highly productive industry which could contribute significantly to the national economy. This contribution to the economic growth, together with a system of quota fees payable by resource users, is essential to achieve the basic objective of redressing existing economic and social disparities. Although the fisheries resources of Namibia can be seen as being managed under a common property regime, the three-part system of allocating long-term rights of exploitation, quotas, and licenses provides private entrepreneurs with the incentive to invest in and develop the fisheries sector.

The fact that Namibia at independence inherited depleted fish resources and that its own industry operated at levels much lower than the potential was a blessing in disguise. It affords Namibia the opportunity to in a sense start with a clean slate, which allows it to avoid the mistakes that have been made in most other fishing nations. The pitfalls of overcapitalization and wasteful practices such as dumping less valuable catches that so often are the results of open-access to resources are avoided by the system of allocating quotas and licensing individual vessels. This system allows for enough flexibility to react to changes in the markets as well as to fluctuations of the resources.

Namibia, as a coastal state, has sovereign rights over the living resources within its Exclusive Economic Zone. This principle as provided for in the United Nations Law of the Sea Convention is fundamental to the policy of establishing TACs and the allocation of quotas. Through independent funding of the required scientific research and the participation of representatives of the different sectors of the economy in the decision-making process, sustainable utilization is ensured. Open-access systems as the one in operation in the pre-independence years have proved to be unsustainable and are therefore rejected.

Last but not least, the fisheries resources of Namibia are common property. The state as guardian of these assets has to ensure the sustainable utilization now and in future. To our minds, the only way to ensure this domestically is by means of constitutional provisions, followed by the required sector-specific policies, legislation, and regulations.

NOTES

1. See Fuggle and Rabie (1983) for a brief history of marine resource exploitation.

2. *Redondo v. The State*, 18th June 1992, 20–26 (an unreported judgement of the Supreme Court of Namibia).
3. *Van Breda and Others v. Jacobs and Others*, 1921, ad 330.
4. See note 2 above.
5. O'Connell (1976) states, "The range of claimants to the diminished allocable resources in the EEZ has widened. The effect of the EEZ upon states which have habitually fished is thus likely . . . to be more drastic than the concept of preference."
6. Proclamation no. 18 of 1992 insofar as it related to fisheries was amended by Proclamation no. 36 of 1930 and Proclamation no. 1 of 1936.
7. Ordinance no. 12 of 1946 was amended by Ordinance no. 26 of 1967; 38 of 1967; and 9 of 1969. The Ordinance applied to the entire extent of the Southwest Africa coastline including Walvis Bay. See *R. v. Akkermann* 1954, 1, SA 195 (SWA).
8. *R. v. Bester and Others*, 1952, 3, SA 273 (SWA).
9. Act 58 of 1973 was amended insofar as it applied to the territory of Southwest Africa by Act 57 of 1975, Act 22 of 1976, and Act 99 of 1977.
10. The regulations were contained in Government Notice 1912 of 12 October 1973 as regularly amended.
11. See Joubert 1973, and Faris 1973, 61 for an analysis on Act 58 of 1973, and Fuggle and Rabie 1983, 269–273.
12. See Briesch and Powell 1992; *Pineiro and Others v. The Minister of Justice and Others* (an unreported judgement in the High Court of Namibia dated 17 June 1991) 23–26; *S v. Martinez*, 1991, 4 SA 741 (NmHC), 747–750; and *S. v. Redondo*, 1992, 8–10.
13. *S. v. Redondo*, 1992, 11–13; *S. v. Martinez*, 1991, 750; and *S. v. Curraz*, 1991 (an unreported judgement of the High Court of Namibia dated 13 February 1991), 12–13.
14. Act 58 of 1973 was amended by Act 98 of 1977 to insert *inter alia* Section 22A into the principle act but confined the scope and application of the section to the Fishing Zone of South Africa which included the area around Walvis Bay and the offshore islands. The Namibian Act no. 3 of 1990 in turn incorporated the section as it applied to Walvis Bay into Namibian law. See *S. v. Martinez*, 1991, 750; *S. v. Curraz*, 1991, 1–2; and *S. v. Redondo*, 1992, 34.
15. The definition of "Namibia" in Article 1(4) of the Namibian Constitution expressly includes the enclave of Walvis Bay and the offshore islands as part of the national territory of Namibia. The Namibian legislature and the courts are bound by Article 1(4) to exercise jurisdiction over Walvis Bay; see: *S. v. Martinez*, 1991, 750 and *S. v. Redondo*, 1992, 18 and 26.
16. O'Connell (1976, 565) and the definition of optimum sustainable yield in the Fishery Conservation Act, 1976, United States of America.
17. Section 14(2) of the act read together with regulation 2(2).

REFERENCES

Briesch, C. F. P. and D. M. Powell. 1992. "Fishing for Convictions: The Namibian Maritime Zonal Regime and The Incorporation of the Sea Fisheries Act 58 of 1973 into Namibian Law in 109." *South African Law Journal* 129.

Carpenter, Gretchen, 1989–1990. "The Namibian Constitution—Ex Africa a Liquid Novi After All?" *South African Yearbook of International Law* 21.

Erasmus 1989–1990. The Namibian Constitution and the Application of International Law. *South African Yearbook of International Law* 21.

Fuggle and Rabie. 1983. Environmental Concerns in South Africa: Technical and Legal Perspectives.

Joubert. 1973. *The Law of South Africa*, 10, (Sea Fisheries), edited by J. A. Faris.

O'Connell. 1976. *The International Law of the Sea*, 1.

Stuttaford. 1994. *Fishing Industry Handbook*, twenty-second edition.

Szasz. *Succession to Treaties under the Namibian Constitution*.

COURT DECISIONS

S. v. Curraz, 1991 (an unreported judgement of the High Court of Namibia dated 13 February 1991).

S. v. Martinez, 1991 (4) SA 741 (NmHC).

S. v. Pineiro and Others (an unreported judgement in the High Court of Namibia dated 17 June 1991).

S. v. Redondo, 1992 (an unreported judgement of the Supreme Court of Namibia dated 18th June 1992).

R. v. Bester and Others, 1952 (3) SA 273 (SWA).

Van Breda and Others v. Jacobs and Others, 1921 AD 30.

TROND VEDELD   CHAPTER 18

State Law Versus Village Law: Law as Exclusion Principle under Customary Tenure Regimes among the Fulani of Mali

Property rights regimes determining the access and use of African range land resources generally depend on the customary tenure regimes and the states' laws, policies, and practices.[1] The property rights regimes interact with the market forces to create the overall conditions within which individuals or groups act. These regimes have very diverse structures of authority, management, and institutions.[2]

A main aim of this paper is to present the diversity, complexity, and dynamism exposed in one particular common property regime, and indicate some of the dilemmas policy- and law-makers are faced with when analyzing what type of governance structures and institutions would best serve sustainable, efficient, and legitimate utilization of the resources governed by such regimes.[3] The discussion is based on observation from the enforcement of law and property rights for access and use regulation of floodplain range lands of the inland Niger Delta of Mali. This is reckoned to be among the most sophisticated customary regimes of pastoral Africa (Vedeld 1997).

A key dimension of the institutional crisis affecting management of range lands in Africa is conceived to be the lack of compatibility between formal law and institutions, largely transplanted from outside and embedded in post-independence state culture, and customary law and institutions embedded in local custom shaped through history (Swallow and Bromley 1995).[4] Empirical work among pastoral groups, including studies in the delta, indicates that major threats to efficient and sustainable property rights regimes often arise more from factors *external* to the local resource setting, rather than from *internal* factors related to rapid growth in human and livestock populations. External factors are, for example, inappropriate state law enforcement and intervention, weak infrastructure and market integration, and increased encroachment by crop cultivators. If such findings are accepted, research should be focused on *how external institutional arrangements can enhance capabilities of resource users*— individually and collectively—to re-invent more robust property rights regimes. Research should go beyond the narrow focus on tragedies resulting from "prisoners' dilemma" games and internal coordination failures, as suggested in Hardin's early metaphor (Hardin 1968).

In the delta, legal practice represents ambigous interpretation by state and customary elites of contradictory historic layers of law— customary, Islamic, French colonial, and Malian state law. Contradictions and conflict result in high transaction costs of enforcement, frustration and social tension, and disintegration of state and customary institutions. In serious cases, conflict leads to loss of lives.[5]

Some of the dilemmas (not all!) faced by policy and law makers might be approached through the introduction of procedural law (rules of procedure), rather than substantive law (rules of right which the courts are called upon to apply) (Vedeld 1993 and 1997). This institutional arrangement would be embedded in custom and provide village and customary structures a more firm mandate in property regimes. For the national law and policy makers a main parameter of choice regarding new regimes would be the *degree and character of excludability for diverse user groups* to be introduced.

Legislation and Tenure Policies in Mali[6]

The legislation enacted by the Malian state ("state law") is based on French colonial law and Islamic law (*Shari'a*). It draws very little on customary or pre-Islamic rules and institutions. Today, the principal law texts regulating property rights to land and natural resources have been gathered in one main Code Domanial et Foncier (CDF) from 1986. The CDF does not recognize local rights to range lands. On the contrary, the only regulation to this end says: "All pastures, transhumance corridors and animal water points are properties of the state." Only by developing the land (*"mise en valeur"*) can a more firm property right be claimed. State ownership of land is the general principle of the law.[7] The state property includes all water resources. This formal ownership of all land by the state reflects influence from both Roman law and the Islamic *Shari'a* (adopted by the French colonial administration).[8] Common property rights to range land have, however, weaker support than individually held crop land—both in Islamic law and in Malian state law. Reflecting this influence of Islam, it is fair to say that customary law—the way it is practiced—often gives weak support to range land and pastoral use (with certain important exceptions).

Historic Evolution in Customary Common Property Regimes

The customary tenure regimes of the Fulani, which still largely control the rich floodplain resources of the Inland Niger Delta, have intrigued several researchers.[9] These tenure regimes have a history of several centuries. They have evolved to regulate access to pastoral, crop, fish, and wild land resources between various producer groups. The customary tenure regimes are adapted to extreme climatic variability in time and space and high diversity in ecology and production potentials.

Historically, these regimes were particularly efficient during the period of the Dina state formation from 1818 to 1862.[10] The Dina laws and institutions basically reflected the interests of the main Fulani clans (the conquerors or founders) and theirs was not an egalitarian system. But it represented a fairly complex common property regime for management of resources. It functioned through relations of interdependence (for example, the slave economy) and reciprocity, backed up by a system of beliefs that accorded first comers and founders of village communities the right to manage (founding lineage). The central administration as well as the other levels in the governance structures guaranteed the appropriators a legitimate claim to the stream of benefits arising from the use of the resources.

The Dina "worked to allocate resources between co-owners of a defined territory and manage access to non-owners, broadly in line with the physical and technical attributes of the resources they used..." (Moorehead 1991, 166).

With the French colonial rule (1893–1960), the hegemony of the Fulani was gradually weakened. This undermined the governmental basis of both the central state administration of the Dina and the governance structures of the Jowro (customary "masters of pasture").[11] The French system of indirect rule implied that the chieftains were made the lowest level in the French administrative hierarchy, with the head of the chieftains being appointed "*chefs cantonnements.*" Unlike the chiefs, the *Jowro* had no official duties under the French administrative system. But he maintained his role in range land management. The colonial administration of the delta, which did not rely on the area for what it produced, introduced land tenure legislation based upon French legal concepts that took little account of customary rules. It denied the rights of local producers to resources.

The French colonial administration in various ways facilitated the opening of the delta for outsiders (for all citizens of Mali). All land which was not "developed" (*mise en valeur*) (i.e., not cleared and used for agriculture), was declared to be nobody's property. This implied a "nationalization" of range lands previously under customary tenure.[12] By maintaining customary chiefs as local administrative rulers (*chef cantonnements*), and obliging them to provide forced labor and army recruits, the colonial administration further discredited their authority and legitimacy as customary leaders. A policy of extracting labor and wealth from the delta and the integration in a wider, monetized market economy led to increased pressure from these outsiders for the state to ignore customary systems of exclusion and to allow them access to resources. The result was increased pressure on local resources (Moorehead 1991; Vedeld 1997).

The post-independence period (1960 to present) implied no major changes in economic policies, laws, and enforcement of tenure regimes. The socialist regime of President Keita did, however, provide a certain support for subordinate groups, like Rimaybé ex-slaves (Vedeld 1997). "Feudal" Fulani chiefs and leaders were imprisoned and suppressed (prior to 1968). A main change was the introduction of new administrative boundaries, including the grouping of a number of villages into arrondissements lead by a government official (*chef d'arrondissement*). He was often of military origin in the French tradition, and became a key actor in local tenure regimes. The village chiefs were appointed the lowest level in the state administration and maintained as tax collectors heading the new village councils. But while tax to the Dina was spent locally, little was now returned for local development. Following the authoritarian regime of President Traoré (1968–1991), a close association developed between the state administration and the only political party (UDPM). The local Fulani elites, through control of the village councils and local party structures, reconfirmed their local political hegemony, allowing them to deny ex-slaves and agro-fishermen access rights to land. But their power often relied on maintenance of ties to the state administration and the party. Leadership was often based on sources of power "from above," not from legitimate support of wider community groups. The increase in power of village councils and chiefs undermined the customary authority of the *Jowro* (customary "masters of pasture"), who have longstanding rights to manage range land, coordinate grazing, and settle conflicts according to customary law. The power of both chiefs and *Jowro* was gradually eroded under the stern rule of the *chef d'arrondissement*. The state administration made access to the delta resources easier for "all Malians" through the structural reforms. Various "outsiders" started to invest in livestock raising, irrigated crop cultivation, fishing, and charcoal burning and selling. These "outsiders" with close ties to the state—urban traders, government officials, or other investors—being foreign to local communities often have less interest in long-term resource management and more in short-term gains (Vedeld 1997).

Population and Resource Use Systems of the Inland Delta

The inland delta is a complex ecological system of great but uncertain productivity, which depends crucially on rainfall (300–500 mm) and annual flood levels of the Bani and Niger Rivers. Today, only about one-third of pre-drought area is flooded (7,000 square kilometers), reflecting large and stochastic fluctuations in pasture (and crop) production (CABO 1991). More than one million cattle and two million small-stock utilize these areas for dry season grazing seven to eight months per year. The total population numbers around three hundred thousand (1.5–2% annual growth/5–10 persons/square kilometer). The field work for this paper was mainly carried out among the Jalloubé Fulani, who are settled in the north-central parts of the delta (arrondissement of Dialloubé: fifteen to twenty thousand people). Dialloubé is the center of a powerful Fulani chieftain-ship, which controls access and use of the largest of the thirty-one pastoral territories (*leyde*) of the delta floodplain pastures (about 2,700 square kilometers). There is a mosaic of different social groups with different interests in local resources—wealth, status (caste), and power. These groups include noble Rimbé pastoral and agro-pastoral groups such as clergymen, pure pastoralists, and traders ("free men"); subordinate Rimaybé cultivators (ex-slaves); and Boso agro-fishermen. The two latter groups are considered of lowest rank according to Fulani identifications. There is an increasing group of absentee investors in livestock and agriculture, such as (urban) traders and government officials. Management and use of resources between groups can be complementary, but increasingly entail resource conflict as a result of drought and external pressure (see Vedeld 1997 for details).

Drought-Induced Resource Conflict

Resource conflict and dynamic changes in communal tenure regimes reflect the thirty years of drought in combination with demographic, technological, and political economic pressures at community and state levels. Political turmoil and insecurity in the northwestern regions have put restrictions on former transhumance patterns (the "Twareg problem"). To varying degrees, the local producers are caught in a protracted crisis which affects resource utilization and survival strategies, and leads to degradation of resources. The drought has led to drastic reduction in land-based production, and has made the inhabitants more dependent on the market and nonland-based activities for the provision of basic needs (Vedeld 1997; de Bruijn and van Dijk 1995; Turner 1992; Moorehead 1991, 1989). The market integration has mostly made the customary tenure regimes more open to influential outsiders with close relations to the state authorities at local and central levels. The customary regimes have gradually been eroded of their economic bases and required authority for resolving conflicts and maintaining access rules.

The drought has resulted in an accelerated conversion of floodplain pastures for crop fields by cultivators (while rainfed fields are abandoned). This process represents an "enclosure" of the most valuable and critical common pool resource for the pastoral Fulani of the delta. The move to crop cultivation is often a necessary and logical response to resource pressures, since productivity per hectare is so much higher than for pastures. Rice is basic for household food security, in a situation of drought and loss of animals.[13] But enclosure processes follow different patterns in different villages and locations as shown below.

Resource conflict and degradation is not only or mainly a result of problems of coordinating use under common property regimes. Resource-enhancement is severely undermined by weak infrastructure and lack of access to appropriate technology and markets under all types of regimes: communal, private, and state property. There are serious problems related to degradation of individually held crop land (for example, fertility depletion and weed infestation) and degradation of common pool resources such as trees (charcoal, browsing), fisheries, and wildlife.[14]

Range Land Ecology: Implications for Law and Policy

New evidence from research on range land ecology has implications for the analysis of legal and institutional concerns related to range land management regimes.[15] Recent work by a few researchers within and around the delta supports the new non-equilibrium ecological theories for the functioning of drier range lands (Turner 1992; Hiernaux 1993). Non-equilibrium theory suggests that strict regulatory measures are unrealistic and probably unnecessary under the drier range lands with ecological conditions of great variability (less than 3–400 mm), which we find in the Mopti region outside the inland delta (Behnke and Scoones 1992, 22; Behnke et. al. 1993; Behnke 1994). Despite high animal densities, grazing has had little negative impact on the productivity of perennial pastures of the dry season zones (Hiernaux and Diarra 1986; CABO 1991, vol. 2)—or on annual rainy season pastures outside the flood plain, where "...the net effects of historic rainy-season cattle actions on overall production could best be characterized as subtle or non-existent" (Turner 1992, 396). Rainfall has been the dominant factor affecting vegetation, not the number of animals. If the local pasture ecology is characterized by non-equilibrium conditions, destocking and removal of grazing pressure will not necessarily lead to restoration of vegetation. Reflecting the variable productivity of range land in time and space, local tenure regimes allow high animal movement and reciprocal arrangements. Customary regimes are more focused on regulation of access rights to key resources and conflict management rather than on controlling resource utilization and stocking rates. Rules for internal use regulation of pastures are difficult to set given the variability and are often unnecessary. Overall, there is little evidence of "overgrazing" and "desertification" being a major *environmental* problem, except for the destruction of trees and bush vegetation for goats. But pastoral resources are declining, and there are regular periods of fodder scarcity (due to drought and low flood). These economic and political management problems require regimes that can minimize conflict and economic loss and distribute rights in a manner perceived fair locally. Under wetter range lands, like what is found on the southern fringe of the Mopti region (500 mm)—as well as in relation to the delta floodplain pastures—the regulation of stocking rates according to available grass and estimated carrying capacities of the range can be more of a concern. Here, more clear rules for use also exist.

Characteristics of Community Arrangements of the Fulani

Empirical and theoretical work on pastoral societies in Africa recognize the wide diversity, adaptivity, and dynamism of land use and property institutions (Scoones 1994). More recently, the weakness of customary regimes is also being revealed in relation to access and use regulation.[16]

The resource regime of the Fulani consists of "bundles of rights" divided according to the resource and according to status-positions and management functions. The rights in a particular resource system may even change through the season; cultivators first have a right to produce rice, then fishermen catch fish, and finally pastoralists graze their animals in the same zone.

Resource regimes to pastoral and agricultural resources range from situations where specific resources for all practical purposes are open for anybody to use (open access) to situations where resources are managed by individuals or local groups as if they were ordinary private property (controlled access) (Table 1). The dominant form is invariably access-regulated common property embedded in community membership, status, and authority, with fairly explicit access-rules and defined boundaries. Internal use-regulation is coordinated through less explicit rules and more subtle status-based systems and custom, where coercive methods come to play. The

TABLE 1
Range Land and Crop Tenure in the Niger Delta

Potential production per hectare	Land type	Tenure system	Access control
High	Flood-plain fields near village	Inheritable/not transferable	Total
	Flood-plain fields outside village	Leasehold with *Jowro*/not transferable	
	Dryland crop fields	Inheritable/not transferable	
	Flood-plain pastures	Inheritable/not transferable	
	Dry land forests	Few restrictions	
Low	Dry land pastures	Few restrictions	Open

internal access- and use-regulated common property regimes would, historically, be for resources such as the perennial floodplain pastures, village floodplain fields, irrigated fields, important dryland pastures, crop residues, fisheries resources, wells, certain trees, wild grain, and wildlife. More open-access (nonproperty) regimes would be for dryland range lands and forest/trees in the rainy season pastures outside the delta (Moorhead 1991, Vedeld 1997). Although the property rights regimes largely follow the rules and regulations outlined above, there are significant changes going on. A hypothesis worth pursuing is that with higher perceived *potential* production benefit per hectare, and lower perceived transaction costs of policing per hectare *and* per produced unit (milt or grain), other factors remaining constant, the more likely there is a move from relatively open access to more strict control over the resources. "Potential" refers to risk and uncertainty being integrated concerns of most producers in the delta. There are risks related to both ecological and socio-economic factors. A more strict regime control represents either a move towards more private control, by individual households and *Jowro*, or a move towards greater state control (under irrigation schemes). Schematically a simplified illustration of this hypothesis is presented in Table 1. Since property rights are so much determined by status and power, it is not likely, however, that one finds a systematic correlation between an increase in the value of a resource and exclusive ownership (Vedeld 1997).[17] In Table 2, I present the hierarchy of property rights associated with different status positions

TABLE 2
Bundles of Rights to Common Pool Range Lands and Flooded Crop Land Associated with Status-Positions in Fulani Villages

	Legal owner	Proprietor (quasi-owner)	Claimant	Authorized user	Unauthorized user
	State	Founding Lineage (village leadership)	Noble pastoralists	Women (Rimaybé Boso outsiders)	Poachers of grass
Access	x	x	x	x	(x)
Withdrawal	x	x	x	(x)	
Management	x	x	x		
Exclusion	x	x			
Alienation	x	(x)			
Inheritance	x	x	x	x	

and functions. *De jure* the state holds all rights. *De facto* it cannot use these rights fully, except by use of force. A striking feature is that all groups have access and usufruct rights to at least some crop land. This right is inheritable. Several groups, in particular the women—subordinate Rimaybé and Boso cultivators—experience limitations on their land rights. Women's rights to land are through marriage (or relatives). They cannot inherit land. All men, women, and children have ownership rights to animals. Products from the married women's animals, such as milk, butter, sale, or wool is her own property and cannot be demanded by the camp chief or the husband. A man will normally leave the right for women to sell milk from his cows in order to buy sugar, spice, batteries, or petrol. He will sell animals when there are particular needs for money: ceremonies, taxes, clothes, voyages (Gallais 1967). The animals owned by the women are rarely touched by the husband. Entrepreneurs among the subordinate cultivators are in some villages denied access to crop land beyond a certain level. They may then need to lease land outside the village commons (for example, from land held by the *Jowro*). Crop expansion is, however, not generally hindered and seems to reflect dynamic response to local market forces (Vedeld 1997).

According to customary tenure rules, the crop land could not be sold, divided, rented, or put in others' charge in any ways by unilateral decisions by the family chief alone. This is still so for crop land of Fulani villages. But not for the crop land developed on former range land rented from the *Jowro*. Today this is often the land that provides the largest share of the household production.

Animal Property and Management of Pasture Rights

The most sophisticated use-regulation system is found in the rank-order each pastoral family holds in the sequence of a given corporate herd (*egguirgol*). There are sixteen corporate herds among the Jalloubé, each consisting of up to a hundred family herds. They are formed to coordinate the transhumance on the delta floodplain pastures, which consist of a web of recognized and fixed cattletracks and grazing zones with the *Jowro* as defined proprietors and managers. This property rights system would be exceedingly difficult to capture in statutory law texts and enforce by central state agencies for several reasons. First, in most cases around the world, property rights to land resources are tied to the ownership of these fixed resources. But in the corporate herd, it is the ownership and management of animals (i.e., the family herd), which decide the family rank-position in this sequence. If a pastoralist for some reason loses his family herd, he may lose his position and access rights altogether. But can an access right or ownership control to range lands be tied to cattle ownership in modern law? The Saami reindeer herders of northern Norway seem to practice a similar system. Secondly, the size of a certain herd obviously determines the amount of grass the owner benefits from. There is no system of quotas for members of the delta communities. Thirdly, the transhumance patterns in important ways decide access rights to different range land areas within and between years. Essential to herd management is the re-grouping of the individually owned family cattle into larger communal herd units (*egguirdi*). These herd units are split according to certain management criteria with the aim of optimizing the use of the pastoral resources and the production output. There are several different types of communal herds which have their own management regime through specific transhumance orbits, the main herd leaving the delta each year for a period of three months. Moreover, one herd unit may have as many as fifty owners since each herd owner prefers to distribute risks on several herds and build alliances (Waagenar et al. 1986).

Processes of Interpreting Access and Use Rights

How are access rights defined and defended today? The point of reference for most local users are the property rights institutions as they stood at the time of the Dina (1818–1862). This rule or "village law" is here denoted: the right of the *founder* (*Dina/Islam*) (Table 3). The founding lineages of a community have legitimate rights to allocate land and to control exclusion, manage, and use resources.

TABLE 3
Mismatch Between Constitutional Rules and Rules-in-Use as Perceived by Groups of Appropriators

Appropriators	Constitutional Rule Claimed	Rules-in-Use
State	Range land: state ownership Crop land: *Mise en valeur* or recognition of usufruct	"All Malians," or first arrival, or founder (Dina/Islam) or ambiguous use of power
Chiefs and members of founding lineage	Community membership, rights of founding lineages (Dina)	Founder (Dina/Islam), all members have usufruct, authority/ambiguous use of power
Jowro ("masters of pasture") (customary leaders)	Customary lineage right Community membership (pre-) Dina/Islam	Customary (Dina/Islam), access for community members. "All Malians" when *Jowro* lease crop land to outsiders.
Noble Rimbé pastoralists clergymen, pastoralists, traders, craftsmen)	Rights of founder (Dina) Rights of last conqueror (divined Islamic right over land and people)	Founder (Dina/Islam) (Supported by the state in the Fulani villages, but often not elswhere).
Subordinate Rimaybé cultivators	First arrivals, or "all Malians", or community members, long-term occupancy	Founder (Dina/Islam) Limited access to CPRs, village law/power, custom
Subordinate Boso/ Somono agro-fisherman	First arrivals, "all Malians" long-term occupancy	Founder (Dina/Islam), limited access to CPRs, village law/power, custom
Women	None	Village law/power, custom

The institutions formalized by the Dina and legitimized by Islamic beliefs and ideologies, separated clearly and formally between insiders and outsiders. Outsiders have to pay a grazing fee (*conngi*), while insiders have free access rights. Such fees are paid for passing the river, for trespassing or pastoral territory (*leyde*) and for camping within a *leyde*. These fees have increased significantly with increasing scarcity of pasture. There are also other important rules from the time of the Dina including the rank-order of the corporate herds; systems of cattle tracks and stop-over points; and inheritance rules for offices, cattle, land, and property. The Dina represented a regime that through conquest overthrew *rights of first arrivals* in certain areas, while providing all groups with new identifications and status positions in society. Groups were conquered, enslaved, or in other ways dominated by the pastoral Fulani and their regimes. In this regard, the rights of local subordinate cultivators (Rimaybé ex-slaves) and Boso agrofishermen were suppressed in the interest of the pastoral Fulani. Moorehead draws our attention to certain nuances in relation to the general picture, and states that different interest coalitions are formed. Moorehead says, "The tendency to defect from customary rules increases the further one moves away from the founding lineage, through consanguine units to resident strangers and seasonal visitors" (Moorehead 1991, 219). This general picture is to some degree supported in this chapter (Table 3). But there is a large degree of fluidity in customary rules and rights reflecting local politics and differential response by different interest groups to broadly similar external circumstances.

In one village of two thousand people (Kakagnan), for example, the expansion in individual cropland at the expense of common pool floodplain pastures occurs through orderly and legitimate processes. Important floodplain pastures remain protected. Here cultivators and ex-slaves conform to the old order and respect the village authority structure. In another neighboring village of four thousand people (Dialloubé), pastures are converted rapidly and disordered. Following the creation of a multiparty system at local levels (after 1991), wealthy local traders sought alliances in tenure disputes with cultivating ex-slaves through new political parties. The two groups established contacts with political party groups in Bamako and the regional capital (Mopti) to enhance local bargaining stands. They challenged the chief and destabilized the old authority structure. The result became a disordered, chaotic, and rapid "rush" for flooded cropland by groups of traders and ex-slaves. Internal conflicts involved military police and state officials.[18] The differences between the village regimes show how values and status positions can change through new political relationships. It also shows how outcomes of property rights conflicts are conditioned by *contextual processes* defining status and power. Village politics matters (Vedeld 1997). Sources of legitimate authority emanate less from legal-rational authority (formal institutions and state law), and more from charismatic and customary authority and power. A constant tension exists between demand for change in the "formal" rules (reflecting demand for access to more crop land) and persistent "informal" or customary status and authority positions (reflecting demand for protection of pastures and a "pastoral way of life").

Below I present a relatively complex typology of state-community encounters which defines the system of property rights in crude terms in the two Fulani village communities (Kakagnan and Dialloubé). The layers of laws and rules are interpreted differently by the different interest groups and bargaining parties to property rights. Table 3 indicates the mismatch between "constitutional" rules as claimed by different appropriators and actual rules-in-use applied by those in power over local regimes (Vedeld 1997). The rules-in-use shift according to local circumstances related to the type of resource-use conflict, power structures, and particularities of the state-locality encounters. This fluidity reflects disagreements over status hierarchies and certain rules (not all), and unpredictability in their interpretation and application. The different legal sources are combined with the accumulated knowledge in oral (and written) traditions from case-to-case decisions over individual tenurial disputes, or cases of tenurial transactions ("case-law"). The knowledge about these cases is embodied in local specialists (marabous, chiefs, elders, Jowro, Bessema [leader of the Rimaybé]). These specialists work as councillors to the chiefs in resolution of disputes. Witnesses are also called. Access to land and security of tenure right is basically maintained through status positions combined with negotiations and political maneuvers. Bargaining involves on-going, open-ended debate over status, authority, and obligation as well as rules and practices. People play several roles, make claims through different channels (administration, court, political party) or historic "layers" of tenure laws and rules. Bargaining stands of individuals are determined through interpretations in different arenas and levels of social organization. Interpretations may change from one arena to another and often from one level of organization to another. Often these different arenas are interwoven and not easy to distinguish from each other. When conflicts arise, coercive methods are often used, especially between groups with different ethnic or social identity and origin. Power is particularly employed by stronger groups against weaker.[19] Both among the Rimaybé, descendants of conquered or enslaved groups, and Boso/Somono there are groups that claim rights to crop land as "first arrivals." They also claim rights of "long term occupancy" and "all Malians."[20] Such rights are overruled by the Fulani elite. Women have no legal principle they can apply to defend rights as proprietors or claimants. Divorced or deserted women must appeal for usufruct rights through kin members.

The Impact of State Intervention in Range Land Management

The state claims *de jure* ownership to all range lands in Mali. In practice, state officials do observe local custom—or rather, the rights of customary leaders to manage and allocate resources. But this is not reflected in formal state law or done in a uniform or strategic manner. State officials, when asked, claim that they adhere to the rule of long-term occupancy (without any reference to what that means). Judges of the court basically claim to judge according to the same rule. But when studying the practice of state officials and judges, another picture emerges. State officials can be observed to use either the rights of the last conqueror (*Jowro*), or founder (Fulani villages), or "long-term" occupancy (in zones dominated by Boso agro-fishermen, for example), or rights of "all Malians" (supporting the entry of influential outsiders). In general, they seem to support the local groups with most power (and wealth), often in rent-seeking manners. This reflects that the state tenure law, in its written form, is basically irrelevant for settling the type of conflicts that occur between local individuals and groups over access to land. There are no set procedures for managing conflicts (except if the penalty law applies). Enforcement of legal regulations or decisions by the administration or the court is mostly weak, non-existent, or unpredictable. Gifts and bribes are common ways of settling disputes. "The one who is willing to pay wins." There are multiple ways of settling disputes or re-opening conflict cases.[21]

There are several additional problems with state interventions. The state gives priority to large-scale upstream irrigation schemes of Mali-south (Office du Niger). This diversion of water lowers flood-peak levels downstream and reduces the area flooded for pasture and flood-fed rice cultivation in the delta and the Mali-north (Turner 1992). The concentration of investments in southern Mali and lack of infrastructural support in the delta is another key issue. Related to livestock and transhumance, the state regulation of dates for entering the delta is a local concern. Due consideration is often not accorded customary leaders' view of optimal dates. Hence, these dates are often not respected. The state taxation per head and per animal without re-investing in local infrastructure or social services is also perceived as unfair. There is the use of paramilitary groups to enforce ambiguous decisions and fining in resource use conflicts involving local groups. Local state officials also engage in arbitrary fining of local people due to "illegal" fuelwood collection or over fishing. Local leaders and communities are not generally able to manage the resources in their own interest. The Malian state has failed to provide an equitable, legitimate, and effective alternative management regime to replace the customary one (Lane and Moorehead 1993; Vedeld 1997).

Local Leaders in Range Land Management: Chiefs and *Jowro*

The *Jowro* and the village chiefs hold key positions in the property regimes defined by historic and ascribed rights. There are several reasons why they are accepted as leaders and custodians of communal resources. Especially the *Jowro*, as customary leaders with given genealogical status, often carry prestige and respect. The "natural" authority they possess bargains for a certain solidity and continuity of a given resource regime. They can activate social and moral norms and therefore lower enforcement costs of a regime. They live close to the resources they are set to manage and have intimate knowledge of their use. The village chiefs, due to close ties to the state, often carry less respect (Vedeld 1997). The Fulani political elite, including the chiefs and the *Jowro*, always refers to the "rights of the founders" of the village in relation to land claims. This right combines rights of long-term occupancy and rights as last conqueror. The village chiefs, although they have formal ties to the state structures, do not, however, belong to the bureaucratic elite or the Malian middle class (Le Roy 1985). In some ways they represent a "hybrid" social group or class different from other aristocrats and the Islamic Imam or marabous. A main personal goal is for them to become re-elected (election every fifth year). They are elected as bearers of customary values and behavior, including Islamic virtues. But they are sanctioned and supervised by state officials at higher levels. They maneuver between the state laws and support from the state officials and customary law with support from among their own councillors, the local nobility (or the local farmers if convenient). Often they manipulate for their own personal benefit. Since the state law is often irrelevant or contradicting customary law, they have to improvise in decision making. Their interests, both ideological and sociopolitical, make them create new institutions in land use and tenure conflicts.

The *Jowro* or "masters of pasture" were historically custodians of communal pasture. In contrast to earlier periods, they today actively facilitate crop expansion in the floodplain pastures. They tend to keep most of the rather significant benefits from the grazing fees (*conngi*) and sale of crop land for themselves and spend it among the close family. The *Jowro*, being dispossessed of animals and rather poor, are prone to pressures from cultivators and agro-pastoral elite groups. These groups are more numerous, better organized, and have better access to state officials. The *Jowro* are no longer performing their delegated authority as "pasture managers," to the benefit of the wider Fulani pastoral community (Vedeld 1997).

Legal Pluralism: State Law or Village Law

The paradigm of "legal pluralism" (in anthropology) takes as a point of departure that law is not representing the ultimate (or even the main) source of order in society. Law is conceived as one among many structuring institutions for behavior (Spiertz 1995). As indicated above, the property rights regimes to range and livestock among Fulani are not regulated by formal state law or by law alone. But law clearly plays a key role when contracts are executed according to custom (for example, rank order, access rights according to status, women's land rights, and entry dates). Conflicts over conflicting historic layers of rights are not and cannot, for example, be clarified by reference to law or legal rights alone, but must involve political and administrative decisions and reforms. The structuring properties that create social order of Fulani society arrive from several types of institutions and actions of agents at different levels of social organization. In Africa in general, the security of range land tenure is mostly not guaranteed by formal state law, "but must be maintained through negotiation, adjudication, and political manoeuvre... If rights in land are defined through ongoing, open-ended debate over authority and obligation as well as rules and practices, the security of farmers' rights (or pastoralists' rights—my addition) depends on the terms in which they participate in such debates and in the domestic, judicial, and bureaucratic arenas in which they occur" (Berry 1994, 11).

There is clearly a certain "legal pluralism" operating at local levels among Fulani. Tenure rights are determined in different sociocultural and political arenas of society. Local users, chiefs, state officials, or entrants make use of more than one normative or cognitive repertoire to rationalize and legitimize decisions or behavior. It is fruitful to distinguish actions and rationalizations (but not always easy outside anthropological work). Focusing on "law" as exclusion principle in property regimes, there are different interpretations by agents at different levels of social organization. One can distinguish between local customary law, which is basically embedded in history and custom, and "village law" (folk law). Village law is basically the form of social control systematically enforced by authority structures of the society at the village or community level, embedded in customary status, meaning systems, and power. The village law has two dimensions. First, it encompasses "customary law" (which may or may not be written, recorded, or clearly codified). Secondly, there is a world of paralegal or quasi-legal forms of rules and regulations of groups operating outside regular systems of courts and law, but embedded within the state systems (Allott and Woodman 1985). Chiefs and local officials in their exercise of (state) power related to land use and tenure conflicts tend to combine state law and various dimensions of village law to produce a third legal system or everyday practice: "local law." Local law represents an attempt to "balance the convergence of central and local general needs" (Le Roy 1985; 257). Local law represents a blend between Fulani custom (village law), Islamic law, and state law. Focusing on law as a means of creating social order, the legal pluralism in the local context can be illustrated, crudely, as follows:

Customary law - Village (Islamic) law - Quasi-legal systems - Local law - State law

Are Country-wide Tenure Reforms Required?

Revisions of state law, and more important, the building of efficient organizational structure for law enforcement at all levels of social organization, are essential for the construction of improved property rights regimes to protect Malian range lands. So far formal state law in Mali, as in most Sahelian countries, disregards pastoral access rights and customary tenure law. Mali nationalized all land after independence in an attempt to simplify the law and promote production and investment. "In practice, however, such legislative initiatives have done little to suppress land speculation or exhaustive commercial exploitation of the natural resources, and they have often intensified the multiplication of claims by incorporating conflicts over land into ongoing struggles for power" (Berry 1994, 10).

Regarding range land management, various beef ranching models have been tested involving private or collective title deeds to a limited range land area for an exclusive group of individuals. Privatization also takes the form of "land grabbing" by influential individuals with ties to the state in many Sahelian countries (for example, Mali, Somalia, Ethiopia, Tanzania, Kenya, Uganda, Mauritania, and Niger). Such privatization of range land or crop land has met with little success in Africa (except possibly in more limited large-scale commercialized farm or ranching areas in Zimbabwe, Botswana, or South Africa, for example). Private titles have neither improved small farmers' or pastoralists' willingness to invest in productive measures/ conservation, nor improved access to credit, which are often the main economic arguments for introducing private titles. In reality, customary tenure arrangements have proven dynamic in response to changes in factor prices—and changes in local power structures. At this stage of development, it seems that country-wide land registration and titling programs are premature on economic grounds—and will be controversial on political grounds. But the pastoral tenure issue, due to the various pressures on pastoral resources, weak positions of pastoralists, and severe conflict over tenure between pastoralists and farmers, is a particular problem that requires urgent attention.

Problems of Legal Reform and National Integration

The problems faced by Mali and its political elite in relation to the delta is larger than making decentralized and community-based resource regimes work in a robust and fair fashion through legal and administrative reforms. Fundamentally, it concerns how to achieve a better integration between the northern multi-ethnic society (dominated by Fulani) and the southern Bambara-dominated society. Since colonial times, the delta societies have been largely overlooked and, in part, the object of manipulation and domination. Political reforms to meet needs in the North are likely to meet resistance among political elites of the South. The shaping of more accountable and efficient regimes for resource management at the community level can be, however, a first step towards facilitating such national integration. Law is at the core of such processes of institutional reform and change. But the introduction of substantive law and legal rules at national level is no guarantee for change of behavior at the local level. The relationship between formal law reform and constitution of meaning, behavior, custom, and power relations is a complex matter. "Jurists and anthropologists are not agreed on the relationship between custom and law. Law grows with societal complexity, with the decline of the importance of primary groups, such as family, with breakdown of organized religion, with industrialization and bureaucratization" (Lloyd and Freeman 1985, 878).

Efficient "top-down" law reform requires a minimum of reflection of new values, norms, and rules carried by the law reform—both related to those who are to abide by the law and those who are set to interpret and execute the new law.[22] As long as there are large discrepancies in state law and customary legal systems, there is room for ambiguous behavior—by all groups. The result is likely to be fluid and inefficient—from any perspective. As indicated by Douglas North, law has the highest adaptive efficiency when "written in the hearts of people" (1990). Customary law, including Islamic law, is based on religious values and patrimonial status-managed authority systems, not rule-governed regimes.[23] Legal-rational behavior is not well developed and requires time to evolve among judges, state officials, as well as local people. If property rights institutions basically reflect constitution of meaning and status in the rural periphery, new regimes will only evolve through slow processes of political motivation, new consciousness, and joint bargaining, not by "top-down" law reform by central state authorities (Goheen 1992). But if "personal rule-ship" continues to dominate over bureaucratic ethics and legal-rational ideals embedded in the "rule-of-law," legal instruments will remain coercive tools in the hands of state officials and local leaders. Nation building becomes dynasty or empire building. The notion "we are all Malians," used by central elites as a way of creating open access for all citizens of Mali to common pool resources of the delta, is a typical value-rational claim. It makes no distinction between different claim right holders to resources, promoting "open access" situations to meet demands of potential powerful political supporters.

Rights of Long-Term Occupancy

Problems of management arise, however, more due to how the law is interpreted and enforced by state agencies than the presence of the principle of "dominium" of the state. Neither in state law nor in customary law, for example, is the important "rights of long-term occupancy" interpreted. What sort of use and how many years of use with acceptance of the proprietor should provide a "right of long term occupancy"?[24] What, more precisely, should a claim right entail in terms of development (*mise en valeur*)? The Fulani have historically used these CPRs annually for grazing for centuries. Hence, they should be able to claim a certain "right of long-term occupancy"— even to land close to Boso agro-fishermen or Rimaybé cultivating villages. These rights are now challenged because the flooded pastures have become more suitable for rice cultivation (due to lower flood water). The land has often not been taken into use until recently by the Boso (as "first arrivals" in the delta).

Legal Dilemmas

For the Malian law makers—and policy planners—there are several dilemmas to face when assessing decentralization, alternative exclusion principles, and future property rights regimes within more liberal market-based economic policies. A key general question is: Who shall decide what in relation to whom? There are many complementary dilemmas: What mandate should be accorded each level of social organization? What degree of decentralization of resource regime provides the most robust, cost-efficient, sustainable, and equitable solution? What role for common-property regimes and communal authorities? How to ensure a legitimate initial distribution of access to multiple users with different historic rights, status positions, and usufruct rights to different resources (in many cases to the same resources at different periods of the year or in different years)? How to balance water rights between upstream and downstream users of river water? How shall criteria for access by outsiders or by others at some later time be determined (third party rights, split rights, inheritance, transferability)? How to establish regimes of enforcement? Who should carry the costs of new regimes? Only a few of these dilemmas will be addressed here.

It is first of all necessary to develop *procedures* for how the more precise content of customary laws can be synthesized, and later how new rules can be enforced through new institutional arrangements. Conflict management institutions should be in focus. Today, there exist multiple and contradictory procedures and channels for settling tenure disputes. New regimes must and will naturally take place within "co-management" frameworks, involving state agencies in different ways and to different degrees. In Malian law and practice, there is only an indirect recognition of customary authorities and communal rights. The establishment of state ownership to all land under customary tenure has not been followed up with any sort of legitimate contracts between the state as the legal owner—and the local appropriators. There are no relevant regulations that address the degree and character of access, management, excludability, and ownership to common-pool resources. It is not possible for appropriators to make reference to a specific formal legal rule (e.g., long-term occupancy), and argue that one person's lawful use is in conflict with or subtracts from the lawful use of others. This enhances problems when customary rights are contested. A key problem regarding the shaping of new regimes is what role the *state* and their local representatives should play. Some argue that the history of authoritarian and oppressive interventions by the state prescribes a "minimum" interference (Swift 1993). There are, however, certain tasks the state has *potential* comparative advantages in doing, such as the enforcement of property rights, maintenance of security and order, and support of weaker groups (such as post-independent support for ex-slaves). The state has also a role to play in support of infrastructure and means of communication. Access to better and cheaper communication is a precondition for local leaders to physically meet, exchange views, and get organized around common problems.

Procedural Law: Land Tenure Reforms as a Process

There are ways through which processes of institutional differentiation can be facilitated and speeded up (through social engineering and creation of enabling arrangements). Despite the delta regime having sophisticated elements, the flexibility and complexity of customary law cannot easily be captured and homogenized in national law by using the written Dina taric held up against pre-Islamic custom and other oral agreements. Even if initial agreement could be reached between different parties involved, the codification would rather quickly be outdated due to resource variability. While, for example, the rank-order and transhumance system are robust institutions and should be maintained, the functions of the *Jowro* and the chiefs require more transparency and democratic control. Historical rights and duties of owners are overlapping and defined through ambiguous political maneuvers. A precondition for legitimate tenure institutions is that solutions gradually emerge as part of locally managed political processes, within the broader political economy.[25]

A possible solution to some of the dilemmas is for the government to elaborate and enforce procedural (rules of procedure) rather than substantive law (rules of right which the courts are called on to apply) (Vedeld 1993a, 1997; Scoones 1994). Instead of legislative dictation of detailed and fixed property rights to pastoral or agricultural resources, the procedural law could specify the arrangement within which the concerned parties could legitimately put forward their claims to a certain resource. This would include the identification and building of administrative or legal institutions which would handle such claims, the principles for judging between opposing claims, as well as procedures for enforcement. Prior to any codification of custom must be an assessment of the principles behind customary law, including local judgements of fairness. At the level of the villages, the administration, and the courts that handle tenure disputes, jurisprudence develops. But few attempts have yet been made to systematize legal cases and accumulate knowledge about local conflict resolution.[26] A first exercise would be to establish systems for the publishing of court and administrative rulings and legal analysis.

Recognizing that Fulani pastoralists are often minority groups or weakly organized groups with low representation in political channels, property rights to pastoral resources require particular protection. One measure to support pastoralists and protection of range lands could be to create independent "land tenure boards" at district levels with majority representation from the local (pastoral) communities and/or specialized "land tenure courts," assisted by staff trained both in customary and modern law who could facilitate this process. Pastoral organizations could play a central role in the construction of such structures as an integrated element of new tenure regimes (Vedeld 1997).

Regulation of Access Rights to the Inland Delta

In order to make state tenure laws compatible with customary law, legal persons' access rights to the delta resource system could be defined according to geographic location (proximity) and membership in a social group. Access restrictions should perhaps be as mild as feasible with a sustainable management—based on general goals of equity in distribution as well as the need to maintain flexibility (also to handle new settlers and new generations). Investments properly done in the delta by outsiders can be to the benefit of the local economy. Politically the national government must face the fact that all citizens can no longer have equal access to all resource systems in the delta and Mali.

A legal distinction could be made between "ownership in common" (the access right is inheritable) and "joint ownership" (not inheritable) (see chapter 5 for a definition of concepts). "Ownership in common" would normally entail a more firm motivation for long-term management. "Ownership in common" could be attributed to residents of the delta for resources of critical or high value, such as key floodplain resources, cattle corridors, wells, and village crop land. "Joint ownership" would be granted to both residents and "outsiders" to less critical pasture, forest, and wild-land resources. It would also be important to recognize temporary rights for pastoralists to pass a certain territory, and systems of secondary or tertiary rights (split rights) of access to groups that already enjoy such rights under the customary institutions. Given the relative indivisibility of the delta as a resource system, geographic boundaries around individual resource units should normally be ruled out as a solution. Property rights to land could be registered without boundaries being geographically fixed. Today, the Malian state property law contains no concept of "joint usage right" to land or legal institutions capable of handling these legal issues. This is ironical when most resources are held and used jointly by various user groups. Even if the range lands in important ways represent relatively indivisible resource systems, the bundles of rights defined by the customary tenure regimes reflect ways of dividing the benefits from these resources.

Regarding the proper management of the resource system itself—as in the physical and biological resources—the future may not look that gloomy. Recent studies from other dryland areas, including the Machakos studies (Tiffen and Mortimore 1994), give some hope regarding the possibility of addressing resource degradation as found in the delta (see Boserupian vs. Malthusian views, Boserup 1990; Turner et al. 1993). Within the delta there are signs of more firm resource control around villages and spread of resource-enhancing technology (for example, improved irrigation control; regeneration; and more rational use of pastures, new crops, and manure). With improved property rights regimes, including better infrastructure and improved access to markets, the productivity of the delta resource system can be improved substantially and sustain population densities at much higher levels than found today. But due to failures of existing property regimes (and drought), the near future might entail further resource conflicts and enclosures of common property range lands, increased land degradation, reduced forest and vegetation cover, and loss of biodiversity. Some of these effects are inevitable consequences of demographic change and development and should be perceived as larger-scale landscape changes. But many of the adverse effects could be minimized through more effective and fair property rights regimes.

Customary Leaders and Institutions: Evolve or Dissolve?

When arguing for community-based solutions it is important not to be "romantic" about the role and capacities of customary institutions and leaders in terms of sustainable resource management. Customary leaders represent tribal, feudal, and hierarchical traditions. They lack basic capabilities and skills for management. They are partial. They are easily coopted by state officials. Being poor, they are prone to self-enrichment. They may, for example, accept outsiders to graze on village commons or relatives to obtain access to crop land. They lack formal education, exposure to the outside world, and visions required to balance interests of wider community groups as dynamic and modern leaders. They are normally not accountable to the local people, nor to the government, in a "democratic" sense. Women have, for example, few customary rights to resources. Decentralization to local, ethnic elites and organizations carries seeds for progress, but also the potential for conflict if elites do not take broader societal responsibilities. Societies often develop through solving conflicts. But recent events in Africa and other places (such as Eastern Europe) show that there are potentials for dangerous conflicts when authoritarian states withdraw and leave more power to local leaders or unregulated market forces. Mali has great ethnic diversity and a long history of tribal rivalry. These conflicts may grow worse if the right balance is not found in the sharing of rights, duties, and powers between state laws and policies, market agents, and customary elites and resource users. A major responsibility for achieving such balance rests with the Malian leadership at local and state levels.

NOTES

1. "Property rights regime" is defined as follows: "A legitimate and coherent system of formally or informally enforced rules and practices used for everyday appropriation of culturally necessary means of subsistence" (Godelier 1984, 71–121). Property is not to be understood as an object but rather as a social relation, a "benefit (or income) stream, and a property right is a claim to a benefit stream that some higher body—usually the state—will agree to protect..." (Bromley 1992, 2).

2. "Management" is here defined as the process of deciding what a collective will do and how it will do it. Structures of authority are created to execute management or maintain order. Institutions enable or constrain what the structures of authority may do to the members of the collective (Bromley 1989; Swallow and Bromley 1991). Institutions also condition meaning, identity, and status positions among members of a community. Property rights regimes can be more or less conditioned by different institutional pillars, such as ascribed status, norms, rules, power, and enforcement systems.

3. "Common property" is a complex constellation of rights, rules, conventions (norms), and contracts (Swallow and Bromley 1991, 3). Common property is one type of land tenure that governs access and use of common-pool resources. It can be defined as a joint management regime controlling assets and allocating rights among co-owners or members (Bromley 1989, 1992,14).

4. "Law" can be defined as "a body of rules of action or conduct prescribed by controlling authority, and having binding legal force" (*Black's*

Law Dictionary, sixth ed., West St. Paul, 1990). With reference to its origin, law is derived either from judicial precedents (e.g., case law), from legislation (by state), or from custom (e.g., established practices). The state law is to be found in its statutory and constitutional enactment, as interpreted by its courts, and in rulings of its courts (case law). Law also embraces long established local custom which has the force of law (customary law). Hence, law is closely related to custom. Rather than discussing the relationship between the two rather abstract terms "state law" and "village law" (custom), I focus on the ways "law" and "custom" interact and materialize through political processes and concrete decisions by "legal" authorities related to property rights disputes. The study of the substantive legal concepts becomes of secondary importance (Lloyd and Freeman 1985). That law and justice relate to "western" ideas about legitimacy, authoritative, and permanent "knowledge" in institutions. These concepts become difficult when "everything is negotiable," as in the context of Fulani society (Moore 1992; Berry 1989a and 1989b).

5. During my visit to the delta (December 7, 1993), a clash between two Fulani groups resulted in twenty-nine people killed and forty-two injured in a conflict over tenure rights to pasture.

6. Focus here is more on land tenure legislation and less on general state regulations related to devolution of power and decentralization within the Malian state administration.

7. Although recent administrative reforms, following the coup d'etat in 1991, recognize in texts and practice a certain authority for village councils in land management within the village territory (*terroirs de village*).

8. In Roman and Continental law all bundles of rights are normally gathered on one hand (the state), ref. the concept "dominium." Similarly, in Islamic law the head of state is conceived to represent the Islamic community, and hence, "the ultimate source for ownership of land" (Park 1993, 1). Islamic law evolved in arid countries of the Middle East. Here, land itself was of secondary importance to, for example, water. Hence, water often became the main object of (private) ownership. Islamic law contains certain general principles regarding water rights (based on agreed saying of the Prophet (Tiffen 1985). Though important, the *relative* importance of state law reforms for development of property regimes may be debated in the Sahelian context where "everything is negotiable" and adherence to the "rule-of-law" is an exception rather than the norm. Law *enforcement* is of greater concern (Toulmin 1991; Vedeld 1997).

9. Gallais 1967, 1984; Ba and Daget 1962; CIPEA 1983; Lewis 1981; Swift 1988, 1989; Moorehead 1991; Turner 1992; Cissé 1991; Vedeld 1993b, 1997.

10. "Dina" means "religion," It refers to the political and cultural revolution under the leadership of Sékou Amadou which led to a hegemony of Fulani clergymen over a territory from Macina in the South to Toumbouctou in the North (1818–1862). Elements of earlier and later systems (e.g., the French colonial system from 1893, independent Mali from 1960) overlap to produce a complex layer of rights, rules, and practices. The Dina governance structures and institutions were largely based on the Islamic

law (*Shari'a*), adjusted according to customary institutions and interests of the clergymen.

11. This right to be a *Jowro* is passed from father to oldest son—often in direct succession since the Dina (1818-1862). Many families held these titles even long before. His jurisdiction is limited to the sub-*leydi* of his hegemony and to the organization of the corporate herding unit (*egguirgol*) which has priority access to the sub-*leydi* under his control (Gallais 1967).

12. In 1955 the French further strengthened the access rights of cultivators to crop land through a government declaration (*Décret foncier et dominal*) (Riddell 1982). This facilitated the break between the Rimaybé rice cultivators and their Fulani masters in many communities.

13. Even so, the livestock sector contributes by far most to regional monetary income. Rice cultivation brings low cash returns both at household and regional levels (CABO 1991).

14. The delta has been recognized as a zone with international conservation values attached to it, both as an important ecological zone and due to its rich biodiversity (World Bank 1996).

15. Ellis and Swift 1988; Behnke 1992; Behnke and Scoones 1992; Behnke 1994 and 1994b; Scoones 1994; Swallow and Bromley 1992; Moorehead 1991; Behnke et al 1993; Lane and Moorehead 1993; Swift 1993; Cousins 1993; Bonfiglioli and Watson 1993; NOPA 1992; Shanmugaratnam et al 1992; Vedeld 1992 and 1997.

16. According to Swallow and Bromley (1991, 9) "The lack of legitimate and powerful governmental organization makes efficient implementation of range land property rights the exception, rather than the norm." They claim that there are not many pastoral common pool resources that are completely regulated by self-enforcing institutions which coordinate access among co-owners as in the assumptions by the assurance problem school of property rights theory (Runge 1981, 1986).

17. Such systematic correlation is found by Behnke among Bedouins of Libya, Karamajong of East Africa, Boran of Ethiopia, and Berber groups of Morocco (Behnke 1994b).

18. It seems that even when the pastoralists constitute a major part of the village population, as in Dialloubé, the pastoral interests are often outmaneuvered. The pure pastoralists are less educated, less organized, and have less influential political alliances outside the village, compared to compact agro-pastoral elites among the Fulani clergymen and traders (Vedeld 1997).

19. As observed by Spiertz (1995), in many life situations resource users, leaders, and state officials make use of more than one normative (rule-based) repertoire to rationalize and legitimize their decisions and behavior. He claims that people orient themselves after local knowledge, perceived context of interaction, and power relations.

20. The rule "all Malians" is referred to by the state as all citizens of Mali having access to state property range land. The concept of "mise en valeur" means that a certain investment in the land, e.g., land preparation, must be carried out before a proprietor can claim a usufruct right or ownership right to the land.

21. To some degree, the interventionist approach by state officials has been moderated following the fall of the one-party system as local politicians have become active in tenure conflicts and question acts by authoritarian state officials.

22. Incompatibility between formal law and local behavior is observed in different forms across the world and often constrain law enforcement. The adoption of "liberal" ideas about the concepts of property, for example, through "top down" privatization creates legitimacy and efficiency problems if reforms are not sensitive and adaptive to local custom (Swallow and Bromley 1995 on range land reforms in Africa).

23. Since Islamic laws and rules are religiously based, i.e. based on local values and ideas, they are slow to change and not susceptible to change by national legislation (Tiffen 1985).

24. In many countries across the world a right of long-term occupancy is obtained after thirty to forty years of use without interference or claims by other users or owners.

25. The "negotiability of rules and relationships in Africa may be seen as an opportunity rather than an impasse" (Berry 1994, 2). It reflects perhaps that land in general is less of a constraint to development than labor and capital, and implies that the scope for dynamic change is maintained.

26. An increasing number of research works have, however, started to raise these problems in Mali and Sahel: CIPEA 1983, Cissé 1985, 1991 and (ed) 1995, Tiffen 1985, Bruce 1988, Rochegude 1990, Hesseling and Coulibaly 1991, Coulibaly and Hesseling 1992, Le Bris et al. 1991, Kintz 1990 and 1992, Le Roy (ed) 1992, Le Roy et al. 1996, Vedeld 1997.

REFERENCES

Allott, A. and G. R. Woodman, eds. 1985. *Peoples Law and State Law*. Bellagio Papers. Dodrecht, Holland/Cinnaminson, USA: Foris Publications.

Ba, A. H. and J. Daget. 1962. *L'Empire Peul du Macina* (1818–1853). Abidjan: Les Nouvelles Editions Africaines.

Behnke, R. H. 1992. *New Directions in African Range Management Policy*. Pastoral Development Network, ODI. UK.

Behnke, R. and Scoones, I. 1992. "Rethinking Range Ecology: Implications for Rangeland Management in Africa." IIED Dryland Networks Programme Issues Paper no. 33. Woburn, UK: IIED/ODI.

Behnke, R. H., I. Scoones, and C. Kerven. 1993. *Range Ecology at Disequilibrium: New Models of Natural Variability and Pastoral Adaptation in African Savannas*. London: ODI/IIED/Commonwealth Secretariat.

Behnke, R. 1994a. "Natural Resources Management in Pastoral Africa." *Development Policy Review*, 12, 5–27. London: ODI.

Behnke, R. 1994b. "Land Tenure and Range Management Institutions in the Context of Commercialisation." RPRRP, Working Paper no. 9. London: ODI.

Berry, S. 1989a. "Social Institutions and Access to Resources." *Africa*, 59(1), 41-55.

Berry, S. 1989b. "Access, Control, and Use of Resources in African Agriculture: An Introduction." *Africa*, 59(1).
Berry, S. 1994. "Resource Access and Management as Historical Processes: Conceptual and Methodological Issues." Paper for seminar on "Access, Control, and Management of Resources in Sub-Saharan Africa." International Development Studies, Roskilde University, June 1–2, 1994.
Bonfiglioli, A. M. and C. J. Watson, eds. 1993. Proceedings from NOPA workshops on Pastoral Development in Africa (East and West); UNICEF/ UNSO Project for Nomadic Pastoralists in Africa, NOPA, Nairobi.
Boserup, E. 1990. *Economic and Demographic Relationships in Development*, edited by T. Paul Schultz. Baltimore and London: John Hopkins University Press.
Bromley, D. 1989. *Economic Interests and Institutions: The Conceptual Foundation of Public Policy*. New York: Basil Blackwell.
Bromley, D. ed. 1992. *Making the Commons Work: Theory, Practice and Policy*. San Francisco: ICS Press.
Bruce, J. W. 1988. "A Perspective on Indigenous Land Tenure Systems and Land Concentration," in *Land and Society in Contemporary Africa*, edited by R. E. Downs and S. P. Reyna. University Press of England.
CABO. 1991. "Compétition pour des ressources limitées: Le cas de la cinquiéme région du Mali." Rapport 1–5. Wageningen. Pays-Bas et ESPR, Mopti, Mali.
CIPEA. 1983. "Recherche d'une solution aux problèmes de l'élevage dans le delta intérieur du Niger au Mali." Vol I-IV, CIPEA/ODEM, Addis Ababa/ Sevare, Mali.
Cissé. S. 1985. "Gestion integrée des ressources primaire de l'élevage dans les systemes pastoraeaux du Delta Interieur du Niger." CIPEA, Prog. Doc. no. AZ 148, Addis Ababa.
Cissé. S. 1991. "Concurrances spatiales et nouvelles cohabitations foncières en 5e Region: La dynamique des leyde." Observatoire du foncier au Mali. ESPR/ODEM, Sevare, Mali.
Cissé, S. ed. 1995. "Caracterisation fonciers et exercise des actes de droit: Prêt, location, métayage et vente de la terre." Bamako: Ministère du Developpement Rural et de l'Environnement.
Coulibaly, C. and G. Hesseling. 1992. "Note sur la Problematique Foncière et la Decentralisation au Mali, Etu de sur un Programme de Recherche-Observation préparée à la Demande de la Banque Mondiale." IMRAD/ Centre d'Etudes Africaines, Bamako/Leiden.
Cousins, B. 1993. *A Political Economy Model of Common Property Regimes and the Case of Grazing Management in Zimbabwe*. Pastoral Development Network 34b, ODI.
de Bruijn, M. and H. van Dijk. 1995. *Arid Ways: Cultural Understanding of Insecurity in Fulbe Society, Central Mali*. CERES Series no. 1. Amsterdam: Thela Publishers.

Ellis, J. E. and Swift, D. M. 1988. "Stability of African Pastoral Ecosystems: Alternative Paradigms and Implications for Development." *Journal of Range Management*, 41(6).

Gallais, J. 1967. "Le Delta Interieur du Niger: Etude de gógraphie régionale, Mémoires de l'Institut Fondamental d'Afrique Noir." No. 79, IFAN, Dakar, Vol I et II.

Gallais, J. 1984. "Hommes du Sahel, Espaces-Temps et Pouvoirs, Le Delta Intérieur du Niger 1960-1980." Paris: Flammarion, Collection Géographes.

Godelier, M. 1984. *The Mental and the Material*. London: Verso.

Goheen, M. 1992. "Chiefs, Sub-Chiefs and Local Control: Negotiations over Land, Struggles over Meaning." *Africa*, 62(3).

Hardin, G. 1968. "The Tragedy of the Commons." *Science*, 162, 1243–1248.

Helland, J. 1992. *Controlling Access or Controlling Numbers: An Issue in Local Resource Management Systems in the Sahel Region*. Forum for Development Studies, NFU/NUPI.

Hesseling, G. and C. Coulibaly. 1991. *La Legislation et la Politique Foncieres au Mali, Rapport dans le Cadre du Schéma Directeur de Développement Rural, Africa-Studiecentrum*, 60. Centre d'Études Africaines (ASC)/IMRAD, Leiden/Bamako.

Hiernaux, P. 1993. *The Crisis of Sahelian Pastoralism: Ecological or Economic?* Addis Ababa: ILCA.

Hiernaux, P. and L. Diarra. 1986. "Bilan de cinq annees de recherches (1979–1984) sur la production vegetale des parcours des plaines d'innondation fleuve Niger au Mali Central." Document de Programme az 142. Addis Ababa: ILCA/CIPEA.

Kintz, D. 1990. "L'Amant Blessé ou une Discussion Peule sur le Pluralisme Juridique." *Politique Africaine*, no. 40, 42–51. Paris: Karthala.

Kintz, D. 1992. Le Foncier Rural au Mali, Bilan et Recommandations pour un Observatoire du Foncier (OFM). Min. de l'Agriculture, de l'Élevage et de l'Environnement/Caisse Centrale.

Lane, C. and Moorehead, R. 1993. "New Directions in African Range Management, Natural Resource Tenure and Policy." Workshop theme paper, May 1993, Woburn.

Le Bris, É., E. Le Roy, and P. Mathieu. 1991. *L'Appropriation de la terre en Afrique noire, Manuel d'analyse, de décision et de gestion foncières*. Paris: Karthala.

Le Roy, É. 1985. "Local Law in Black Africa: Contemporary Experiences of Folk Law Facing State and Capital in Senegal and Some Other Countries," in *Peoples Law and State Law*, edited by A. Allott and G.R. Woodman. Dodrecht, Holland/Cinnaminson, USA: Foris Publications.

Le Roy, É. ed. 1992. *La Mobilisation de la terre dans les strategies de developpement rural en Afrique Noire Francophone*. Paris: APREFA/LAJP (Laboratoire d'anthropologie juridique de Paris).

Le Roy, É., A. Karsenty, and A. Bertrand. 1996. *La sécurisation foncière en Afrique. Pur une gestion viable des ressources renouvables*. Paris: Karthala.

Lewis, J. D. 1981. "Range Use and Fulbe Social Organization: The View from Macina," in *Image and Reality in African Interethnic Relations: The Fulbe and Their Neighbors*, edited by E. A. Schultz. Publication no. 11, St. John's University.
Lloyd, Lord of Hamstead, and M. D. A. Freeman. 1985. *Lloyd's Introduction to Jurisprudence* (Fifth Edition). London: Stevenson & Sons.
Moore, S. F. 1992. Treating Law as Knowledge, *Law and Society Review*, vol. 26(1).
Moorehead, R. 1989. "Changing Takes Place in Common Property Resources Management in the Inland Niger Delta of Mali," in *Common Property Resources, Ecology and Community-Based Sustainable Agriculture*, edited by F. Berkes. London: Belhaven Press.
Moorehead, R. 1991. "Structural Chaos: Community and State Management of Common Property in Mali." Thesis. Development Studies, University of Sussex.
NOPA. 1992. "Pastoralists at a Crossroads: Survival and Development Issues in African Pastoralism." UNICEF/UNSO Project for Nomadic Pastoralists in Africa. Nairobi: NOPA.
North, D. 1990. *Institutions, Institutional Change and Economic Performance*. Cambridge: Cambridge University Press.
OCDE/CILSS. 1990. "West African Systems of Production and Trade in Livestock Products." Issues Paper d/90/351.
Ostrom, E. 1990. *Governing the Commons: The Evolution of Institutions for Collective Action*. Cambridge: Cambridge University Press.
Park, T. 1993. *Risk and Tenure in Arid Lands: The Political Ecology of Development in the Senegal River Basin*. Tucson and London: University of Arizona Press.
Riddel, J. C. 1982. "Land Tenure Issues in West African Livestock and Range Development Projects." Land Tenure Center Research Paper 77, University of Wisconsin-Madison.
Rochegude, A. 1990. "Le droit de la terre au Mali, un aspect juridique du développement économique." Thése pour le doctorat. Univèrsité de Paris I, Panthéon-Sorbonne.
Runge, F. C. 1981. "Common Property Externalities: Isolation, Assurance, and Resource Depletion in a Traditional Grazing Context." *American Journal of Agricultural Economics* 63, no. 4.
Runge, F. C. 1986. "Common Property and Collective Action in Economic Development." *World Development* 14, no. 5, 623–635.
Sandford, S. 1983. *Management of Pastoral Development in the Third World*. London: John Wiley & Sons and Overseas Development Institute.
Scoones, I. 1994. "Living with Uncertainty: New Directions for Pastoral Development in Africa." Overview Paper of the Workshop on New Directions in African Range Management and Policy. Woburn, UK, June 1993, ODI/IIED/Commonwealth Secretariat.
Schlager, E. and E. Ostrom. 1992. "Property-Rights Regimes and Natural Resources: A Conceptual Analysis." *Land Economics*, 68(3):249–62.

Shanmugaratnam, N., T. Vedeld, A. Mossige, and M. Bovin. 1992. "Resource Management and Pastoral Institution Building in the West African Sahel." World Bank Discussion Papers. Washington D.C.: AFTAG.

Spiertz, H. L. J. 1995. *State and Customary Laws: Legal Pluralism and Water Rights*. 13:1-7. The Farmer-Managed Irrigation Systems Network, International Irrigation Management Institute.

Swallow, B. and D. Bromley. 1991. "Rethinking the Conceptual Foundations of African Range Tenure Policy." Workshop on New Directions in African Range Managment Policy. Matopos, Bulawayo, Zimbabwe, January 1992.

Swallow, B. and D. W. Bromley. 1995. "Institutions, Governance, and Incentives." *Common Property Regimes for African Rangelands, Environmental and Resource Economics* 6:99–118.

Swift, J. 1988. *Major Issues in African Pastoral Development*. FAO, Rome: Institute of Development Studies, University of Sussex.

Swift, J. 1989. "Land Tenure and Pastoral Resource Conservation." *The IUCN Sahel Studies 1989*. The IUCN Sahel Programme, IUCN.

Swift, J. gin. "Fulbe Herding Practices and the Relationship Between Economy and Ecology in the Inland Niger Delta of Mali." Thesis, Energy and Resources. Berkeley: University of California.

Tiffen, M. 1985. *Land Tenure Issues in Irrigation Planning Design and Management in Sub-Saharan Africa*, Working Paper no. 16. London: ODI.

Tiffen, M. and M. Mortimore. 1994. *Environment, Population Growth and Productivity in Kenya: A Case Study of Machakos District*, IIED Issues Paper no. 47, London: IIED Dryland Networks Programme.

Toulmin, C. 1991. "Natural Resources Management at the Local Level: Will This Bring Food Security to the Sahel?" IDS Bulletin 22, no. 3. Sussex: IDS.

Turner, M.D. 1992. "Life on the Margin. Fulbe Herding Practices and the Relationship between Economy and Ecology in the Inland Niger Delta of Mali." D. Phil. in Energy and Resources, Berkeley: University of California.

Turner, II, B.L., G. Hydén, and R.W. Kates, eds. 1993. *Population Growth and Agricultural Change in Africa*. Gainesville: Florida University Press.

Vedeld, T. 1992. "Local Institution Building and Resource Management in the West African Sahel." FORUM for Development Studies, no. 1. Oslo: NUPI/NFU (also in Pastoral Development Network Paper no. 33c, London: ODI).

Vedeld, T. 1993a. "Rangeland Management and State-Sponsored Pastoral Institution Building in Mali." Case study for Research Workshop on New Directions in African Range Management and Policy. Woburn, UK.

Vedeld, T. 1993b: "Enabling Pastoral Institution Building in Dryland Sahel." Paper presented to the UNSO Donor/Agency Consultation Meeting on Pastoral Resources Management and Pastoral Policies for Africa, Paris.

Vedeld, T. 1994. "The State and Rangeland Management: Creation and Erosion of Pastoral Institutions in Mali." *IIED Dryland Network Papers*, no. 46. London: IIED.

Vedeld, T. 1997. "Village Politics: Heterogeneity, Leadership, and Collective Action among Fulani of Mali." Dissertation presented in partial fulfillment of the Doctor Scientiarum degree, thesis 1997:13. Department of Land Use and Landscape Planning, Aas: Agricultural University of Norway.

Wagenaar, K.T., A. Diallo, A.R. Sayers. 1986. *Productivity of Transhumant Fulani Cattle in the Inner Niger Delta of Mali*, ICLA Research Report no. 13, Addis Ababa.

World Bank. 1996. "Towards Environmentally Sustainable Development in Sub-Saharan Africa." A World Bank Agenda. Washington: World Bank.

AUDUN SANDBERG   CHAPTER 19

The Analytical Importance of Property Rights to Northern Resources

The North

For opportunistic harvesters, settling within the North had several advantages—one was the short distance to the rich fishing and hunting grounds and the opportunity to constantly monitor the changes in fish and wildlife and over time build a "folk science" related to fluctuations in natural resources.[1] The first settlements of the North are therefore to a great extent the history of primary occupation of favorable places (*primi occupantis*); when these were settled, the larger area was in fact occupied at the prevailing level of resource utilization. The main disadvantage for northern settlers has always been the huge distance to the more densely populated areas of the continents, which meant that exchange of goods, barter, and marketing was—and still is—a difficult and sometimes risky operation. With opportunities for flexible resource utilization, the conditions for self-sufficiency were favorable, while the power over world markets for specialized surplus products has remained negligible.

Another advantage of settling within the North is the fact that permanent residence is an important element in the long process of establishing socially acceptable property rights to northern resources. Initially we shall here use the term property rights to describe all forms of relationships between northerners and northern resources, in the legal heritage of the North these were categorized as: *primus occupat, possessiones, usum fructum,* and *dominium* (Stephanus 1629).

In a world of freely migrating humans, there are always two sides to resource-endowed regions. The seemingly great advantage to resource management in the North and its sparse population relative to its richness in natural resources was also its great disadvantage; the scarcity of defenders of the "favorable places" and the inability to defend the rest of the resource base—the "outer and the upper." Most northern societies were able to accommodate—and even welcomed— a trickle of poor migrants from the South. But faced with various kinds of intruders, from plundering armies to large-scale state settlement programs, the societies of primary and secondary occupants showed their vulnerability. Despite the vast open areas of the North, the number of favorable places was limited and in the case of the intruders' occupation of these, the result was often starvation and poverty for the original inhabitants. Although full state sovereignty was imposed relatively late in the "Far North" (some observers hold this to be from the 1930s onward) it can be instructive to explain some of the longer lines in the development of the "internal colonization" (Young 1992).

At the height of the Viking Age, around A.D. 1000 to 1250, the sea between Norway, Spitzbergen, Greenland, Iceland, and Scotland/ Ireland was a Norse "inland sea" and numerous Norse "northern settlements" were established along the northernmost coast of Norway, along the northern river valleys of Sweden, on Iceland, Greenland, and even on Labrador. This expansion ceased in the 14th century. Both the increased power over trade routes by the Hanseatic league, the "little ice age," and the demographic effects of the Black Plague in 1349 are believed to be important factors here. After the 15th century we no longer find Norse colonies in Greenland and Labrador.

After the rediscovery of America there was again new European expansion towards the North. The first known "biocide" of the North was in the 16th century, when Dutch whalers and walrus hunters adhering to the newly won "freedom of the seas" doctrine depleted huge herds of fat-bearing sea mammals and birds (e.g., the *Geirfowl*). All they left were their names: The "Spitzbergen," the "Barents Sea," and the "Jan Mayen."

The overriding question here is who has the rights to northern resources. This does not only depend on which nation state claims the land or sea area, but also on who has lived there before and who lives (permanently) in this area today. It is a big question in the international arena—between the nation states in the UN or law of the sea framework (e.g., the cod wars between Iceland and Great Britain and between Iceland and Norway). It is also a big question within each of the nation states: as to which categories of citizens have what kind of rights to northern resources—all nationals, all resident northerners, all rural northerners, or only resident aboriginal northerners (NOU 1997, 4). On the international level there is mounting agreement that it is unrealistic to regard the North as a global commons; the jurisdictions of the eight "Arctic states" have become too entrenched for that (Young 1992). This debate is now reduced to the question of whether parts of the North should be considered a "shared resources region" where resources can best be managed by two neighboring nation states or a "regional commons" where, for instance, international institutions or supranational organizations like the European Union would manage a section of the European Arctic.

On the national level it is much less clear who has rights to the resources of the North. On the European and Asian continent, the distinction between the "aboriginal" peoples of the North and the "natives" of the colonizing states is today often blurred; in many instances they have been living in the same geographical areas for several thousand years. However, their different adaptations to ecological and man-made niches, and their different cultures have existed side by side for hundreds of years despite frequent intermarriages and heavy pressure on the minority cultures to "go Russian," "go Finnish," or "go Norwegian." In the northern areas of America, the distinction between aboriginal people and colonizing people—or "native peoples" and "immigrant peoples" (Sproule-Jones 1993)—is easier to draw and is also much more an integral part of "northern politics." However, the vivid but long-drawn political processes of aboriginal (native) land and water claims in Canada and Alaska have to a large extent influenced the public debate on aboriginal and local land rights and water rights in northern Europe. With the formation of an association of the "small peoples" of Russia and Siberia, there are also signs that such a debate gradually is in the coming in both the European and Asian part of Russia. Both the recommendations of the United Nations Conference on Environment and Development (UNCED) agenda 21, and the emphasis on "subsidiarity" within the European Union contribute to this.

In this respect the political situation of the whole of the northern circumpolar region has become more similar in the last twenty years. This should be of benefit to the policy analyst as it permits more meaningful comparative analysis (for example, analysis of the managing of similar resources in two or three areas with distinctly different systems of nation state supremacy). The grand experiment of the North that was initiated by the European nation states has now been running for about three hundred years and there is sufficient evidence to start an evaluation of it at a larger scale.

In the final analysis the question of northern politics is often boiled down to this very simple question of who has the rights to the resources of the North. But this form of rhetoric often conceals the analytically more interesting questions of how property rights to resources are created and maintained. These questions should therefore be made more complicated by rephrasing them and treating them with sharper analytical tools:

- Do we find elements of social contracts in northern politics, where the nation states continue to have ownership rights and access to the northern resources in exchange for northerners' continued access to the state treasury?
- What happens to such social contracts if the welfare benefits to northerners diminish or when the state loses or gives away centralized power to local communities or to institutions based on aboriginal identity?
- How are property rights in the North affected by decentralization of political authority on one hand (devolution) and on the other hand the slow legal/political process of transferring more resource governing authority to local authorities or aboriginal associations in the North?

We shall here mainly use northern Norway as a case to present some preliminary hypotheses about such relationships and try them out on six different types of resources—some major resources and some resources of minor importance. However, the reader should bear in mind that these questions can be raised at a generalized level in all the northern circumpolar areas, and that comparative studies in two or three northern areas could bring new insights which a one-shot approach does not give.

The Puzzle

In a modern "welfare state" like Norway, one might think that the links between society members and natural resources were of minor importance, that a benign state caters to the livelihood and happiness of all its subjects irrespective of their inherited or achieved rights. It is a puzzle that this is not the case and that resource rights again have entered the public realm. In fact, the questions of both individual and collective, local and regional rights to resources even increased in importance as, for instance, the question of the relation of the nation state to the greater European Union moved up on the political agenda. In addressing this puzzle, it is therefore important to keep in mind that even a seemingly solid institution like the welfare state exists only as long as it is maintained by its members. Following the tradition of Aristotle, Tocqueville, and Weber, the position taken here is that if individual members do no longer act in support of social institutions, these will wither away, break down, or change their form (Kaminski 1992).

The welfare of northerners has always depended on their relations to natural resources, on their command of access to resources, and on their ability to exclude others from extracting these resources (Dacks 1981, 1990). These kinds of relationships between individuals, groups of individuals, communities, and states with respect to resources are what we commonly call property rights. In the political sphere such rights carry heavy symbolic values and "our rights to resources" are frequently used to create identity and to rally followers for or against a particular solution. One basic argument in this article is that more sharply defined concepts of property rights will be useful tools for analyzing the interdependence between human and economic enterprise and the governing of resources. Further, it is argued the full potential of property rights concepts has not been realized due to lack of definitional clarity, ideological bias in Scandinavian social sciences, and sloppy operationalizations by scientists.

In a mixed economy like the Norwegian economy, markets have usually been governed by legal frameworks initiated by political bodies or by institutional arrangements agreed upon by way of negotiations between organized interests. The high transaction costs and loss of efficiency from cumbersome market regulations have in recent years set in motion a prolonged process of deregulation of markets, where the explicit aim of government economic experts has been to allow the market mechanism work wherever this is more efficient than other means of aggregating decisions. With less government interference in the commodity markets, the product markets, and the labor markets, the underlying structures of ownership and user rights will resume some of their importance and again bear influence on the distribution of income- and benefit-streams and on the overall distribution of welfare. One of the reasons for the renewed academic interest in the study of origin and maintenance of property rights is therefore the fear that deregulation might thus lead to increasing inequalities, social misery, and political unrest. Through refinements of the art of crafting institutions for property rights–based management regimes, such unwanted side effects of deregulation can hopefully be avoided.

Northern areas are basically resource-based regions and the number of resources in northern areas are large. Both nonrenewable resources and biologically renewable resources are the backbones of northern economies and societies. It is therefore impossible in one study alone to analyze all the changes in resource relations that result from greater northern self-consciousness and from changes in the role of the nation state. It is necessary to select some of the most typical resources and carry out the analysis with a scope that is wide enough to encompass both the individual and household level, the local community level, the intermediate level the nation state level, and the international or "regional" level. In this respect, some resource governing systems give a higher "analytical payoff" than others as they reveal more of the underlying social processes in contemporary society. In using northern Norway as a case, six different kinds of resources stand out as having a rich analytical potential. As will be seen from the brief exposition of the governing challenges facing each of these resource types, they are comparable on some dimensions, while they differ greatly on others. The governing systems for these six resource types also interact with each other in a number of ways and such interactions might increase or decrease the challenges of governing that particular resource.

Birds' Eggs

Birds' eggs are analytically important in spite of their negligible economic importance. The eggs of sea gulls and certain wild ducks have for thousands of years been collected during the spring by fisher households and farmer households as subsistence food. Later in spring the down of certain ducks, notably eider ducks, was also collected and treated for use in downs and pillows. The nesting environments of "down-birds" were enhanced by simple constructions of protecting stones, wrecked boats, and the like, while no enhancement was usually undertaken in relation to "egg-birds." Traditionally, spring was the time when the stores of cereals and meat were run down after a long winter, when fish was hung to dry for cash, and when fresh eggs represented an important source of protein for all members of the household. The eggs were only collected from those kinds of birds that have the capacity to lay additional eggs if some are lost or stolen. For obvious reasons, collection took place while the eggs were still fresh, thus giving the birds a chance to add new eggs. Strict rules of how many eggs can be removed from the various kinds of nests and at what times by which households have maintained this sustainable resource management system through centuries (Vold 1981). The only threat to this kind of viable egg collecting system seems to be external factors such as over-fishing of certain key species of fish which are crucial food for the newborn chicks and international regimes, notably EU regimes, which indiscriminately ban all egg collecting from wild birds in order to protect some endangered species of birds.

Because of their minimal economic importance, the egg-collecting institutions of northern Norway have in most places been untouched by interventions from government or big business. Thus we here find design principles that have evolved gradually and have been continuously refined since the Viking Age. In most respects these principles can serve as a baseline for comparison with the design principles of newer resource governing systems.

Wild Sea Fish

Wild sea fish are economically the most important renewable resource in northern areas—which is, in fact, mainly northern waters. Because of this, sea fisheries and ocean fisheries are entangled in state regulations, international regulations, and the continuous games played by the organized interests of fishers, fish industries, and fish exporters. These are not only games played against the nation states or weak international supervising bodies, but to a large extent games where one group of organized interest uses the apparatus of the state or the "union" in its gameplay against another group of organized interest. Apart from a few exceptions, the organization of the Lofoten Fisheries being one, several local fjord fishing regimes in Finnmark being another, there is only a small potential analytical contribution to the body of knowledge of design principles from "naturally evolved" resource management systems in Norwegian fishing. However, the analytical importance of wild fishery governing regimes is significant in dealing with the role of the nation state and the role of provinces, counties, and associations in resource management and in dealing with the workings of international governmental organizations (IGOs) and international nongovernmental organizations (INGOs) (McGinnis and Ostrom 1993). However, some of the effects on local fisheries from the incentive structures provided by the government-designed governing institutions can provide insights that are analytically important. Crucial factors here seem to be:

- The lack of correlation between rights and duties for Norwegian fishers
- The negative effects on recruitment of young fishers of the increasing exclusiveness in the authorization of fishers
- The decreased flexibility of coastal fisheries and the social and political costs of a high-mobility ocean fishing fleet (Sandberg 1993a)

Recently there have been several moves by local communities, coastal fishers, politicians, and academics within the northern areas to modify the extremely complex and cumbersome government and international regulatory systems for harvesting wild fish. The idea has been to replace these systems with simpler systems of fishers' co-management regimes or territorially based management regimes that utilize the capacity of the northerners to exercise self-discipline and self-control. A convergence of folk knowledge and modern multispecies management tools has made these ideas politically more feasible (Eikeland 1993). Recent experience has shown that neither national nor regional (European) management regimes can manage the highly efficient and highly mobile international fishing fleet that belongs in the big fishing nations, but increasingly fish under various flags of convenience on the high seas (where it is not bound by quota agreements between the "responsible fisheries nations"). Mounting difficulties in monitoring this international fleet and its adherence to agreed quotas might in the long run result in a total closure of the global commons called the "high seas"—a preliminary test of this is the recent agreement of United Nations on the management rights and responsibilities of coastal states regarding "straddling stocks of fish." With a further enclosure, the design of institutions benefiting the coastal fisher as the main category of saltwater fisher would be more feasible. Recent studies have also shown that when confronted with streamlined and simplistic government resource management regimes, coastal fishers have the capacity to design their own supplementary rules within a short time in order to avoid gear collision and secure a reasonable harvest for all participants in practical fisheries (Bjørnaa 1993).

In line with this kind of thinking and with findings like these, fisheries management could increasingly be handed over to coastal fishers themselves or their coastal communities in the form of:

- "Producer organizations" (POs—a tolerated exemption to EU principles)
- Coastal territorial "boxes" (an EC invention)
- Fjord basin "management boards" (a local demand)
- Aboriginal councils (a Saami demand)
- Local government (a provincial demand in northern Norway)

What primarily carries analytical importance here are the actions of the nation states and the interest groups connected to the various proposals, and the reactions to these proposals of the different parts of national governments and international governmental organizations, notably of the environmental sections' opposition toward positions of the fisheries and export sections.

Coastal Environment for Aquaculture

Coastal environment for aquaculture is the basic resource in the development of aquaculture along the northern coasts. The gill-net "pens" that are the present basis for salmon farming interact directly with the flowing sea water, and the health and rapid growth of the farmed salmon is highly dependent on a pure and healthy coastal ecosystem. This is important even at the level of the individual aquaculturalist; the externalities (pollutants and contaminating agents) produced by one firm are not likely to affect only the neighboring aquaculturalist, but also the firm itself when the tide turns. Thus there should be very strong incentives in aquaculture firms to either be far away from each other, or if that is not possible, to internalize all or most of the externalities. The effect of both strategies is basically to treat the clean and healthy coastal environment as a crucial production factor that must be maintained, even if a cost is incurred to do so.

In reality the world is different: aquaculturalists have the same tendency as other enterprises to be free riders on what they consider to be a public good—the clean and healthy coastal ecosystem. The time horizon is often short ("this crop of salmon—while the price is good"), and the fisherman's cultural paradigm of chance luck prevails.

Also the government licensing system with size limits for each firm gives adverse incentives to stock too many fish in a small volume, thus exceeding the environmental carrying capacity of a certain location. In addition to this, the aquaculturalists' demand for clean and healthy coastal environments and their subsequent consumption of these has to compete with all sorts of needs and uses of coastal ecosystems for sewage disposal, recreation, sport fishing, spawning areas, and fry feeding areas for wild fish.

With a situation where the national government is administering the licensing system, while the local government has the responsibility for sewage and recreation, the end result has often been a "right" of the aquaculturalists to consume the local coastal environmental resource by "central government authorization." At the same time as this "right to degrade the environment" is contrary to the objective interests of the aquaculturalists as a collective, it also renders the central government vulnerable. This awkward position of a central government comes as a result of historical/political ambitions to "manage" the growth of a new trade so as to avoid "boom and bust" situations. Experience has shown that this is virtually impossible to achieve; the enterprising individual aquaculturalists will always outsmart the government while the government's attempted governance will eliminate any self-discipline and self-restraint that the aquaculturalists might have had at the outset. It therefore seems necessary to replace management of the aquaculture sector with governing of the aquaculture resource itself—the clean and healthy coastal environment.

The future governing of coastal ecosystems of the North therefore needs to take into consideration all the different uses of coastal environments and to place the coordinating authority at a level of governance where efficient deals can be struck between competing uses. Coastal zone management by local government, with mandatory adherence to strict environmental standards, would make the health of the coastal ecosystems the prime objective, thereby also benefiting the aquaculturalists as a group in the longer run (Sandberg 1993b).

If decisions on this kind of multipurpose use of fragile ecosystems are taken to the local level, property rights have a tendency to assume importance. Depending on the particular history of a community, fjord, or archipelago, individuals or collectives will claim rights to islands, sounds, and shores; to river mouths and the runways of salmon, sea trout, or arctic char; or to traditionally good fishing places for marine species.

Wild, Migrating Salmon

Wild, migrating salmon is a very special kind of resource and maybe one of the most valued resources among northerners. In a resource management context it is special because as an anadromous fish, it moves between the open ocean, the coastal waters, and up the numerous rivers surrounding the northern oceans. Along its route, the salmon encounters widely different property rights systems which are as difficult to traverse as the waterfalls of the most rapid rivers. On the high seas no one owns the salmon, but it roams widely and is hard to get. Downstream from these vast feeding grounds, the salmon become more concentrated closer to the coast, where it used to meet a virtual fence of drift nets. The salmon then enter the sounds and fjords, where shore owners have their traditional fixed-net-set places (*kilnot*), the rights of which date back to medieval ages. Here the sport fishers also troll for salmon. In the fjords the salmon also encounter runaway distant relatives from numerous fish farms who disturb its homing instinct and mating behavior. When the remaining salmon finally reach the river of their origin, both the shore and the river channel itself are someone's private property, usually the property of the farmstead bordering the river. Fishing is done here usually by rod and can only take place with specific permission from the owner. In some rivers of the North, farmers still practice traditional salmon "harvesting" with traps or nets. In other rivers, the river owners have pooled their property rights together and sell fishing permits that give access for the general public to numerous private shores at a reasonable price. With secure property rights to salmon rivers, the owners are usually eager to enhance the river environment to achieve a greater spawning success, in order to supplement the fish stock or to restock the rivers after hydropower development, road construction, or attacks from the *gyrodactilus* parasite. Most of these activities are undertaken as co-operative effort between the owners, often with assistance from the local government or from the local sport fishers' association—in a co-management arrangement.

The analytically challenging aspects of wild, migrating salmon shows the relative importance the different kinds of property rights take on when action has to be taken on the dwindling stocks of river salmon. The suspension of the coastal fishers' right to fish ocean salmon with drift nets has not helped the stock of wild salmon and stronger measures are under consideration.

An even more difficult debate is a current discussion over the rights of river owners to have "their" salmon and rivers protected from genetic contamination of runaway farm salmon and to have the runway of their salmon—all the way from the free ocean to their private river—protected from "foreign pheromones" (the smell substance that is the basis for the homing instinct). The "protection zones" necessary to give optimal runways to all the salmon rivers of northern Norway would render large areas out of bounds for aquaculture and significantly hamper the commercial development of new forms of farming the seas, notably sea ranching and marine transhumance (NOU 1990:22). The traditional property rights of river owners to "their" genetic brand of salmon are reasonably strong when confronted with the newly acquired rights of aquaculturalists to farm salmon in certain locations or the more diffuse rights of sportfishers to troll in the fjords. Adding to this is the division of licensing authority at both the central government and regional levels. A Directorate of Nature and Wildlife has become the agent of the river owners, environmentalists, and sport fishers, while a Directorate of Fisheries has become the agent of the aquaculturalists and ocean fishers.

Forests, Berries, and Pastures

Forests, berries, and pastures represent as a resource the whole of the uninhabited lands and mountains of the North. Although nobody lives in these areas, that does not mean that they are not used by anyone or that they are useless. In Norwegian these areas are traditionally called *utmark* (the outer fields), which means that they were used by the nearby farms or villages as additional fields in order for inhabitants to make a living in a harsh environment. Wood and timber were collected from the forests, wild berries were picked during the autumn season, cows and milk goats were grazed in natural mountain pastures in the transhumance system, and sheep and reindeer were left to roam freely in the mountains. Instead of expensive herding of these animals, farming communities and Saami communities have increasingly relied on predator control in what they consider "common property lands." Recently, international commitments have forced the nation state to take over most of the management of predators in order to preserve a viable national stock of wolves, bears, and lynx in the North. This does, however, clash with the present processes of extensification of animal husbandry in the same areas. Partly these extensification processes are state-induced through the various support programs for district agriculture and reindeer herding.

In addition to these farm-related activities, hunting and the fishing of fresh water fish have been major sources of food and cash. Although the areas look much like wilderness to the untrained observer, the entire uninhabited area of the North has been utilized by northerners for thousands of years. Lakes have continuously been restocked with fish by certain families, clans, or villages; game has been managed in different ways in defined territories (*vald*); and snares have been set for grouse in mutually respected places. Thus there are bundles of more or less visible rights tied to these areas.

It is in the case of forests, berries, and pastures that the role of the nation state as a holder of vital property rights in the North is most marked. The state does not merely exercise jurisdiction over this territory—for a greater part of the northern lands, the state is *de facto* owner (see chapter 10). Because of the special history of the North, its rich resources, and its sparse and in many cases nomadic population, the nation states could in most cases colonize the North and make all uninhabited lands state property without much opposition from the indigenous population. For instance, in Norway the whole northern part is exempted from the laws that govern the operations of commons in rural areas (*statsallmenning* and *bygdeallmenning*). In northern Norway and northern Sweden and to some extent also in northern Finland, such state-owned land was crucial for the massive resettlement programs in the eighteenth century that induced people to move from the South to the North.

The question of whether the state also is *de jure* owner (a socially accepted owner) of these northern lands is extremely complicated and related to fundamental constitutive processes in the whole society. As in most western societies, this kind of definition of socially acceptable rights goes on as a "double path" process—sometimes constitutive rules are changed by the political system, sometimes by the judiciary system, and sometimes by both (Ørebech 1991). In Norway the powerful coalition of hikers, hunters, and sport fishers; the tourist industry; the state; the state corporations; and the urban and southern "public" has been challenged by land claims from the Saami, claiming aboriginal rights to land, forests, rivers, and lakes in the "core area" of the Saami. A Royal Court Commission (*Samerettsutvalget*) has been reviewing these claims for the last fifteen years (NOU 1997: 4). The state ownership has also been challenged by claims of northern farmers to have restored the old "common property rights" in the outer fields that the state has "stolen" from the local villages. A royal court commission (*Utmarkskommisjonen*) has for the last ten years been reviewing the rights of various northern villages to forest products, unrestricted pasture, fish, and game. This commission has reached several verdicts, some of them confirmed by the Norwegian supreme court, that legally clear the way for the politicians to introduce laws that give state lands a "commons" status also in northern Norway.

This also shows that the *de jure* rights of the state to "state lands" are not so fundamental that they cannot be changed by ongoing constitutive processes. The property rights questions involved in these kind of legal and political battles therefore point to their fundamental importance in analyzing the role of the state in a resource-endowed region.

Water Power

Water power is maybe the most important resource in explaining the transition of the European northern societies from traditional farming, herding, and fishing societies to modern industrial societies and subsequently to "welfare" societies. Access to cheap hydroenergy was a significant comparative advantage to many towns or regions in the North, whether its primary resource was fish, iron ore, timber, imported bauxite, or simply the nitrogen in the air. The entire industrialization process of the North and a number of the urban conglomerations of the North are products of proximity to abundant hydropower resources.

The property rights of the energy in waterfalls (*fallrett*) were bought by industrialists, municipalities, or the state at an early stage of hydropower development. In most cases they were bought cheaply as the farmers who owned the rivers saw no immediate value in the masses of falling water. To the traditionally inclined farmers it was the salmon that was the real value of the river, and for their own mills they preferred to use the smaller and more manageable streams. The development-oriented farmers were also easy to persuade that hydropower development would benefit the local community and "put lightbulbs in every home." To facilitate this, the property rights of the whole river often had to be gathered to one source. Thus the property rights of water power were separated from the other property rights in most communities at an early stage.

After almost eighty years of hydropower development and production in the fragile northern environments, experience has shown that although water energy is a renewable resource, it is not without effects on the local environment. The massive multiyear water storages tend to change the local climate; in many instances the vegetation of the area deteriorates—these processes affect wildlife, game, and tourism. The fisheries of regulated lakes and rivers have gradually deteriorated and the disappearance of the spring flushing has gradually affected the ecology of the fjords and near coasts, making them less favorable to the free salmon. Thus this kind of energy is not totally "clean." While the owners of northern hydroelectric power plants earn good incomes from their property rights, the holders of other property rights and the local population of northern communities have experienced significant environmental deterioration, particularly in their mountain and river areas.

Until recently, the deal has been somewhat fair—in return for some lowering of environmental quality, the northern communities have had secure jobs guaranteed by cheap and "clean" hydropower. However, with the recent liberalization of the European energy market and new institutional arrangements that lower the transmission costs dramatically, the favored position of northern "energy-communities" has changed. If outside interests (or the state), acquire ownership of the power plant, the proximity to the waterfall is no longer advantageous and only the disadvantages of the energy/environment deal are left to the local community. In a future where the ownership of waterfalls is becoming more significant, this means that there are continuous opportunities for renegotiations of the property rights for water power, with a scope for local government, local industry, multinational industry, or the state to change the constitutive rules regarding water power.

Developmental Strategies

Our societies are not governed merely by tangible resources, harvest technologies, and the more or less appropriate institutions we design to organize the social and economic life in a resource-based region. They are also governed by the ideas we nourish about the correct and incorrect paths towards a future which in the modern age has come to be viewed as an "open future." It is therefore customary that the debates and the formation of "schools" relating to development strategies occur in legislatures, administrations, and institutions of higher learning and research.

In the brief history of northern Norwegian academic institutions dating back to the 1960s, there have been numerous schools of thought relating to development strategies of the North. Especially in relation to rights to natural resources and ways of organizing resource users, these philosophies have differed substantially over a short period of time.

- The idea that traditional northern societies contained special qualities of peasant economy and regional self-reliance that made them robust in relation to fluctuating resource basis and to changing markets. The flexibility of northern households with regard to means of livelihood and sources of income was the backbone of this "robustness." These qualities were being eroded by the attempts by the national government to "modernize" the North after World War II (Brox 1966).
- The idea that local resource dependencies were to a large extent overcome by "regional integration." This meant that improved communications, commuting, enlarged labor markets, and growth in public sector investments and employment, together with improved education, to a large extent had modernized the North (Brox 1984). Increased mobility and increased public investments had to some extent facilitated a kind of development that maintained some of the inherent "robustness." The real challenge to this was the internal processes of specialization and increasing rigidity within the fisheries sector and within the agricultural sector.
- The idea that northern Norway has lost control over its own resources and that the correct development path lies in regaining control over resources and the generation of knowledge. Therefore it is necessary to initiate a total reconstruction of the North, where local solutions to resource control and development strategies are emphasized. A grand scientific program, "The Project New Northern Norway" (PRONOR) was considered essential to provide Northern Norway with the scientific base for this reconstruction (NAVF 1990).

- The idea that the "tragedy of the commons" was the main obstacle to development in the resource-based North and that privatization was the only way to achieve accelerated economic development and responsible resource maintenance. In fisheries this meant that transferable quotas should be introduced in order to create a favorable "incentive structure" (Hannesson 1990). In reindeer herding areas, fencing and privatized range management techniques were also introduced.
- The idea that resources held as "common property" were not doomed to tragedies, but were both traditionally (and would be in the future) the most efficient, just, and legitimate way to govern resources. The theories of co-management advocated a reduced role in resource management for the nation state and an increased role for local communities (Jentoft 1991).
- The idea that in the new European postindustrial era there will be industrial regions with a potential for "flexible specialization" and flexible adaptation of households that will do well. This means that the resource-endowed North, with its traditional cultures of flexible resource users, will have a competitive advantage. This would call for a halt to the recurrent government attempts to modernize the North and destroy the inherent robustness (Nilsen 1992). It would also need a smallholder greening instead of the gradual withering away of smallholder culture (Netting 1993).
- The idea that greater regions with similarities in resource base and strategic market position can benefit from concerted action in research, production, and marketing. Recent European experience has shown that "the industrial regions" of Toscana, Rhone-Alps, and Westfalen have achieved a high degree of competitiveness based on certain structural and cultural characteristics of those regions together with active networking among small- and mid-sized enterprises. In the North, the creation of a "Barents region" including areas of Norway, Sweden, Finland, and Russia, should thus be seen more as an attempt to create—by political action—such a successful industrial region rather than an attempt to create a "shared resource region."

None of these ideas or strategies were "right" or "wrong," but some worked better and some worked more poorly in the discourse of the time when they were formulated. This means that these various ideas have influenced the course of developments in the North—some to a great extent, some to a lesser extent. Some of the ideas have contributed to an opening of the "public realm," some have contributed to a narrowing of the relevant "models of a future society." By discussing development strategies in the universities and administrations of a particular region and by teaching them to students, this means that the conceptual heritages are present among those who remember and can be mobilized for or against any solution at any point. This was more than amply shown by the debate in the North on whether Norway, Sweden, and Finland should join the European Union or remain outside it. From the popular base the often defunct theories raised their head and surprised the well-trained but ignorant young professionals.

Still, one thing that all—both "living" and "dead"—northern theories of development have in common is the relationship between the northerner and the natural resources of the North as the focal point. Even for resources with very small economic significance such as cloud berries and birds' eggs, their value as symbols of a particular northern culture tends to increase in times of insecurity about the exact role of the nation state.

In order to analyze a possible new role for the state and the prospects for private enterprises and collective action in the resource-favored North, it is therefore necessary to sort out the fundamentals of resource relationships. What are we really talking about when we argue for a need to manage or to govern a resource—be it whales, cloudberries, herring, or mountain pastures for reindeer?

The Analytical Importance of Property Rights

Property rights are the links between the social world and the biophysical world. In brief codes often interpreted as institutions, property rights systems constitute the basis for most resource management (North 1990). However, the question is not the nature of the relationships between the physical properties of a resource and the way it is governed. Neither is it a question of one person's direct relation to the physical resource itself—a stock of fish or a pasture area. The main question is about the relationship between people and their respective resources. Such relations are the real content of property rights—claims to a future stream of benefits. This kind of property relations can be direct and personal as in face-to-face encounters. They can be indirect and personal as in local resource governing collectives. Finally, they can be indirect and impersonal as in the case of intervening states. The governing of resources is therefore mainly about relations between humans and between humans and their collective institutions (Bromley 1991).

This way of analyzing property rights follows a Kantian interpretation where man has never lived in "natural conditions" as Locke assumed, and where Hobbes could be believed to take his point of departure. There has always been some sort of social contract which comes before real property rights. If we follow Kant's definition of property rights, it therefore follows that all property rights are derived from collectives, which again are based on shared communities of understanding and common knowledge associated with the use of language. Even if property rights are often communicated to succeeding generations as cultural heritage, all property rights are basically instrumental variables that are designed or organically evolved to suit particular needs of the collective. They are means to reach certain ends and can be changed if this is considered to be desirable for the collective or those with influence within the collective. Expropriation, nationalization, and privatization are examples of a collective changing basic property rights in order to reach specific goals. Such changes usually take place at a constitutive level of the society and usually after some crises or as part of a lengthy political or legal process. Variations in forms of property rights can thus be analyzed comparatively as solutions to different kinds of resource management challenges posed to societies, communities, or collectives. And property rights can be consciously designed in order to achieve specific objectives (as in crafted to make up appropriate institutions with built-in incentive systems that will work in more or less predictable ways in deregulated markets). In this respect, property rights are also useful concepts for various schools of "new institutionalists."

As mentioned above, some of the greatest limitations to workable systems of governing resources lie within the academic communities and in government departments. Imprecise and blunt concepts and frequent abuse of these by fashionable academics and "experts" seriously limit the effects of resource governing systems, constrain the modification of existing systems, and preclude the design of new and more workable systems. The notion of "Global Commons" is only one example of such imprecise concepts. Some more familiar examples will make the need for sharper concepts more clear:

- In the Norwegian resource management debate, the term "common resources" is used indiscriminately. But the term "common" has here been used to define fundamentally different kinds of property relations.
- Resources owned by no one (*res nullius*) which have completely open access, have often been termed "common resources" in the public debate (*tjod/allemannsrett*).
- Resources owned by the state (*res publica*) in northern Norway, where the state alone has all rights connected with ownership, have been termed "common resources" (*statsgrunn*).
- Resources owned by the king (state) and the local community together, but where the members of the local community had clearly defined commons rights—user rights and governance rights at the exclusion of others—have been termed "common resources" (*statsallmenning*).

- Resources owned from ancient times in common by a local community—a true commons (*res communes*) where all members of the community originally had commons rights—have also been termed "common resources" (*bygdeallmenning*).
- Resources owned in common by owners of private portions of a resource deciding to pool these together in a common pool for operational convenience have been termed "common resources" (*realsameige*) (Stortinget 1992)

When the concepts used in science and in public debate become so blunt as these misconceptions of "common resources" indicate, the analysis based on these concepts and the related sciences and political discourse loses its ability to be analytically precise and to work out suitable institutions for specified tasks. Norwegian academics and government economists have thus uncritically imported the term "the tragedy of the commons" and used it indiscriminately on all these different Norwegian forms of property rights systems. Quotas in fisheries (semitransferable) were introduced specifically to avoid further tragedy in the "common fish resource." The same kind of "enclosure thinking" has also led to unnecessary fencing and extensification of resource use in the reindeer grazing areas.

Around the world there are abundant examples of such self-fulfilling tragedies resulting partly from sloppy analysis and abuse by power holders of blunt academic concepts paraphrased as slogans. Both predatory and benevolent states tend to take over the responsibility from complex but transparent and well-functioning local common property governing systems (grazing systems, local fishing systems, forests, or irrigation schemes), because they foresee a "tragedy" dynamics. In most cases some mixture of private ownership and a protection of collective rights would be the natural response to the challenges posed by modern technology, improved communications, unemployment, and exposure to the world markets. But what often happens is that the state—and its advisers—have limited capacity to think of all the detailed institutional arrangements necessary for governing a resource properly. If these are not in place in time, the resource degenerates into an "open access" resource and the tragedy becomes a reality. The state's desire to govern is usually far greater than its ability. In panic, the benevolent state then privatizes the resource to get rid of the problem; in many cases enclosure and privatization are offered as solutions to problems that do not even exist. More often than not, this leads only to even larger problems of landlessness, unemployment, social misery, and political unrest.

In order to avoid further tragedies like these, it is necessary to develop more precise concepts regarding different kinds of property rights, especially concepts that can better distinguish between what is private or corporate; what is common to a more or less defined group; and what is public, state, or global. This is because the concepts of property rights seem to be fundamental in any system of governing resources and their design thus decisive for the level and distribution of welfare of large groups in resource-dependent or resource-endowed regions. This is because with lesser state intervention and continued deregulation of markets—also for fish and agricultural products—property rights will assume new and increased importance.

To aid us in an analysis of northern resources, we shall employ an analytical framework that breaks down the right/duty correlates that make up property rights into units that can be analyzed in relation to empirical, real life situations. One way to do this is to define bundles of property rights cumulatively so that an individual is a part- or full-owner according to the number of bundles of property rights he or she possess. Thus only a full set of property rights qualifies one as a full-owner. Such an analytical framework has been developed by Edella Schlager and Elinor Ostrom, but has only been tested out for analytical power in the Maine lobster fishery (Schlager and Ostrom 1992) (see Figure 1).

In such a framework it is useful to distinguish between property rights on an *operational level*—where things happen, and property rights on a *collective level*—where things are decided. Property rights at the operational level are weak, while the power to participate in the shaping of future possibilities and to craft future property rights on the operational level, makes property rights on the collective level strong. A deeper *constitutive level*, where designs of all property rights are laid down—and intermittently contested—is also a necessary part of the analytical framework, but is not included in Figure 1.

On the operational level we can distinguish between the property rights as the *right of access*, to enter a defined, physical area, and the *right to subtract*, to take away or harvest the products of a particular resource. On the collective level we can distinguish between the *right to manage*, to regulate internal patterns of use and to transform the resource through improvements or negligence; the *right to exclude*, to decide who shall have rights of access and how these rights can be obtained, lost, or transferred; and the *right to alienate*, to decide to sell or hire out one or both of the other property rights on the collective level.

These five different property rights make up various bundles of rights that make it possible to distinguish analytically between different ownership positions. A full owner has all five rights, a mere user has only access and subtraction rights. In between these positions we can find all the known forms of property relations to resources. Compared to the analytical framework proposed by Schlager and Ostrom, this has here been expanded with one position: the unauthorized user with only a right of access, because this is a category that is analytically useful in northern areas.

This kind of subdivision of property rights has great analytical advantages. At the same time it keeps the concepts of various property relations away from being muddled, it also enables the classification of specific resource management systems according to the empirical distribution of these five different kinds of property rights. Also, in real life these five different categories of ownership positions can easily be recognized in relation to northern resources.

The "unauthorized user" in northern Norway and northern Sweden is given access to most wilderness resources under the legal categories of "everyone's rights" (*allemannsretten*), which has a correlated set of duties concerning proper conduct in protected nature or in landowners' forests or mountain areas. This kind of right of access gives the public a general protection from charges of trespassing, although for the protection of nesting birds and breeding animals, temporary limitations to the right of access can be introduced. If an unauthorized user starts to harvest other products from the resource than those granted in the "tourist laws" (mainly noncommercial species of wild berries), he or she is a thief. The implicit "harvest" contained in this kind of modest property right is the recreational value, and the improved health and mental balance of the public resulting from experiencing untouched nature. So far, this has been viewed as a genuine public good where one person's "harvest" does not represent a real subtraction from the resource.

The "authorized user" has both right of access to the resource and is granted the right to subtract from the resource—a right given by the owner or by the holder of management rights over the resource. For instance, northern Norwegian coastal fishers formally have no more than this kind of user rights to the state's fishing grounds, although this clashes with local perceptions of property rights and is contested. This character of the property rights of northern Norwegian fishers to a large extent explains the amount of energy spent on obtaining—and keeping for themselves—the state authorization as bona fide fishers.

FIGURE 1
Bundles of Rights Associated with Ownership Positions

	Owner	Proprietor	Claimant	Authorized User	Unauthorized User
Access	X	X	X	X	X
Subtraction	X	X	X	X	
Management	X	X	X		
Exclusion	X	X			
Alienation	X				

Source: Schlager and Ostrom 1992

In addition to rights at the operational level, the "claimant" also has rights in relation to management of the resource. These kind of rights at the collective level enable the claimant real participation in the formulation of rules for harvesting or subtracting from the resource, but not in exclusion decisions. Norwegian fishers have wanted to have such rights for a long time, and are given limited management rights through participation in the Fisheries Regulatory Council. This organization does not, however, make the final decisions on controversial management questions. The best Norwegian example of genuine claimant management rights for fishers is the old Lofoten management system (Jentoft and Kristoffersen 1989).

"Proprietors" have *de facto* property rights and participate fully in management and exclusion decisions. Most members of genuine common property regimes should be classified as proprietors, as they cannot sell out the resource—neither individually nor collectively. Analytically they have all the property rights except the right to alienate the resource. This means that in most cases they are protected against themselves through a legal bonding. Aboriginal rights are also believed to belong to the same category of ownership (including all property rights except the right to alienate the resource). The right to exclude others gives the commoners the necessary tools to avoid "the tragedy of the commons" and usually the proprietors' property rights are socially accepted as long as the resource is managed in an ecologically sustainable way.

A full "owner" has a full set of all five property rights and is a *de jure* owner usually justified by a title-deed. Only a full owner can alienate the resource and become landless. Compared to ancient common property regimes, the full owner is the making of the western romanistic maxim, "No one can be held in co-proprietorship against his will" (Grossi 1981). But in the final analysis also the five owners' rights are socially derived rights, granted and sanctioned by the collective, and they can thus be nullified if the resource is severely mismanaged or neglected. In principle this should also apply when the state is the holder of a full set of property rights. Interesting forms of ownership are created when owners by voluntary decision pool their portions together and agree to submit to binding rules for operational convenience (*sameige*). However, as such portions in principle can be alienated, the holders of pooled portions are owners in this framework.

What new insights will current analysis of these varying bundles of property rights to northern resources give in relation to self-governing capacities, the role of the state, and the level of welfare among northerners? We shall briefly apply these analytical tools on the above mentioned resources: birds' eggs; wild seafish; wild salmon; forests and grazing land; waterpower; and coastal marine environments for aquaculture, protection, and recreation. From this we shall conclude whether property rights are analytically important.

Property Rights of Eggs, Fish, Forests, and Coastal Waters

The egg-collecting institutions of northern Norway present a multitude of institutional variety tailored to the local ecology and the demographic processes of the birds, but in some cases also tailored to the demographic processes of the households involved. In most cases the nesting islands or rocks (*eggvaer* or *dunvaer*) are owned by two or more of the households that make up a fishing hamlet. But they own these only as residents or household members or heirs related to community members—they are proprietor rights. Usually the egg-collecting rights are constructed in such a way that they belong to the plot in the hamlet and cannot be sold out or separated from this. The right to graze sheep on the islands and around the homestead is sometimes the link between property rights in the fishing community and the egg-collecting rights. Still, the limitations on alienation of these kind of rights do not make them common property rights to the whole hamlet; on a certain island there are usually only two or three specific families of co-owners. But with a multitude of islands around a hamlet, most households used to have egg-collecting rights on several of these islands, thereby increasing the chances of finding eggs in a poor year.

The egg-collecting institutions are very different from one community to another, as there has been no government effort to standardize them. By comparing different institutional designs in different ecological settings and different social dynamics, it is possible to reach some conclusions regarding the tendency for self-evolved governing systems to combine the different kinds of property rights in particular ways. This involves the practicalities of taking turns collecting eggs, the techniques for fencing off neighbors and distant relatives by generous distribution of egg gifts, the generation of tacit agreements on quantity restrictions, and the monitoring of environmental changes that invoke a need for change in management decisions. All these rules for exercising egg rights make up an array of different egg-collecting institutions in northern Norway. The analytical importance lies in the way the different property rights combine to make up different institutions. The practical importance of this is that these institutions taken together make up a "palette" of institutional solutions to general resource governing problems. Thus studies of seemingly insignificant birds' eggs can be very valuable in the future crafting of governing regimes for economically significant resources like wild sea fish or water power.

The property rights of wild sea fish are almost opposite to those of birds' eggs. Since A.D. 872, the real owner of wild ocean fish in Norway has been the king, when King Harald (the "hairy fairy") took as his property "all lands, all seas, and all lakes." Eleven hundred years of contestation, colonization, uprisings, and wars have not changed this basic property right of the "king's estate"—in this way state property rights have a tendency to be very "sticky." Still, only the state has the full owner's right to alienate wild fish in the sea, as it does in negotiations with other states. The state also takes all vital management decisions relating to wild fish and decides who shall have access to the fishing grounds. This sovereign ownership position of the state has been constantly contested by fishers and by coastal communities, claiming "ancient commons rights" to their nearby fishing grounds, dating back to the era before the formation of the kingdom. Apart from a few supreme court decisions granting aboriginal common property rights to Saami fjord fishers, the state has been successful in maintaining its ownership rights (Salten 1994). However, it would be erroneous to consider the state as a purposive actor making only rational choices to maximize its assets. More often the state is an arena where competing groups of powerful actors bargain over the future direction of public policy (Young 1982). This implies that powerful groups of ocean fishers and fish exporters have a strong interest in the continuation of the state ownership position, provided that private ownership of the fish (individual transferable quotas) are politically impossible to achieve. Recently marine biologists have shown that coastal cods in the fjords of northern Norway are separate stocks of cod—and not part of the shared resource of Norwegian/Russian (Arctic) cod. This kind of research finding spurs new discussion of management and exclusion rights to these "local" cod stocks. Fjord fishers and coastal fishers have already claimed local management rights to these stocks; ocean fishers and mobile offshore fishers will argue strongly for a continuation of state managed general quotas for cod.

The amount of self-government on the part of the collective of fishers and fishing communities has actually decreased in recent years, as the management decisions of the state regarding issues such as quotas, fishing periods, and gear use have become increasingly specified as formal control efforts are stepped up. This weakening of traditional claimants' management rights has resulted in the gradual breakdown of informal control mechanisms among fishers with regard to over-fishing, high grading, false reporting, and black market sales. Local actions to gain more control by closing certain fjord basins to shrimp trawlers or herring purse seiners have often been overruled by an insensitive state.

The coastal fisher has now only the right to access and the right to subtract a specified amount of fish; he or she has moved from a position of a "claimant" to a position of a mere "authorized user." Fishers must therefore spend considerable effort towards the state in order to maintain their authorization, but are simultaneously tempted to spend as much time and energy on cheating the owner and the owner's increased formal control efforts. In addition, there is no excitement in fishing anymore—one can no longer strike luck with a big catch, but is merely a "contract harvester for the state sea lord." In this respect, the increasing control problems in Norwegian fisheries can be explained by recent changes in property rights and subsequent changes in incentive structures.

Property rights in coastal ecosystems have had little attention during the twenty-five years of aquaculture development in northern Norway. The development of legal and practical property rights categories are therefore lagging behind the rapid developments in aquaculture and related sea-cultivating activities. Private shoreowners traditionally had management rights only as far as the maximum extent of the tidal zone—outside this zone is "the king's sea." Although most fishing communities can point to traditional ownership to certain fishing grounds or certain sounds, the state ownership has implied that the fishing hamlet, the larger coastal community, or even the local government (municipality) has no legal management rights in these "public waters." Legally, the coastal population is merely an "authorized user" of the state coast. But everyone can "subtract" from the healthy coastal ecosystem without any duties towards it.

Several local governments have felt the need to manage fjord basins, archipelagos, and sounds, and have tried "coastal zone planning" in order to bring together all the different rightholders, user groups, and interest groups in a coastal area. These have failed on the whole because certain groups—notably mobile fishers—have central government authorization and the legal right (from the Salt Water Fishing Act) to move freely and thus disregard local attempts to regulate the use of the coastal zone (Nilsen and Emmelin 1992; Sandersen and Buanes 1995). This is also one of the overriding questions related to the development of stock enhancement or sea ranching of marine or anadromeous fish species, and one of the main reasons why this has not developed in Norway as it has, for instance, in Japan. However, there is a deep reconstitutive process going on regarding property rights in coastal environments and the outcome of this might well be that the state will be forced to give up its management rights and its sovereign right to make decisions about access and exclusion, and hand these over to local government.

The property rights connected to wild, migrating salmon also offer numerous interesting avenues for analysis. The river owners usually hold proprietor rights to the salmon on their stretch of the river. This means that this kind of right is tied to the farmstead and cannot be alienated from this—although long-term and very lucrative leases are quite common. Because the salmon move along the entire river, river owners usually pool their management and exclusion rights into an "association of river owners" that agrees on license policies and enhancement efforts. These associations cooperate with government natural resource managers through "freshwater fish management boards," which are one of few examples of "co-management" in Norway.

However, when salmon migrate to the fjords and to the ocean, and then return along the same runway, they are outside "the management room" of the river owners and the freshwater management boards. As activities here over the years have increasingly affected the migration of the salmon, the owners feel a desperate need—in order to save the remaining salmon stocks—to extend the management type property rights beyond the rivers and into the coastal zone. Contrary to most other interests in the coastal environments, the salmon river owners have had some success in this —a ban on coastal fishers' drift nets was accomplished mainly through lobbying. And in the ongoing constitutive battle of property rights in coastal areas, the river owners are clearly "contracting for property rights" of the management type to fjords and streams at the expense of the aquaculturalists. However, this has not helped the wild salmon, whose stocks continue to decrease in most salmon rivers.

The development of property rights to forests, berries, and pastures also offers interesting insights into the ways in which composition of property rights determine the governing of a resource. Parallel to a number of fishing communities in northern Norway, the Saami and the farming communities in northern Norway also want to hold property rights on the collective level in relation to the forests and mountain areas they have been using for hundreds of years. This question was made even more acute by the questions raised in the negotiations over entry to the European Community and the practical arrangements necessary to accommodate the freedom of movement and freedom of enterprise without disrupting the fragile northern environments.

At one level of analysis this can be seen as part of the circumpolar process of "devolution" or decolonization of the North (Young 1992). As such, it is of basically the same character as the granting of "home rule" to Greenland and resource governing rights to Nunavut in northern Canada. In Scandinavia this kind of deep process involves both very complicated and long-standing legal questions and very sensitive political questions. It therefore moves on rather slowly. But for the trained policy analyst, the positions taken by the various organizations and political parties, the timing, the argumentation and use of symbols, and the coalition building and legal initiatives all offer a rich ground for study of a very deep constitutive process that eventually will affect the property rights on the collective level (for example, management and exclusion rights) in a fundamental way.

But part of this process is also the practical feasibility of crafting workable resource governing institutions based on such "local ownership" of northern resources. If the sustainability of alternative resource management systems is not convincingly argued, this is most likely to slow down the devolution process or maybe even reverse it, thus strengthening the role of the nation states or supranational organizations in the governing of northern resources. Therefore, the nesting of institutions works both ways; the available options at the collective and operational levels influence the deeper processes at the constitutive level.

In claiming reindeer pasture management rights for the Saami parliament, the actual management institutions were not worked out in great detail. They could be thought of as the traditional familial/territorial units of the *siida* type or they could be other types of institutions that provide the incentives necessary to remedy the immediate overgrazing of the Finnmark mountain plateau. The national parliament did not want to give the Saami parliament resource governing powers at the time of its establishment and the claim was thus temporarily shelved by the state (awaiting "legal clarification of the Sami right to graze reindeer on non-owned lands" [St. meld., 52 1992–93]). On the legal track of the constitutive process, however, the property rights of the Saami to the material base for maintaining their culture (lands and waters) have a strong standing both in the Norwegian Constitution and in international law. The process along the political track will therefore resume in 1999 when the hearing of the Commission on Saami Rights to Lands and Waters is finalized and the government starts work on legislation for land and water rights (NOU 1997:4).

By claiming the introduction of institutions close to the old commons (*statsallmenning*), the farmers of northern Norway are in reality also claiming property rights on the collective level (management and exclusion rights). The claim was politically justified by arguing that this would give farmers in the North the same kind of property rights as farmers in the southern part of the country. The northern Norwegian farmers have strengthened their position by pointing to institutional designs that can be utilized without modifications—the "Mountain Governing Council" (*fjellstyre*) in the existing "mountain law." This law also provides for the inclusion of members on the council from reindeer herding communities in mixed farming/reindeer mountain areas. Although opposed by both Saami and state forest interests, these claims have a substantial support in the legal track of the constitutive process and a growing support in the political track. The pace of this process has accelerated in response to both the apparent success of the Saami claims and to the prospects of a changed role for the nation state in relation to local government. This has caught the national government by surprise, as they had hoped to settle the Saami land claims first before dealing with the northern Norwegian farmers' claims. This shows that a deep constitutive process cannot be planned from the top; actions will usually spur reactions until the matter is resolved by all parties agreeing to lay down new fundamental rules for the distribution of property rights. These will then become the new constitutive rules, which again are likely to be changed by a new turn of the ongoing process. But in systems based on a dual track mechanism for constitutional change, this is—and is supposed to be—a very slow process.

Conclusion

There are various ways to analyze these developments in the governing of major—and minor—northern resources. These cases from northern Norway have aspects that are common to most northern resource governing systems, and should therefore be of interest for comparative studies throughout the circumpolar region.

The crucial point in the cases briefly presented here has been the analytical importance of the concepts of property rights and the relationship of these to the collectives we call community, local government, or state. This implies that a certain distribution of property rights is not given for "all times." As analytical concepts, property rights can also be institutional instruments, design principles that can be utilized to make institutions work better—for instance, work for a more sustainable governing of northern resources. It is therefore not so much a question of "having the rights," but of what distribution of the rights will work towards a stated goal.

It is the use of such analytical concepts that can enable us to carry out studies with an analytical depth that is greater than the level achieved through the inherited jurisprudential concepts of *possessiones*, *dominium*, and *usum fructum* or through a plain comparison of official state policies or state adherence to international treaties. The way the different circumpolar states handle their internal fights between state, local government, and northern communities over various kinds of property rights to northern resources would then be a fertile ground for such comparative studies of the "Northern Problem." It is towards such a greater northern research objective that this study is a contribution.

NOTE

1. An initial paper outlining the preliminaries of these questions was presented at a colloquium held at the Workshop in Political Theory and Policy Analysis, Indiana University, in September, 1993. Based on comments and suggestions, the paper was substantially expanded and revised and appeared

as a report in the series "Los i Nord-Norge" (Los-Notat no.18). In preparing the present version, the author is appreciative of the valuable comments from Thrainn Eggertson, Vincent Ostrom, and Hans Sevatdal. The author is also appreciative of the support received from the Royal Norwegian Research Council grant no. 530–93/034.

REFERENCES

Andersen O. J. and H. T. Sandersen 1992. *"I markedets tjeneste—en studie av høgskolemiljene i Bodø og Narvik,"* NF. Rapport nr. 24/92–50.
Bjørnaa, H. 1993. Hovedoppgave. ISV, University of Tromsø.
Bromley, D. W. 1991. *Environment and Economy: Property Rights and Public Policy*. Cambridge, Massachusetts: Blackwell.
Bromley, D. W. 1992. "Common Property as an Institution," in *Making the Commons Work: Theory, Practice, and Policy*, edited by D. W. Bromley. San Francisco: ICS Press.
Brox, O. 1966. *Hva skjer i Nord-Norge?* Oslo:Pax forlag.
Brox, O. 1984. *Nord-Norge—fra allmenning til koloni*. Oslo: U-forlaget.
Brox, O. 1989 *Kan bygdenæringene bli lønnsomme?* Oslo:Gyldendal Forlag.
Daly, H. E. and J. B. Cobb. 1989. *For the Common Good: Redirecting the Economy Toward Community, the Environment and a Sustainable Future*. Boston: Beacon Press.
Dacks, G. 1981. *A Choice of Futures: Politics in the Canadian North*. Toronto: Methuen.
Dacks, G., ed. 1990. *Devolution and Constitution: Development in the Canadian North*. Ottawa: Carleton University Press.
Eikeland, P. O. 1993. *Distributional Aspects of Multispecies Management of the Barents Sea Large Marine Ecosystem—A Framework for Analysis*. Paper presented at a MAB Conference on "Common Property Regimes: Law and Management of Non-Private Resources" in Nyvågar, Lofoten.
Eriksen, E. O. 1993. *Statens Rolle i Nord-Norge*. LOS i Nord-Norge, notat nr. 4.
Fiskeridepartementet. 1992. Mandat for utvalget til å utrede juridiske spørsmål i forbindelse med Havbeiteprogrammet PUSH.
Gaski, L. 1993. *Utnyttelse av utmarksressurser; endringer i samhandlingsmønstre og kulturell betydning*. Hovedfagsoppgave, Planlegging og lokalsamfunnsforskning. ISV, Universitetet i Tromsø.
Gordon, H. S. 1954. "The Economic Theory of a Common Property Resource: The Fishery." *Journal of Political Economy*, 62.
Grossi, P. 1981. *An Alternative to Private Property, Collective Property in the ... Nineteenth Century*. Chicago: University of Chicago Press.
Hannesson, R. 1990. En samfunnsøkonomisk lønnsom fiskerinæring: Struktur, Gevinst og Forvaltning. Sektoranalyse for prosjekt "*Effektiviseringsmuligheter i offentlig virksomhet*. Administrasjonsdepartementet.
Hanssen, J. I. 1995. "Cultural Contrasts in the Norwegian Welfare State." Mimeo. Nordland Research Institute (forthcoming).
Havforskningsinstituttet. 1993. *Multispec-Nytt*. Vår—sommer.
Jentoft S. 1991. *Hengende snøre, Fiskerikrisen og framtiden på kysten*. Ad Notam, Oslo.

Jentoft, S. and T. Kristoffersen. 1989. "Fishermen's Co-Management: The Case of the Lofoten Fishery." *Human Organization* 48, no. 4.

Kaminski, A. Z. 1992. *An Institutional Theory of Communist Regimes: Design, Function and Breakdown*. San Fransisco: ICS Press.

Keohane, R. M., McGinnis, and E. Ostrom. 1993. Proceedings of a Conference on "Linking Local and Global Commons." Harvard University, April 1992.

McGinnis, M. and E. Ostrom. 1993. "Design Principles for Local and Global Commons," in proceedings of a Conference on "Linking Local and Global Commons." Harvard University, April 1992.

NAVF. 1990. *Nybrott og gjenreisning* Ny kunnskapspolitikk for Nord-Norge. Innstilling fra et utvalg oppnevnt av NAVF's styre. Oslo.

North, D. 1990. *Institutions, Institutional Change and Economic Performance*. New York: Cambridge University Press.

Nesheim, J. and K. Rystad. 1990. *Om eiendomsretten til moltebæra på privat grunn i Nord-Norge*. Hovedoppgave NLH Inst. for Planfag.

Netting, R. McC. 1993. *Smallholders, Householders:* Farm Families and the Ecology of Intensive, Sustainable Agriculture. Palo Alto, California: Stanford University Press.

Nilsen, R. 1992. "Look to Toscana og Sunnmøre, Fleksibel spesialisering og distriktsutvikling." *Plan og Arbeid* nr. 3/4.

Nilsen, T. and L. Emmelin. 1992. *Næringsutvikling i Skjerstadfjorden* Miljø—og ressurshensyn som grunnlag for næringsutvikling i kommunene Bodø, Fauske, Saltdal og Skjerstad. UNIT, University of Trondheim Rapport nr. 3/92.

NOU 1997: 4. *Report from the Saami Rights Commission on Saami Rights to Land and Water*. Oslo.

NOU 1994:10. Lov om havbeite (Proposal for Law of Sea Ranching). Oslo.

NOU 1993:34. *Retten til og forvaltning av land og vann i Finnmark*, Bakgrunnsmateriale for Samerettsutvalget. Oslo.

NOU 1990:22. LENKA (Suitability Analysis for Aquaculture on the Norwegian Coast). Oslo.

Pipe, R. 1974. *Russia Under the Old Regime*. New York: Charles Schribner & Sons.

Salten, Herredsrett. 1994. Dom i sak nr. 93-00233A. Fagervik og Olaisen vs. Staten.

Sandberg, A. 1993a. *Merker for Institusjonsutforming i Ressursforvaltning*, Norges Forskningsråd (Avd.NFFR) Seilingsmerker for Fiskeripolitikken. NF Særtrykk nr. 1/93.

Sandberg, A. 1993b. *Lokal forvaltning av oppdrettsressursene*. LOS i NORD-NORGE Notat nr. 11, Tromsø.

Sandersen, H. T. and A. Buanes. 1995. *Lokal miljø- og ressursforvaltning i kystsonen. Nye roller for kystkommunene* ? MILKOM-notat no. 8/ 95. NIBR, Oslo.

Sandvik, G. 1993. Statens grunn i Finnmark—et Historisk perspektiv Vedlegg 1 i *NOU 1993:34*. Rett til og forvaltning av land og vann i Finnmark—Bakgrunnsmateriale for Samerettsutvalget.

Schlager, E. and Ostrom, E. 1992. "Property Rights Regimes and Natural Resources: A Conceptual Analysis." *Land Economics*, August 1992, 68(3).

Sproule-Jones, M. 1993. *Government at Work, Canadian Parliamentary Federalism, and its Public Policy Effects.* University of Toronto Press.
Stephanus, S. H. 1629. *De Regno Daniæ et Norwegiæ Tractatus Varij.* Leiden.
St.meld. nr.52. 1992–93. Om norsk samepolitikk. (Norwegian Government white paper on "Norwegian Sami Politics").
St.meld. nr. 33. 1986–87. Norges framtidige energibruk og—produksjon.
Stortinget. 1992. Innst. O.nr. 67, Innstilling fra Landbrukskomiteen om A: Lov om bygdeallmenninger. B: Lov om skogsdrift m.v. i statsallmenningene C: Lov om opphevelse av og endringer i gjeldende lovgivning om allmenninger m.v.
Stortinget. 1992. Lov om Sameiger med endringar, sist ved lov av 26. juni 1992 nr. 86.
Vold, H. A. 1981. "På våres måte og etter våres regla" En undersøkelse av eiendomsforhold og eggsanking på Bleik, Andøya. In *Norveg, Årg. 24.*
Weber, M. 1968. *Economy and Society.* New York: Bedminster Press.
Young, O. 1982. *Resource Regimes Natural Resources and Social Institutions.* Los Angeles: University of California Press.
Young, O. 1992. *Arctic Politics: Conflict and Cooperation in the Circumpolar North.* Hanover: University Press of New England.
Ørebech, P. 1991. *Om allemannsrettigheter* Særlig med henblikk på rettsvernet for fiske ved igangsetting og utøving av petroleumsvirksomhet. Vettre Osmundsson Forlag.

The Contributors

TORGEIR AUSTENÅ is professor of law in the Department of Land Use Planning at the Agricultural University of Norway and holds a Dr. Juris degree from the University of Oslo. He has served as a member of a number of government committees and has been a member of the Comité Europan de Droit Rural and the Saami Rights Commission. His research interests include land law and minority rights issues.

BERTIL BENGTSSON is a retired supreme court justice. He has a doctor of law degree from Uppsala University, and has been a professor of private law at the University of Stockholm and Uppsala University. At present, he is temporary professor at the universities of Lund and Luleå. Bertil Bengtsson has published twenty books and a large number of papers related to tort, insurance, contract, family, real estate, the environment, and constitutional issues. He has published several articles on the rights of the Saami people.

ERLING BERGE is professor in the Department of Sociology and Political Science at the Norwegian University of Science and Technology in Trondheim. He holds a cand. polit. degree (sociology) from the University of Bergen and a PhD in sociology from Boston University. He has previously held positions at the Central Bureau of Statistics of Norway, the Institute of Applied Social Research, and the Agricultural University of Norway. He was a visiting fellow at the University of Essex during the winter of 1986-1987. He has published on population issues, urban and regional problems, and land-use theory. His current research is concerned with cultural and organizational aspects of resource use and rural development.

HANS CHRISTIAN BUGGE is a professor of environmental law at the University of Oslo. He holds a doctorate in regional planning and economics from the University of Paris II and is Dr. Juris of the University of Oslo. He held various senior positions in the Norwegian ministries of the Environment and Finance from 1972 to 1982. From 1982 to 1991 Hans Chr. Bugge was secretary general of the Norwegian branch of Save the Children, a nongovernmental organization with major development programs in Asia, Africa, and Latin America. From 1986 to 1987 he acted as state secretary in the Ministry of Development Cooperation and also served as a personal advisor to Mrs. Gro Harlem Brundtland in her work as chair of the World Commission on Environment and Development (the Brundtland Commission).

THRÁINN EGGERTSSON is jointly a senior fellow at the Max Planck Institute for Research into Economic Systems in Jena, Germany and professor of economics in the Department of Economics at the University of

Iceland. He was educated in England and the United States and holds a PhD in economics from Ohio State University. He has worked as an economist in the Central Bank of Iceland, taught economics, and acted as dean to the Department of Business Administration and Economics at the University of Iceland. He has been a visiting scholar to universities in the United States (Washington University in St. Louis in 1984 and 1992, Indiana University from 1993 to 1995), an honorary professor at the University of Hong Kong in 1992, and from 1995 to 1997 was a visiting fellow at Stanford University's Hoover Institution. Eggertsson has carried out research and published on inflation and on labor market issues, and on the economics of institutions. He is the author of *Economic Behavior and Institutions* (published by Cambridge University Press in 1990), and he has edited, with Lee Alston and Douglass North, *Empirical Studies in Institutional Change* (Cambridge University Press in 1996).

THOR FALKANGER holds a Dr. Juris degree from the University of Oslo and has been professor of law in the Faculty of Law at the University of Oslo since 1970. He is attached to the Scandinavian Institute of Maritime Law and is responsible for teaching property law, bankruptcy law, law of mortgages and secured transactions, and maritime law. He has published a number of books and articles on various subjects including property rights and maritime law.

ALF HÅKON HOEL holds a cand. polit. degree (political science) from the University of Oslo. He is associate professor of international fisheries at the College of Fisheries Science at the University of Tromsø. His main research interests lie in international relations and fisheries management.

HEIKKI J. HYVÄRINEN holds a master of arts degree in law and is juridical secretary of the Saami Parliament in Finland. He has published several reports on the legal situation of the Saami in Finland and has drafted proposals for new bills.

KAISA KORPIJAAKKO-LABBA is a senior research fellow in the Nordic Saami Institute situated in Kautokeino, Norway. Her most important scientific work is her doctoral thesis dealing with the rights of the Saami in Sweden-Finland (published in 1989; published in Swedish in 1994). She has also held various positions in the universities of Helsinki and Rovaniemi. After moving to the northern Finnish Lapland, she became a member and secretary in several local associations attending the interests of minority and Saami rights and reindeer herding. She has also served as secretary in governmental committees, preparing bills concerning fishing rights in the North and the whole land use problematic of the Saami in Finland. She has often visited the Finnish Parliament as an expert on northern questions. She is a docent of legal history in the University of Helsinki.

GARY D. LIBECAP is professor of Economics and Law, director of the Karl Eller Center at the University of Arizona, research associate with the National Bureau of Economic Research, and co-editor of the *Journal of Economic History*. His major academic interests are in the fields of American economic history, industrial organization, natural resource economics, and law and economics. He has been involved in research on common property issues with the World Bank, National Science Foundation, and Sea Grant. He has published widely on property rights and regulatory issues in the United States and developing countries. His recent work includes the books *Contracting for Property Rights* (Cambridge University Press, 1989); *Titles, Conflict and Land Use: The Development of Property Rights and Land Reform on the Brazilian Amazon Frontier*, with Lee Alston and Bernardo Mueller (University of Michigan Press, 1998); *The Federal Civil Service System and the Problem of Bureaucracy: The Economics and Politics of Institutional Change*, with Ronald Johnson (University of Chicago Press, 1994); and *The Political Economy of Regulation: An Historical Analysis of Government and the Economy*, with Claudia Goldin (University of Chicago Press, 1994).

ELINOR OSTROM is Arthur F. Bentley Professor of Political Science and codirector of the Workshop in Political Theory and Policy Analysis at Indiana University. Her current research is centered on institutional analysis and design. She has studied institutional arrangements related to governance of natural resource use in the United States and in developing countries. She is an elected fellow of the American Academy of Arts and Sciences. She has published extensively on urban and political issues in the United States and on common property resource issues in both America and in developing countries. Her recent books include *Governing the Commons: The Evolution of Institutions for Collective Action* (1990), *Crafting Institutions for Self-Governing Irrigation Systems* (1992), *Rules, Games, and Common-Pool Resources* (1994) with Roy Gardner and James Walker, and *Local Commons and Global Interdependence: Heterogeneity and Cooperation in Two Domains* (1995) with Robert Keohane. She was the president of the International Association for the Study of Common Property in 1990 and of the American Political Science Association in 1997.

ROBERT PAINE holds a doctor of philosophy degree from Oxford University and has been professor, head of department, director of sociological research, and Henrietta Harvey Professor of Anthropology at the Memorial University in Newfoundland. He is now Emeritus.

He has also held research and teaching positions at the various universities in Canada and Scandinavia, and at the Hebrew University of Jerusalem. He is a Fellow of the Royal Society of Canada and of the Norwegian Academy of Science and Letters. His principal research work has centered on the coastal Saami, Saami reindeer pastoralists, and Saami ethnopolitics in the north of Norway. He has published extensively on these topics since 1955. His books include *Coast Lapp Society I and II: A Study of Economic Development and Social Values* (1957 and 1965); *Dam a River, Damn a People* (1982); *Herds of the Tundra: A Portrait of Reindeer Pastoralism* (1994); and *Politics of Reindeer Wealth* (forthcoming).

PIERRE ROUX has been the chief legal advisor in the Office of the Attorney General in Namibia since 1991. He has previously been an administrator and deputy head of finance and administration within SWAPO of the Namibia Election Directorate.

BJØRN SAGDAHL holds a cand. polit. degree (political science) from the University of Oslo. He is an associate professor and elected head of the Department of Social Science at the Nordland College, Bodø. He has been a visiting research fellow at the University of Exeter and at the University of Western Washington. His main research has been on policy formation and implementation in the field of fisheries, fishery resources, and the impacts of the oil policy for northern Norway. He has edited and contributed to several books including *Fiskeripolitikk og forvaltnings-organisasjon* (*Fishery Policy and Management Organization*). Currently he is doing research on the management of marine resources.

AUDUN SANDBERG holds a cand. polit. degree (sociology) from the University of Bergen. He is associate professor in the Department of Social Science in the Nordland Regional University, Bodø. He has also been head of several research projects on common property management regimes in northern Norway at the Nordland Research Institute since 1992. Audun Sandberg has a wide range of experience from research work in developing countries—in particular, Tanzania, Kenya, and Sri Lanka. During several periods he has been employed as advisor or consultant by the Norwegian Development Agency on issues related to integrated rural development. He has been a visiting research fellow at the University of Alberta; Pacific Lutheran University, Tacoma; Indiana University; and the University of Sussex. He has published on fisheries organization, aquaculture development, and coastal societies in northern Norway; on management of fishery resources; and on common property issues related to forests, mountain pastures, and wildlife.

GUDMUND SANDVIK was professor of legal history in the Department of Public and International Law at the University of Oslo from 1975 to 1995. He has previously held senior teaching positions at the Section norvégienne du Lyce Corneille in Rouen, the Université de Paris, and in the Faculty of the Humanities at the University of Oslo. His published works concern, among other issues, the history of law with emphasis on Roman Law and European legal history in the Middle Ages. His most recent publication is concerned with the legal status of the Saamis. He has been chairman of several Norwegian government committees and was a member if the Saami Rights Committee from 1980 to 1997. Gudmund Sandvik is a fellow of the Norwegian Academy of Science and Letters.

CARL HERMAN GUSTAV SCHLETTWEIN has been permanent secretary in the Ministry of Agriculture, Water, and Rural Development since 1990 in Namibia. He has previously acted as senior hydrologist in the same department and also has intermittently acted as permanent secretary in other departments.

HANS SEVATDAL is professor of cadastral law and land consolidation in the Department of Land Use and Landscape Planning at the Agricultural University of Norway.

NILS CHRISTIAN STENSETH is professor of zoology at the University of Oslo. He is chairman of the National Norwegian Committee of the "Man and Biosphere" program under UNESCO and a member of the Norwegain Academy of Science and Letters. He has been attached to the Norwegian Institute of Nature Research with particular responsibility for research activity on wildlife management in developing countries in a broad multidisciplinary perspective. He is, and has been, a committee member of a host of national research programs within the environmental science field. His research work in zoology has focused on evolution, landscape ecology, population ecology, and behavioral ecology. He has a particular research interest in developing models that integrate ecology and economy, pastoralism, and uncertainty issues. He is co-editor of a number of international journals and has published extensively internationally and in Norway.

GEIR ULFSTEIN is associate professor in the Department of Public and International Law at the University of Oslo. He has held teaching positions at the University of Tromsø and in the Scandinavian Institute of Maritime Law at the University of Oslo. His academic field of interest is mostly focused on international law, law of the sea, and international environmental law and he has published widely on these topics. He has also served as a consultant to the Norwegian Development Agency on various evaluation and advisory missions.

TROND VEDELD is research officer at the Norwegian Centre for International Agricultural Development (NORAGRIC) at the Agricultural University of Norway (AUN). He holds a Dr. Scient. degree from AUN. His dissertation focused on the heterogeneity, leadership, and collective action among the Fulani of Mali. He has experience in the Norwegian Agency for Development Cooperation (NORAD), the International Fund for Agricultural Development (IFAD), the African Development Bank (AFDB), and various international consultancies on dry land management in Africa, among others with the World Bank. His current research is on state and common property regimes related to range land management in Mali, West Africa.